Country Life Diary

Three Years in the Life of a Horse Farm

❖

By Josh Pons

❖

Illustrations by
Ellen Blackwell Pons

THE BLOOD-HORSE, INC.
Lexington, Kentucky

ISBN 0-939049-49-X

Printed in the United States of America
First Edition: November 1992
1 2 3 4 5 6 7 8 9 10

Book design by Suzanne C. Depp
Cover design by John D. Filer
Hook rug by Mary Jo Pons

To Ellen.

❖ CONTENTS ❖

Year Three

Foreword

"Ponsie," my father said, "we have to get the kids involved. They are the future of our game." The statement was made in response to a query as to why a busy man would spend time judging a minor junior horse show in the farther reaches of Maryland; the questioner was Finney's good friend Adolphe Pons, grandfather of the author of this chronicle.

That conversation occurred more than 50 years ago, at the office of the Maryland Horse Breeders Association, in what otherwise would have been the front second-floor bedroom of our house on Highland Avenue in Towson. I had not thought of Finney's statement in nearly that length of time, but it came to mind the other day when Josh Pons honored me with the request that I write this foreword.

Whatever else each may or may not have achieved, Adolphe Pons and Humphrey Finney certainly got their kids involved. "Ponsie," as Finney affectionately called him, is remembered as a small, stocky man, always nattily presented, with a stylish mustache and a precise and colorful mode of expression. In time he was succeeded in stewardship of Country Life Farm by the duo known to us as "the Pons boys," and to the author as "Uncle John" and "Dad," who in turn have been relieved of that stewardship to a considerable extent by the third generation of Country Life Ponses. The apple doesn't fall far from the tree, however; absent the mustache, but with the conformation, twinkling eye and fluent turn of phrase, the author is more than faintly evocative of his grandsire.

When Ed Bowen (to whom we give thanks) fortuitously asked Josh to undertake a Maryland farm journal in 1989, he was asking to retrace footsteps Finney had taken when I was a yearling, but the path was over very different terrain. Finney was paid to oversee a fifteen-hundred-acre spread for a wealthy patron who did not expect the farm to feed the family, in an environment free of encroaching hostile development, and in a time when qualified labor was readily available. Pons is an entrepreneur, operating with not unlimited family capital a high-risk enterprise, the success of which is of vital importance to him and his family, in an extremely difficult labor market on a small parcel of ground constantly in danger of being overrun by suburbia.

The external pressures confronting Finney and Pons differed significantly, but the fundamental challenges of the primary farm activity—breeding, raising, and preparing Thoroughbred horses for racing—have changed practically not at all in the intervening half-century, or indeed since the breed began.

It is the thoughtful and artfully expressed evocation of these fundamentals which so enriches the daily reports in this work; the reader gleans an understanding of and appreciation for the cadences of horse farm life, dictated by Mother Nature and necessarily respected by successful Thoroughbred breeders of every generation. The externals may change through time, but the joy and cruelty of nature's process are immutable. This truth is superbly vignetted by Pons, as is the abiding faith in and affection for the animal which characterize the men and women who elect to spend their lives in this pursuit.

Country Life and its generations of stewards are a charming anachronism in today's world, one which all who harbor affection for the Thoroughbred and the life surrounding his production would like to see go on forever. Few scribes are presented with such a rich lode to mine; fewer still have the prescience to marry a gifted illustrator. Josh Pons has chronicled with love and precision a segment of this charming anachronism; wife Ellen has skillfully illuminated it. May their efforts afford you as much pleasure as they have brought to me.

> *John M. S. Finney*
> *Newport, Rhode Island*
> *October 1992*

Preface

Write while the heat is in you.

Those words of advice, and a large family who took over my farm chores so that I could write while the iron was hot, are the reasons this diary came to be. The idea, however, belonged to Edward L. Bowen, senior editor of *The Blood-Horse* magazine, who summoned me to his office four years ago this autumn with the notion that it might be time to update Humphrey Finney's *Stud Farm Diary*, published in *The Blood-Horse* in 1935.

The prospect of once again writing for *The Blood-Horse* thrilled me. I owed this institution a game effort. The magazine was my first real employer outside of Country Life, when then-editor Kent Hollingsworth took a flyer on a sophomore English major from the University of Virginia. For three years after college, I worked full time. Ed and Kent both encouraged me to apply to law school at the University of Kentucky in Lexington. Once again, I became summer help, and once again, *The Blood-Horse* paid me to write, thus enabling me to pursue an education in law, an insurance policy against the ups and downs of the horse business.

With such motivation, and much to my surprise, I never felt the task of writing the diary to be a burden. Rather, it was a form of continuing education—graduate school in the real world. I started writing drafts the night I arrived home from my meeting with Ed, on the floor of a house not yet finished, by a creek whose waters I could hear from the basement window where I worked. Thereafter, each month from 1989 through 1991, *The Blood-Horse* carried the Country Life Diary in serial form.

This diary was a jealous mistress, but my wife Ellen never objected to the time I spent trying to get the words right. She might as well have hung a Do Not Disturb sign on the door of my study every evening. I am so very grateful to her for her understanding. My mother, meanwhile, often edited the monthly installments before I forwarded them to the magazine. It is appropriate that her hook rug, sewn on many evenings in front of the fireplace after she had put five children to bed, graces the cover of this book.

From the beginning, I took Thoreau's advice, and wrote while the heat was in me. In the end, I hope my efforts warm you.

Josh Pons
Country Life Farm
October 1992

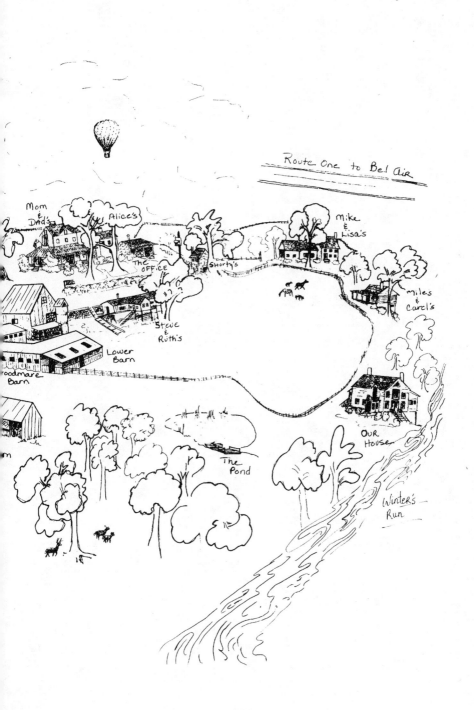

❖

Year One
THE STAGE IS SET

Sunday, January 1. While the world sleeps off the New Year's Eve cele-bration, we are quietly plying our trade, greeting yesterday's weanlings with newfound respect. At midnight last night, they became yearlings. We almost feel like throwing a huge birthday party today—the uniform date for *all* Thoroughbreds to officially become one year older.

I paused to assess our brand-new yearlings as we turned them loose at their gate. The three fillies will be separated from the five colts in another month. For now, though, all eight are learning to run in a herd, no different from what they will experience as racehorses. A nutritionist regularly reviews our yearling feed: sixteen percent protein pellets fed twice a day, free-choice mineral block sup-planting trace deficiencies in the pastures, and alfalfa hay. I feel comfortable with their condition. They have a lean, athletic look—growing *up*, not out.

"Happy New Year, kids. Good luck in the future," I shout, as I clap and shoo them from the gate, sending them off to run wild on this first day of a new year.

Monday, January 2. Spent the morning at the Maryland Horse Breeders Association office, a valuable asset to our state's industry. The office not only disburses the Maryland-Bred Fund, but also headquarters the Maryland Million program, and publishes the monthly *Maryland Horse* magazine. With the help of assistant editor Marge Dance, whose father, Humphrey S. Finney, founded the *Maryland Horse* and wrote a *Stud Farm Diary* more than sixty years ago, I put the finishing touches on this year's farm brochure.

Tuesday, January 3. Every day starts with the stallions. We spend a great deal of time schooling them, and no one person ever handles the studs alone. Like policemen, we always have a backup, an experienced person to follow the stallions as they are led to their paddocks. I've attended seminars on equine behavior modification, but armed only with a meager carrot or a sugar cube, we gently coerce obedience out of headstrong stallions. If you reward these big boys for good behavior, they seldom are bad.

Wednesday, January 4. The local county paper came out today. Our little classified ad sold a retired racehorse: "Thoroughbred Filly Four Yrs Old. Bay with Blaze. Show or Hunter Prospect. Make offer."

The alternatives for this filly were not promising. To simply break even on a horse at auction, that horse must be worth a thousand dollars: The entry fee, the sales company commission, and the bloodstock agency commission each are three hundred dollars, and the van ride to the sale adds another hundred dollars. Auctioning culled-out horses for burned-out owners is an exercise in futility. Yet I remind myself of what John Finney, Humphrey's son, cautioned about becoming a collector of horses: They *simply* accumulate.

At the sales now, a dour man with a cigar sits in the front row and buys every horse failing to exceed his opening bid of three hundred dollars. He loads these mares into gooseneck trailers and drives them off to the meat-packing plant in Connecticut to be slaughtered, quartered, frozen, and exported to France. This is not a pretty business sometimes. Hence, the classifieds: Much better for a young girl to get a riding horse, than to watch the man with the cigar buy our filly.

Thursday, January 5. The County Executive said he had a few surprises for developers if the Five-Year Comprehensive Rezoning Plan wins approval. He assured me the farm across Route 1 from Country Life would not be rezoned from agricultural to commercial.

"Citizens make government work," he said. "If you organize your community, you will have a profound impact on the county commissioners."

The farm across the street has been sold to a fellow from Arizona. He plans to erect a two-story mall with restaurants overlooking our front fields. (He suggested the name "Country Life Cafe," he has such gall.) He showed me his beautiful view of our farm, a view he expects to show potential investors in his mall. Horse farms enhance the value of surrounding real estate, but I do *not* enjoy enhancing this fellow's mall site. He bought the property with agricultural zoning; he should be stuck with that zoning. I wish he would go back to Arizona.

The development of horse farms is a concern to horsemen everywhere these days. When Man o' War Boulevard cut through the Bluegrass region around Lexington, Kentucky, an editor commented: Should have named the road "Upset Boulevard," for the only horse to finish ahead of Big Red, and for its impact on legendary farms of the Turf.

Friday, January 6. Went to a party at an old friend's house. Only one person in the throng had an interest in horses. He was *not* an ambassador of good will for the game.

"I bought three broodmares in partnership a few years ago," he moaned. "Those three mares have produced nothing but trouble every year. Now I have nine horses on my monthly board bill, and I couldn't get two thousand dollars for the entire lot."

His partner keeps the horses at her farm. She has no impetus to sell the horses for peanuts when she can send him a board bill every month.

"How do you force a partner to give the horses away? I wanted to be in the horse business, but hell, I didn't want to ruin my other business in the process."

Saturday, January 7. Spent the morning in a goose hunting pit on the Eastern Shore. I can't hit the bed with my hat, so the geese on my side were safe. Nevertheless, a trio of shareholders from Louisville, Kentucky, insist that I accompany them on their annual pilgrimage to the beautiful estuary that is the Chesapeake Bay. These fellows own shares in Assault Landing, a young stallion whose first crop performed well at the track as two-year-olds last fall.

"Why don't you raise his stud fee?" they asked.

Country Life primarily serves the five-state region of Maryland, Virginia, New Jersey, Pennsylvania, and West Virginia. One hundred and seventy-eight Thoroughbred stallions stand in Maryland. That's a lot of stallions for a little state. The key to attracting mares is to provide value to their owners. So I think we'll leave the stud fee where it is, send the mare owners a good message.

Long answers to short questions from goose hunters prevents them from resorting to their repertoire of dumb jokes. I think that is why they invite me.

Sunday, January 8. Phoned clients to firm up bidding requests for the January mixed sale in Kentucky. We bred two hundred and fifty mares here last spring, and a half dozen are being sold in foal. Kentucky buyers won't be interested because of the regional appeal of our sires.

Marylanders ship mares to Kentucky to get a better price from other Marylanders. There is a hole in the local auction scene you could drive a gooseneck through.

Monday, January 9. As requests were made today for this spring's bookings, I paused to reflect on the necessary paperwork; to wit, the Stallion Service Contract, the Mare Information Form, the Breeding Shed Form, the Health Requirement Form, the Boarding Policy Form, plus cover letters and brochures and return envelopes. Some days it seems the paperwork rolls on *ad infinitum*, and that the horse business is not the simple game among sportsmen I naively imagine it was in Grandfather's time.

Tuesday, January 10. Ten million Americans served in the Armed Forces in World War II, but only one million saw combat action. The other nine million, I think, must have been in the U. S. Cavalry Remount Service at Front Royal, Virginia, with Dad and Uncle John. Remount horsemen are legion. Claiborne Farm's longtime resident veterinarian Col. Floyd Sager taught herd health to these boys. Finney was in the auctioneer's stand, culling out thousands of horses when the Cavalry was disbanded in 1943. A number of famous Thoroughbred colts served as sires in the Remount. In the office of Gen. Harry Disston in Charlottesville, Virginia, I once saw a photograph of five thousand cavalrymen on horseback, on parade at Fort Riley, Kansas. It is a most impressive scene.

"We know a little something about horses," Gen. Disston said.

The old Remount boys often speak of the cyclical nature of the horse business. I pray they are right.

Wednesday, January 11. A mare shipped in this morning with no warning of impending arrival. At six-thirty, with the temperature at fifteen degrees, my younger sister Alice, home from a teaching assignment in Australia, got out of bed, signed the van driver's Bill of Lading, unloaded the mare, brought the health papers to the office, and left a note instructing me to find the latest arrival watered off and bedded down in the isolation barn.

When you run a family restaurant, it helps when everyone can cook.

Thursday, January 12. Spoke today with R. Richards Rolapp of the American Horse Council regarding racing partnerships.

"Try to get folks to the winner's circle," Rich advised. "You'll be amazed at how excited they'll be, and it will give them an understanding of just how difficult it was to get there."

How many people have come in and gone out of this game over the years, without much success at reaching the winner's circle regularly? The winner's circle is an elusive ring.

Friday, January 13. A veterinarian from Pennsylvania, Dr. John R. S. Fisher, visited this morning. He previously bred his mares in Kentucky, but with stud fees among the lower-echelon stallions reaching parity, he decided to forgo the shipping expense, and instead try Maryland stallions.

We had given the stallions quick baths on this warm afternoon, and outfitted them with new halters. It was a polished presentation. Dr. Fisher appreciated our efforts and booked a mare. After he left, I felt as I used to feel in law school, when I had prepared for class, luckily given the correct answers, then gone out on a warm afternoon to play basketball. A cycle of preparation, a test, then relaxation. That's the way to beat a Friday the 13th.

Saturday, January 14. To Laurel this afternoon to see Mosquito Coast run fourth. She is by Assault Landing out of Trek, a mare bred by the late Stuart S. Janney, Jr., who was a Marine in World War II. He must have assaulted some mosquito coasts on his trek to Japan. Besides, *Mosquito Coast* was a good book and movie.

Naming horses is like solving the Sunday crossword, always a fresh challenge to the imagination.

Sunday, January 15. As we fed the yearlings this morning, from deep in the thick fog we heard the single, lone call of a Canada goose. Suddenly the yearlings snorted and disappeared into the mist, as a flock of thirty geese appeared less than twenty-five feet overhead. They flew in perfect V formation, low under the fog, making no sound save for the whoosh whoosh whoosh of wings propelling fifteen-pound birds through the air. In stealthy fashion, they passed through my Sunday morning.

It is often spectacular to work outdoors, even on this little one hundred-acre farm in suburbia, five miles from Interstate-95, two hundred yards from a Chevrolet dealership. We work outside so we can see the geese, or watch in awe as the Great Blue Heron lifts off the pond with prehistoric grace, rising clumsily to a height where air currents catch him and carry him away with dispatch, down the wooded ravine where the river flows through the farm.

Monday, January 16. "Unemployment in Harford County is only five percent," the Maryland Job Service officer in Bel Air informed me. "We consider this to be *zero unemployment*, since five percent of the population is chronically unable to hold a job."

A labor force of hardworking country boys is a vanishing breed, most local farms having been sold for development. We are heading into the labor intensive breeding season with only the chronically unemployable comprising the talent pool, so I commit myself further to the "breed and return" *modus operandi*, utilizing family farms in adjoining counties to house some mares we expect to breed this spring. Fortunately, we border Baltimore County, whose one hundred and twenty-five Thoroughbred farms are more than any other county in America except those around Lexington, or around Ocala, Florida. We must play to our strengths to survive.

5

Tuesday, January 17. David Harp, a photographer, visited the farm today with an idea to capture the life of a Thoroughbred from birth to the race track. I told Dave *Sports Illustrated* set the standard some thirty years ago, when it followed Iron Liege from the foaling stall at Calumet to the winner's circle of the Kentucky Derby.

"Well, pick the one that wins the Preakness," he quipped.

I suggested we photograph two different mares as they foaled, and we'd pray for a winner. Then I explained some aspects of broodmares to him. Barren mares under lights, for example. Like chickens laying eggs. Photo sensors in their eyes, the lights in the barn advancing the day by half an hour per week, until by February 15th, it will be 11 P.M. before the simulated day is done. Frankly, though, I wished Mother Nature did not have to be so coerced. Foals born in January have meconium balls you can putt; foals born in April benefit from the laxative effect of the green grass the mares have eaten.

The photographer vowed to be prompt when alerted at two in the morning that a mare is foaling. We'll see about that.

Wednesday, January 18. Dr. William E. Riddle, Jr., our farm veterinarian, drew blood from the stallions for their annual Coggins test. The stallions do not like Dr. Riddle. Allen's Prospect cowers when he hears his voice. Bill Riddle to them is a pain in the neck, or a tube down the nose. The stallions are petrified of even the tiniest syringe. Meanwhile, the list of vaccinations grows annually: Influenza, Rhinopneumonitis, Tetanus, Rabies, Potomac Fever, East-West Encephalomyelitis, Streptococcus Equi, Botulism. The studs receive these shots plus annual Equine Viral Arteritis boosters. It is easy to see why they act like such children when Dr. Riddle visits.

Thursday, January 19. To Laurel today to see Hal C. B. Clagett's Bullet Assault break her maiden. In the paddock, Mr. Clagett informed me that the filly's dam, Golden Bullet, died of a twisted intestine several days ago. I figured Bullet Assault would romp today. That's the way the horse business works. In the starting gate, however, the filly next to her flipped over and broke the legs of the jockey and of the assistant starter. Laurel broadcasts the starter's orders over the loudspeakers to enhance the action for the fans, like sideline microphones. Yet we felt like helpless eavesdroppers as we listened to metal banging, jockeys screaming, starters shouting, horses hollering.

One jockey dismounted to help the fallen rider. Soon all the jockeys were off. Their weight on their mounts during the twenty-minute delay might be a disadvantage at the finish line six furlongs away. Betting windows stayed open and wagers were second-guessed, and lead ponies walked with the fillies while jockeys congregated. A lot happened while nothing was happening.

Driving home afterwards, I recalled what a hunter once told me about dangerous game: "Some days you get the bear. Some days the bear gets you." Mr. Clagett and Bullet Assault got the bear today, but a jockey and a starter in a Laurel hospital tonight might feel as if the bear got them.

Friday, January 20. Wonderful rituals are the fortnightly visits from Ben Campanella, a shareholder in the Northern Dancer stallion Carnivalay. He elevates immigrant entrepreneurial instincts to a new high, and runs a handful of Philadelphia businesses—a flower mart, a meat shop, a sandwich deli. When he dons Mom's apron and prepares Philly cheesesteak subs for the farm crew, it is a morale booster surpassed only by a winner.

"I got rid of all my horses," he boasted as he carried bags of bread, onions, olive oil, and sliced steak into the house. I looked suspiciously at the steak, but he assured me it was no horse I knew. We had nagged him to dispose of assorted off-bred beasts ever since he walked into our kitchen five years ago. The market finally shook him to his senses.

Saturday, January 21. Full moon tonight, and I might never see the surrounding farmland appear so beautiful again, especially the farm across Route 1. Spring promises the relentless approach of development. We will not come up for air to observe the changes until August 1st, when the last of the breeding season mares have returned to their year-round homes.

As we walked along Winters Run, the river that forms a natural riparian boundary for our farm, I told my wife Ellen about the county's plan for a public walkway along this creek—a "greenway," they call it. I imagined graffiti on the boulders, beer cans on the bank. A few delinquents can ruin a park for the many who might enjoy it, and so we prepare to fight another battle on another front.

For a few minutes in the moonlight, the walk was no fun. But we shook off fears of the future and enjoyed the here and now, watching the delicate ice floes gliding silently on the creek as the temperature dropped below freezing.

Sunday, January 22. Anticipation of a quiet Super Bowl Sunday was broken by the sound of metal crunching. The new girl drove the club cab pickup into the side of the hay barn. Any person with a driver's license should know how to pull into a sixteen-foot doorway, but the long wheelbase surprised her. Now I have to explain to the Monday crew why the Sunday crew dinged up the new truck.

When the truck hit the barn today, I knew it was not serious. Yet it is a symptom of the greater problem of finding capable farm help.

Monday, January 23. Our weather is mid-Atlantic moderate, but South Carolina must be downright balmy, judging by the short coat on today's arriving mare. Our barns are full of windows almost always open, allowing air to move freely. But some of the ship-in mares are accustomed to a hothouse climate, where barn doors are pulled tight and windows latched all winter. The openside barns of Kentucky point out the value of fresh air for horses. Nevertheless, we must allow nature a grace period to grow a coat on the new mare from South Carolina.

Tuesday, January 24. It is morning on the *Ponserosa*, and the following scenes occur:

Elmotion, a homebred filly who strained her flexor tendon in September, worked a half mile on Saturday, and my brother Andrew, who trains her, says the tendon feels firm, as he knocks on wood. In the yard, bales of timothy and alfalfa flank bright yellow straw as my brother Michael, the farm's business manager, speaks to a hay broker. Sister Norah, who keeps the books, is on the phone, consoling a client whose wife has been very ill. Marva, Andrew's wife, types a cover letter to a new partner in a racehorse, explaining that fingerprinting is a prerequisite to being licensed as an owner.

Dad brings in today's *Daily Racing Form.* No babies by our sires entered, he reports. Mom calls from the *Maryland Horse*, where she is a pseudo Kelly Girl, proofreading the February stallion register. Peter Richards, our broodmare manager and the son of legendary English jockey Sir Gordon Richards, leads two pregnant mares to the foaling barn. I phone the printer doing the farm brochures: "They'll be done by Friday," he says. Right. Dr. Riddle stops in the office and says the colt which underwent the "tieback" operation will be arthroscoped later this week, and might return to training by February 1st.

Meanwhile, relentlessly, the breeding season approaches.

Wednesday, January 25. "I go out of my way to drive past your farm and see the foals. That's what they're called, aren't they? *Foals?*"

Our district councilwoman is sincere when she describes her fondness for the bucolic scenery we provide. It is a sentiment I hear frequently, and it will be an asset as the public hearings on rezoning begin in February. I tell her we had our first foal last night, named it Salvador because the painter Salvador Dali died yesterday, and I suggest she come see the foals.

"Yes, then I can see for myself what kind of impact a mall might have. I'm free next week. I'll bring my rubber boots."

The farm's ledger sheet does not reflect profits from time spent protecting the quality of our life. Yet the satisfaction from energizing politicians to see *our* side of the fence is a rich reward.

Thursday, January 26. Drove three hours to New York City to meet with a client for whom we dispersed fourteen horses in the December sales. I gave him a check for eleven thousand dollars, less than a grand per horse, less than break even. He said, "Well, at least I don't have to feed them anymore."

He gave me directions from midtown New York to New England. Then I drove around Harlem for an hour, looking for the Bruckner Expressway, wondering if I would end up like the characters in Tom Wolfe's *Bonfire of the Vanities*. Finally made it to Providence, Rhode Island, by dinnertime. The couple I was visiting breed mares to our stallions each year. Last spring, Ellen filmed their favorite mare foaling, dubbed in the music of violinist Stephane Grappelli, and presented them with foaling as an art form. These days, anything to enhance the experience of owning horses is worth the effort.

Friday, January 27. To Suffolk Downs today to see a filly who was foaled at the farm, with the mare standing up. We cradled that soaking-wet foal to the stall floor, covering ourselves in afterbirth in the process. We foal about fifty mares each spring, yet I quite clearly recall this filly's birth five years ago. It gave me a sense of satisfaction to enter the forty-stall barn on Suffolk's backstretch and, with no identification on the doors, find her eating hay in a distant stall. Buster's Girl is an old friend. With luck, she'll come back to the farm to be bred someday.

Saturday, January 28. Back home by 4 P.M., greeted by the farm crew finishing up chores in warm winter sunshine. The new foal, Salvador, is leading well now, the result of our putting a halter on a foal the first day. Foals which are allowed to simply follow alongside their mothers soon become rapscallions. Two new foaling mares are under the closed-circuit camera, and the vigil has begun in earnest, not to abate until the last foal is delivered around the first of June.

Sunday, January 29. Had our second foal tonight, after a birthday party for Ellen. The mare stood up immediately after foaling, tearing the afterbirth from the foal but leaving the umbilical cord attached. I hate when this happens. The umbilical cord is the first avenue of entry for bacteria, and I much prefer when it snaps naturally. Now I had to cut the cord a few inches from the stomach. I hope the remaining tissue sloughs off without incident. The alternative was to pull on the cord, but that can tear the foal's stomach lining and result in a hernia, or create a "patent urachus," causing the foal to urinate through his navel. We applied iodine to the navel, and on the foaling report asked Dr. Riddle to pay particular attention to that area when he examines the foal tomorrow.

Monday, January 30. Dr. Riddle pronounced the tieback operation a success, and said the black colt was ready to return to training. This colt is a chip off the old block, a carbon copy of his sire Assault Landing, and even with an impeded windpipe, he ran second in his only start at two.

We have high hopes for him as a three-year-old, but we have been humbled before by unfulfilled expectations of success on the race track.

Tuesday, January 31. Sixty degrees today. The fields are trying to push up grass as if it were April. Salvador is turned out in a four-acre field and appears well pleased. Our two mares both foaled several days early, and it seems Mother Nature knows best when to pop the cork. We will be in high gear in just a few days. The stage is set. The breeding season beckons.

SIGHTS AND SOUNDS

Wednesday, February 1. "We're in trouble here. I've got a head but no feet."

It is 2 A.M. The night girl calls for help, and forces the mare to walk. Nine out of ten mares foal with no intervention required. This is that tenth mare: She is trying to deliver a foal whose front tendons are mildly deformed, contracted. Instead of arriving in the classic diving position, head tucked between outstretched front legs, this foal is coming headfirst. And *only* his head has cleared the mare's pelvis. His front legs and the rest of his body are lodged behind her pelvis, locked in her uterus. His legs are folded up like undescended landing gear. This foal is a victim of intrauterine crowding in the late stages of gestation. The hydraulic thrusts of labor are working against him. We are all in trouble. It's dystocia time at the foaling barn.

A directory of veterinarians is posted in the foaling barn. We dial for help. Then we prepare to search for the foal's legs in the mare's uterus. Two principal factors take priority: The cleanliness of our intervention, as fertility problems are common in mares after difficult deliveries; and ample lubrication of the birth canal. Any lubricant is better than none. Humphrey Finney used olive oil back in the nineteen-thirties. K-Y Jelly is at the breeding shed. Mom's vegetable oil is in the kitchen closet.

Last year, we led a mare to a sloping hillside, stood her facing downhill, and gravity helped pull her contracted foal back into the uterus, making manipulation of the foal's legs easier. This year, we added a block and tackle to the foaling stall, to hoist the mare's hindquarters if necessary.

Tonight, Dr. Peyton Jones took it one leg at a time, starting at the foal's knee, working down to the fetlock, then finding the toe and cupping it gently over the mare's cervical area to minimize trauma to her reproductive tract. We got lucky. We got him out. Now comes the hard part: Keeping this rascal alive.

Thursday, February 2. The three Harford County Council members arrived at ten-thirty this morning to see our panoramic view of the proposed mall site. It was as if we were taking the jury to the scene of the crime. Off we went in the club cab, four-wheeling across the beautiful high ground we hold on the area topography.

Someone said politics consists of choosing between the disastrous and the unpalatable. A mall would be a disaster, we all agreed after studying the site. Yet I don't think I can swallow *any* plan that would destroy the rural beauty of the terraced cornfields, standing like sentinels above the flood plain through which meanders Winters Run.

"What are the alternatives?" they asked.

Can't plant corn on such expensive property, so I suggested the county should issue bonds, buy the land, and twice a year invite the entire town to picnic at Bel Air's version of the Camptown Races—a hunt meet, of sorts. Use a modest admission fee to pay down the bonds. This idea struck a chord with older council members, who remember the Bel Air race track which operated from 1937 through 1961 two miles away. The Harford Mall now occupies the former race grounds, a rusted water tower standing anachronistically among the trees that survived the bulldozers.

The Bel Air track formed a link in the fair circuit of half-mile tracks that once dotted the Maryland countryside: *Cumberland*, a poor man's Santa Anita nestled in the mountains of western Maryland; *Marlboro*, site of a famous match race in 1768 (Selim was defeated by Figure); *Hagerstown*, where races had been held since the early eighteen-eighties; *Timonium*, the sole survivor, still offering ten days of racing during the Maryland State Fair.

Grandfather settled on a very accessible farm—ten miles to a train station, twenty miles to a major city, Baltimore. In the nineteen-sixties, I-95 rolled through, four miles away as the crow flies. There followed the developers, who now outflank us. Our accessibility is both a blessing and a curse. We are too close to hired guns like the Arizona fellow. His plan would take the country out of the land surrounding Country Life.

Friday, February 3. Ellen nicknamed the contracted foal "Rollaids," because his life is a roller coaster, and because we dose him regularly with stomach antacid in defense against ulcers. Dr. Russ Jacobson ministers to his health, while we act as physical therapists, gently stretching his pliable tendons to ameliorate the curvature of his front legs. Rollaids cannot stand alone, though he tries. The exertion exhausts him. To add to his woes, the stress of labor caused his dam to prematurely lactate her vital colostrum. From the freezer, we removed a plastic bag containing sixteen ounces of colostrum stripped from a mare who foaled last year. Thawed slowly under warm water to preserve the sensitive immunoglobulins, the colostrum was tubed into Rollaids before he was twelve hours old.

The next hurdles our young friend will face are likely to be diarrhea and dehydration, but his will to live has endeared him to all.

Saturday, February 4. At nine forty-five this morning, not one person had shown for the first-ever organizational meeting of our unnamed group opposed to the development project. I was crestfallen. By ten after ten, however, forty cars lined the driveway.

"We've never been back here before," was the consensus. "All we've ever seen of the farm is the hills and the foals playing. By the way, what shall we call ourselves?"

From the back of the standing-room-only crowd came the suggestion: "The Winters Run Preservation Association." All in favor said "Aye."

How can we fit our name onto a bumper sticker as we campaign against the mall?

Sunday, February 5. Rollaids' breeder owns a Chevrolet dealership. He boards horses with us. We buy pickups from him. He grew up playing football in the same Bronx neighborhood as Baltimore Colts great Artie Donovan. He knows the value of sportsmanship. I discovered this last year, when I sent this telegram to him in Florida: "Unable to reach you by phone. Bad foaling last night. Mare laid down with foal's legs already out. Broke foal's right front leg. New Bolton vets consulted. Foal destroyed. Please call."

There are a thousand ways to lose a foal, but we were within two minutes of the end of an eleven-month wait. He sent us a second mare, and she produced Rollaids. This gentleman beat cancer twenty years ago. He thinks Rollaids is cut from the same cloth. Maybe he's right. Rollaids is persevering. We put fluids in one end, they come out the other. But the colt is still strong and is responding well to the frequent massaging of his limbs. We may yet put splints of plastic pipe on his crooked legs, but for the moment, he is stressed enough.

Monday, February 6. Grandfather bought this farm in 1933. Three years later, he wrote the following letter to Mr. Harry Scott of Faraway Farm near Lexington.

"Mr. Sam Riddle has been good enough to let me have a complimentary to Man o' War for the season of 1936, which he had promised me while Man o' War was a three-year-old but I never took advantage of until I felt I had a mare worthy of such a horse. As you probably know, I was the one who called Mr. Riddle's attention to the Belmont yearlings, which included Man o' War. I told Mr. Riddle I would send the mare Sunny Ann. This mare is now in foal to Flying Ebony. The produce resulting from breeding this mare to Man o' War, or a Fair Play horse, would be about the same bloodline as Chance Sun, winner of last year's Futurity."

I often reread Grandfather's letters. They tide me over difficult moments, such as when I had to call my Chevrolet dealer, tell him I am zero-for-two in producing healthy foals from his mares.

Tuesday, February 7. In our equine lineup, teaser stallions bat one-two-three before the real studs take a swing at the mares. At the breeding shed reigns a fourteen-year-old Thoroughbred named Dew Burns, who has the manners of a prince but the libido of Henry VIII. His stand-in is Popeye, who did his racing under the name Power of Grandeur—a bit too pretentious for a teaser. A gentler horse than Dew, Popeye serves as a courtly introduction to maidens suspicious of overt male attention.

Finally, there is Jazzman, a rover whose antics on the trail-riding circuit are legendary. He is a shade over fourteen hands. His coat is a burnished liver chestnut. He is my little Man o' War, my "field general." This stocky Quarter Horse never met a mare he didn't like, which is why his former owner took him off the Sunday circuit. Each morning during breeding season, Jazzman is led directly into the fields of mares. New mares soon learn that Jazzman is available. Safe and unthreatened in their fields, mares often show better heat there than they do in a teasing chute. Nor are they generally concerned for their foals. New mothers frequently stand safely in the distance, displaying signs of estrus. Still, an irascible minority of mares simply do not show heat, and they require close veterinary palpation to detect estrus.

Under any circumstances, the process of teasing mares—particularly field-teasing—is inherently hazardous, not recommended for the uninitiated, and *definitely* not for the slow of foot.

Wednesday, February 8. From Kentucky this morning arrived thirteen "dry" mares from Allen Paulson's Brookside Farms, to be bred to Allen's Prospect and Corridor Key. Off the Sallee Vans pranced sleek maidens entering the broodmare phase of their careers. We let them settle into their stalls today

"Rolaids"

after spending twelve hours on a van, and we'll turn them out tomorrow in small paddocks until they become acclimated. Brookside manager Ted Carr has had these mares under lights since the first of December. If our weather and the stallions cooperate, a passel of Maryland-sired babies will hit the ground next January and February in Kentucky.

Thursday, February 9. From upstate New York this morning arrived a maiden mare booked to Assault Landing. Her arrival information was impeccable. The Certificate of Veterinary Inspection necessary for interstate travel bore today's date. A copy of The Jockey Club registration papers supplied us with her lip-tattoo number and date of birth, and indicated she had won a maiden race at Aqueduct. (We like to know if mares have raced, which means they have been broken to the saddle. Two of our stallions will bite mares' necks for purchase while breeding, requiring us to sling a heavy leather breeding pad over the mares' withers. Cinching the girth on a mare unbroken to a saddle can result in a violent surprise.)

A complete vaccination report was included, together with uterine culture results from her last cycle. A hand-written note stapled to her teasing chart explained her activity during her current cycle. A Coggins test and a negative Equine Viral Arteritis test completed the full history on the mare. Her Stallion Service Contract was already signed and on file in our office.

Tonight, a fellow in upstate New York, name of Jack Conner, is taking a great deal of pride in the professional presentation of his mare for breeding.

Friday, February 10. We handed out brochures this morning at the Middleburg Training Center in Virginia in one last flurry of salesmanship before being planted on the ranch for the breeding season. The emphasis on racing in today's market has thrown trainers into the spotlight as advisors to mare owners. If a trainer has success with foals by a particular sire, he continues to buy yearlings by that sire, and he tells his clients to support that stud.

We are swimming upstream on this score. Only one of our four stallions has foals to have raced, so trainers have no line on seventy-five percent of our roster. It is easy to be overlooked in this saturated market. The time on the road is a necessary aspect of marketing.

Saturday, February 11. "Foals who overcome adversity make good race-horses," optimistically stated Rollaids' breeder upon seeing our Poster Child. The foal is developing a million-dollar body on March-of-Dimes legs, but his heart is gold, pure gold. Off all medication now, Rollaids is benefiting from the therapy provided by the night girl, who braces her knee against his knee while he is resting, grasps his toe and his forearm, and firmly flexes his legs against their bowed shape.

Rollaids is becoming spoiled with attention, though, as if he were an orphan foal. Hard to resist at this age, he will be hard to restrain later in life.

Sunday, February 12. The Maryland race tracks in my formative years were decrepit. Apathetic track management cared little for the appearance of the facilities, nor for the comfort of the bettor. I thought all race tracks were run like this until, at sixteen years of age, I saw Belmont Park and Saratoga, at eighteen Keeneland, at twenty-four Hialeah. Now, a modern Maryland racing scene is coming of age. At Laurel this afternoon, the Barbara Fritchie Handicap was contested by the best sprint fillies in the nation. Maryland's top trainers were matched against out-of-town peers. In the Sports Palace, latter-day Diamond Jim Bradys ran the handle up while keeping an eye on the big-screen basketball games. Cashiers politely issued wagering tickets, and the owner of the race track escorted an elderly lady to a reserved box.

I feel fortunate that the progeny of our stallions is coming on line during a resurgent period in the long history of Maryland racing.

Monday, February 13. The breeding season does not officially open until Wednesday, but I did not see a game warden around, and I let Assault Landing take a shot at a maiden mare. The first cover of the season for the stallion is usually a macho adventure—seven months of pent-up testosterone building into a Tarzan call. We will know in fifteen days if we hit the mark, when we ultrasound today's opening-day mare.

Tuesday, February 14. A new farm sign arrived today.
ATTENTION VAN DRIVERS:
1) Please Close Gate After Entering *(real horsemen among the drivers will be sure to close the gate behind them);*
2) Use Phone In Barn To Call Office *(is it a breed-and-return van, an arrival to stay through a heat period, or the UPS guy at the wrong address?);*
3) Absolutely No Loading Or Unloading Without Farm Personnel Present *(careless or exhausted drivers will only use the short board ramp, with no wings on it, instead of the safer, heavy steel ramp).*
This new sign quietly demands compliance with policies I learned the hard way.

Wednesday, February 15. On the tack box outside the foaling stall, Ellen affixed a Dry-Erase Board. The message she wrote—in plain sight of curious employees who otherwise might have approached last night's newborn—prevented an accident: "CAREFUL! This mare is extremely protective of her new foal. Please DO NOT enter her stall!"

It took us twenty patient minutes to catch the mare. A green employee might have been our first workman's compensation claim of the breeding season. Communication, in all forms, makes this little horse factory run. (Darn, I spoke too soon. At feeding time, a yearling kicked the new man in the knee. Probable length of disability? "Uncertain," Marva pencilled on the claim form.)

Thursday, February 16. Our stallion brochure lay open on a table in an Italian restaurant. The restaurant owner, dressed in a finely tailored suit and silk tie, noticed the glossy photos of horses. He sat down.

"I used to own racehorses," he said. "I got out of the business. Their shoes cost more than mine."

I was certain he was going to say "never invest in anything that eats," but instead he gave me an excuse I had never considered.

Friday, February 17. *Arson.* The mere word brings the white heat of adrenaline to a farmer's face.

"I'll hear that siren in my sleep for years, the way it roared down the sandy road from the barn to the highway," Mr. Clagett told me. "This was no accident. The barn didn't even have electricity in it. Just five thousand bales of hay under a roof in the dead of winter."

Despite the senseless crime just committed against his property, he phoned to report on the outstanding foal he delivered at 5 A.M.

"Although I am fatigued, I am still roused with the thrill of having watched this beautiful colt raise up his hind end, stagger from his knees to his front feet, and take those first precious drinks of mother's milk."

Mr. Clagett did business with Grandfather. He has bred *hundreds* of foals, and although he may be fresh out of hay, he has a barnful of reasons to persevere.

Saturday, February 18. Dr. Riddle believes that Thoroughbreds spend their entire lives trying to commit suicide. I spent today with a foal who supported that belief. I am not by nature a cynic, but I feel tonight like H. L. Mencken when he said the sweet smell of flowers reminded him of funerals.

Sunday, February 19. For four consecutive nights last week, the Baltimore *Evening Sun* gave front-page coverage to the sad fate of many Thoroughbreds sold at auction. The gist of the lengthy series is that horsepeople ought to be ashamed to sell their once-pampered Thoroughbreds for slaughter. The articles raised the consciousness of horsemen and laymen alike. Today, our office received a chilling call from a lady who refused to identify herself. She called us killers, said we should suffer the same fate as those poor mares. Then she hung up.

Since we are the lone entry in the Baltimore *Yellow Pages* under Horse Farms, our phone number is public record to outraged readers. Marva looked assaulted when that lady hung up on her. Marva had no opportunity to mention our success in finding good homes for the dozen mares we gave away, mares which boarded here just last spring. As for the mares we've sold at auction? Well, we act as agent for some stubborn mare owners, fellows who refuse to believe that broodmares are perhaps only worth the killer's opening bid—until those girls are appraised objectively through sale at public auction.

It disturbs the quiet of this peaceful Sunday to reflect on the many issues, pro and con, raised by Ross Peddicord's well-researched series so aptly entitled *Last Ride.*

Monday, February 20. Andrew did a masterful job nursing Elmotion back to the winner's circle this afternoon, and when Dad phoned from the track, the entire farm crew shared vicariously in the success. Her saga has been a character-building experience: Broken pastern bone at two, Maryland Million starter at three, valiant return from injury at four.

The day finished strong when the stallions put good covers on two tough maiden mares. High-fives for the breeding shed crew, hot mashes for the studs. We'll all sleep good tonight—unless the mare from South Carolina foals.

Tuesday, February 21. Driving rain all day today, the weather coming from Kentucky, where eight inches fell the other day. The fields take a terrible beating in this weather, saturated topsoil offering no resistance to fifty mares. Yet nature's resiliency will win out by April, when these same fields will be lush with bluegrass.

The mare from South Carolina foaled last night. We were concerned with the transfer of immunities even though the foal nursed well soon after birth. The mare had waxed for several days prior to foaling, milked heavily during labor, and was producing her first foal—red flags for failure of passive transfer. We were relieved when Dr. Jacobson phoned with good results from the foal's blood tests, taken within twelve hours of birth. In a world teeming with pathogens, no better gauge exists for evaluating a newborn's chances than to test his blood, quickly.

Wednesday, February 22. Working through the pouring rain, all we did today was move manure, shake straw, and haul hay. But thirty foaling mares are in their last trimester of pregnancy, and they need clean, dry bedding in this cold, damp weather. Standing in ankle-deep mud as the mares were brought in, the crew resembled the worn-out rangers of the *Lonesome Dove* cattle drive.

As if to reward our efforts, one of the heavy mares was kind enough to start foaling at the civilized hour of 5 P.M. Within minutes, a healthy colt by Allen's Prospect whinnied from his bed of yellow straw in the foaling stall. Made today all worthwhile.

Thursday, February 23. A barren mare with a troubled reproductive history showed heat today. Peyton recommended breeding her.

"It's her first heat period," Peter Richards informed Dr. Jones. "She was clean in November, but we haven't had a chance to culture her yet this spring."

We hate to miss barren or maiden mares early in the season. They will haunt you in April and May when foaling mares need the stallions. But our policy requires a clean culture on such mares prior to breeding. A quick screening test was necessary. Peyton pulled a microscope from his jeep and suggested taking a cytology smear to detect white blood cells. Although not as conclusive as full culture results, the clean smear assured us of negligible risk of infecting the stallion.

It's *Catch-22*: Wait forty-eight hours for a complete culture and miss this heat cycle, or stain the cells from the uterus and make the decision on the spot.

I figure there has only been one immaculate conception, and that was two thousand years ago. We went ahead and bred the mare.

Friday, February 24. The breeder of last night's filly is resting in a hospital bed in Rhode Island, teaching his left hand how to put pegs in a board. His interest in horses defies the effects of the stroke he suffered two weeks ago.

"You had a beautiful filly last night," I told Tom Baylis over the phone. "The first baby by Corridor Key we've foaled. Everyone's excited. She's straight and tall and Ellen has already turned her out for some exercise."

"That's very good news," he said happily. He told me that Buster's Girl won again on Monday. He asked about his two-year-olds in training in Virginia, and reminded me to submit the name Read My Lips for one of his yearlings. I quickly checked *Registered Thoroughbred Names*: It's taken, I told him. They say pets are great therapy for depression, but raising your own racehorses has no equal for recuperative powers. Fellows who own two-year-olds have *this* world at their feet, and are not anxious to enter the next.

Saturday, February 25. Standing with a mare in the veterinary stocks while a Caslick operation is performed is *almost* as boring as holding mares for the blacksmith. Today, however, a barren mare, bred off the van last spring, arrived back here at the farm. She might have produced a foal in Corridor Key's first crop had her vulva been sutured last summer. Fecal material falling into the vulval aperture caused a bacterial infection, killing the developing fetus. Our half of the bilateral Stallion Service Contract had been completed—namely, we had gotten the mare in foal. Even more frustrating, though, our stallion lost a potential winner from his first crop, in a game where numbers are everything. It sustains effort to believe that every foal might be another Carry Back, conceived here in 1957, winner of two legs of the Triple Crown, a millionaire, a Hall of Famer.

Well, no merit in fixing the blame. Fix the problem instead. After all, it's easy to check under the hood in February, quite another in the dog days of July, in the exhausting conclusion of a breeding season. So we stood in the cold and sutured the mare.

Sunday, February 26. Dozens of Canada geese tumble out of the sky like giant oak leaves, pitching into the farm pond. Three young deer feed alongside the barren mares, then ridicule gravity, clearing five-foot fencing en route to cover. A maiden mare succumbs to the stallion. A foal rocks on legs like stilts. A Beagle barks. These are the sights and sounds of a snowy Sunday morning.

Monday, February 27. No better advertisement for stallions exists than to have fields full of their offspring. (No one ever says anything bad about foals, anyway; it is akin to telling a proud father his children are unattractive.) A safari train, as at the zoo, would be most appropriate for our guided tour:

"To your left, please notice the yearlings. The black ones are by Assault Landing, the bays by Allen's Prospect, the blaze faces by Carnivalay. On your right are foals born this spring. The chestnut mare has Corridor Key's first foal at her . . .'ROLLAIDS, GET OFF THE TRACK!'...Excuse me. Straight ahead are the maiden mares from Kentucky. We're entering the stallion complex. Please keep your hands inside the car."

Wishful thinking. Instead, we trudge through wet fields, and prospective breeders never fail to wear casual shoes on the muddiest days of winter.

Tuesday, February 28. All eyes are focused on the gray ultrasound screen as Dr. Riddle probes over the mare's uterus for the telltale black dot indicating the developing embryo. The mare strains against the veterinarian's arm in her rectum. Dr. Riddle is patient, joshing with the new employees who are curious about the ultrasound procedure. He waits for the mare to relax.

If not for the detection of twins, and the remedy of pinching, ultrasounding mares at fifteen days past breeding would not be such common practice. Many old-time horsemen feel that the benefit of ultrasounding is not worth the intrusion at this delicate stage of pregnancy.

Today, the screen stayed gray. My guess is that Assault Landing's first cover of the season was more bark than bite. Jazzman will tease this gal back into heat, and a few weeks from now, Dr. Riddle will perform another ultrasound to see if she has too many biscuits in the oven. We head for March with our work cut out for us.

AN UNPREDICTABLE JOURNEY

Wednesday, March 1. We go roaring into March with a barnful of cycling mares, as the lighting program works its magic. On the books for this morning was the good stakes mare Alden's Ambition—a filly of distinction, as Mr. Clagett says. That certainly is true, judging by our efforts to impregnate her. Heretofore, she refused to enter the veterinary stocks, fought the twitch, sat down on Dr. Riddle's arm while being palpated, resisted the restraint of a leg strap, and blasted Jazzman when my gallant teaser tried to mount her.

Today, though, she progressed with a soft follicle demanding fertilization. Carrots were the way to her heart, as she acquiesced to the twitch. Next, she permitted us to take her right front leg (her near front ankle being fused with calcium from racing). Finally, she succumbed to Jazzman's gentle nuzzling, and Corridor Key wasted no time. The thought of the foal resulting from today's mating sent us marching away on a very good note, indeed.

Thursday, March 2. Rollaids looks as if he could step into a boxing ring with the heavyweight champion, his blue bandages neatly covering his contracted legs. A bevy of stable girls follows Dr. Riddle's every instruction: Vigorously massaging Furacin into Rollaids' tendons, applying a thin layer of cotton, followed by a layer of plastic wrap, then more bandage cotton, and finally the fancy blue leg-wraps. Heat generated from the Furacin keeps the tendons pliable, while the bandaging provides support.

Rollaids is turned out only four hours each morning, so his legs gain strength without stress. Still, I half expect this animated little clown to appear one morning, his name promotionally stitched onto his leg-wraps, shadow boxing, safe in his opinion of himself as the heavyweight champ of this year's foal crop.

Friday, March 3. A homebred filly was claimed today. So, what's the fuss? Horses are for sale in a claiming race. That's the deal. But this was

Elmotion, a runner who filled us with admiration. Moreover, the filly's new owners didn't want Owen, the goat who had accompanied her throughout her racing career. Andrew acquired Owen to calm the stall-walking filly, tethered him to her hay rack, then tolerated the goat's territorial ramming of persons entering her stall. I imagine the scene tonight—Elmotion calling in vain for Owen, pacing her stall, aggravating her old racing injury. Meanwhile, her new trainer can't leave the barn, curses his bad luck with claims as he examines her bowed tendon, and vows to return the nervous hussy to her former owners at the first light of day.

They say horses are like streetcars: Stand on the corner long enough and another one will come by. Perhaps the *same* car will come back by. Better keep Owen handy, just in case.

Saturday, March 4. "Good morning, Mr. and Mrs. Members of the Committee. I'm here today to speak in favor of the off-track betting bill. Why? Because we live in an age of convenience. It's an age of drive-in food, drive-in churches, even drive-in funeral parlors. Yes sir, just pull up and ask the funeral-home guy, 'Who's laid out where?' Pass along your condolences, and away you go. I don't see any reason why betting on the ponies shouldn't be as easy, right here in Maryland."

Such compelling arguments abounded at this morning's public hearing on the satellite wagering bill in Annapolis. To our north, Pennsylvania legislators overrode their governor's veto and implemented off-track betting. To our south, the consortium driving Virginia racing will not overlook gambling dollars pouring like coal down a tipple from the Blue Ridge mountains, hours away from the proposed race track in the Tidewater area.

Preliminary plans in Maryland call for four betting parlors, mini-Sports Palaces, located throughout this tiny, congested state, with a thirty-five-mile blackout area to protect attendance at the two operating Thoroughbred tracks. Unfounded rumors portray the old Bel Air Post Office as a logical local casino. Zoning approval will run smack into the NIMBY factor (the "Not In My Back Yard" folks), but as the man said, we live in an age of drive-ins. Dad has never relished picking up the mail, but the Post Office someday might hold greater attraction for him.

Sunday, March 5. The leather volumes of Dickens, Balzac, Poe, and Conrad stood out in the darkness of Maj. Goss L. Stryker's library. He died in 1971. His widow, Marion, died last month. Their beautiful two hundred-acre farm has been bought for preservation in the Gunpowder Park System. Today, Mrs. Stryker's executrix presented me with a 1920 anthology of horse poems, a few of which Finney recounted in his *Stud Farm Diary* published in *The Blood-Horse* in 1935. Finney admired Rudyard Kipling, and a chill struck me as I read Kipling's poem *The Undertaker's Horse:*

> *I meet him oft o' mornings on the Course*
> *As I canter past the Undertaker's Horse*

the cadence of his hoof-beats
to my mind this grim reproof beats;
"Mend your pace, my friend, I'm coming.
Who's the next?"

And I am spooked the rest of the day.

Monday, March 6. Twenty-three buildings comprise the curtilage at Country Life, and tonight a horse is in most every structure. It is all because Old Man Winter awoke in a spiteful mood this Monday morning in March. The snaps on the gates were frozen, and it is somehow humiliating to be bent over, cupping hot breath against spring-loaded snaps until they thaw. A sheet of ice covered the roadways, forcing us to take the long routes to the bank barns. The walk to the foaling barn will be bobsled-slick tonight.

Tuesday, March 7. The morning "trials" (as Finney referred to teasing) brought signs of heat from two stubborn maidens. Normal progesterone test results indicated the mares' disregard for Jazzman to be temperamental rather than ovarian and, of course, they cycled on the coldest day in weeks. The stallions covered the mares lustily, and returned steaming, like boxers, waiting for the bell to ring again.

Warm water in the stallions' buckets encouraged drinking in this frigid weather, bowel impactions and colic being more frequent when exercise is curtailed and water supplies are ice-cold. A hot bran mash as a laxative is on the menu tonight as well.

Wednesday, March 8. White-hot, five hundred-watt bulbs cross-lit the foaling stall. Oblivious to motor-driven bursts from the fifteen-millimeter camera along the ceiling, the mare pushed through labor. Dave Harp kept firing the impulse to the remote camera while he crouched in the straw, readying his hand-held camera for the foal's emergence. Outside, the wind shook the tree-tops before pushing cold air through the battens of the foaling barn. Inside, steam from the mare's efforts fogged the windows and clouded the cameras' cold lenses. Heat lamps and staged lighting bathed the stall in the sterile glow of an operating room.

A foal's front feet, even when merely grasped in the birth canal, reveal the size of the body to follow. This foal had a handshake like a stevedore's, and we braced for the difficult pull through the mare's pelvis. Three years from now, when photos bookend the life of a Thoroughbred from the foaling stall to the starting gate, we will recall delivering tonight's Amazon Queen.

Thursday, March 9. It's not as much fun to lose money in the horse business as it once was. In 1979, when a rich man lost one dollar breeding horses, his tax bill decreased by as much as seventy cents. Nowadays, that same losing dollar results in a tax deduction of only twenty-eight cents—assuming, of course, that the breeder could prove he materially participated in five hundred hours of concentration in the horse business. Five hundred hours!

So, when breeders testify in favor of an OTB bill that appears to help the tracks, legislators think track owner Frank De Francis controls us like pawns. But if OTB helps the overall betting handle—that umbilical lifeline coursing through the track to the owners to the breeders—then we are all nourished. In Annapolis this week, suspicion exists that the monopoly of ownership of the Maryland tracks cuts too much in favor of one man.

Friday, March 10. Our Friday evening changed from languish to anguish with one ring of the telephone.

"I've got a placenta that looks like a basketball sticking out of this mare, but no foal. Get here quick," demanded our night girl, Sue Clayton.

Sliding down the firehouse pole carries no more urgency than does delivery of a foal whose *after*birth is arriving *before* birth—"placenta preview," the old-timers call it. Tonight, the tough, velvety, bright-red chorion protruded from the vulva instead of the thin amnionic membrane easily torn open in delivery. Time is of the utmost essence. The umbilical cord attached to the placenta often is severed from the foal, depriving him of oxygen. Mere seconds can result in piti-ful foals called "barkers" or "dummies." A wrenching experience for farm per-sonnel ensues as these foals often linger for days.

We have burst the chorion with a blunt pair of scissors before, but tore it manually tonight. The foal gulped for air even before his head fully emerged from the mare. We pulled hard and fast. His nostrils flared. We *had* him. All he had to do was jump from the burning building into our net. But the mare stopped pushing, with the foal's chest not yet clear. Three strong men, feet propped against the mare's buttocks, hands wrapped around the foal's front feet, could not overcome the suction of the uterus. Precious moments passed.

After her rest, the mare resumed contractions, and the foal inched out, only this time not gasping for air—just a limp head, glassy eyes, white gums. Somehow, though, behind his elbow, his heart thumped faintly, repeatedly, against his thin, wet skin.

"Breathe, you little sucker! Breathe!" we shouted into his throat as we tried to resuscitate him, pumping his frail chest all the while. His lungs were impenetrable. Our breath filled him only so far, and not far enough. Dizzy from hyperventilating, we rested in vain. He rested in peace. The fire had gone out.

Saturday, March 11. The photographer wandered unattended down to the foaling barn, his eight-year-old daughter in tow. He read the message on the Dry-Erase Board: "Foal Born 10:10 Friday Night. Foal Died 10:20." He explained to the young girl that things hadn't gone well last night here on the horse farm. Then they quietly walked off, hand-in-hand, towards the field where the Amazon Queen at three days of age cantered in perfect sync with her dam.

Sunday, March 12. Back from the barn for the end of Charles Kuralt's TV show *Sunday Morning*: Bald eagles nesting on the Mississippi River. No voice-over. No explanation. Just footage, just freedom of imagination.

While mud clogged the gates this morning, high overhead in the blue sky, fifty snow geese banked into the morning sun, whooping like Indians. A gaggle of tundra swans, with their immense wing span, followed the same flyway, chasing winter from the farmlands below. These two advance teams cleared the skies for the Canada geese, pumping low over the woods, wave after wave, a herringbone pattern certain in its direction—due north. Mr. Kuralt's quiet camera would find ready subjects in the air above the grounds of a Maryland horse farm.

Monday, March 13. It's a Monday, all right.

"I used to carry a .38 Special with me," the driver of the rendering truck gabbed. "In the old days, we had to put 'em down before we hauled 'em away. But nowadays you need a permit for a handgun, so the vets do the putting down part—what nature herself don't do, of course."

The magnetic sign on the truck door read "Valley Protein," as if the truck delivered nutritious health food. The driver lifted the tailgate to reveal another farmer's weekend misfortune—a mare and foal mortally locked in labor. We had been lucky enough to extract the foal and save the mare. This morning, a neighbor called, said his stallion had a heart attack covering his first mare of the season, asked me who to call to pick up his dead horse.

The breeding and foaling season is a busy time for the innocuous vehicle with the telltale winch on the roof.

Tuesday, March 14. Clearly evident on the ultrasound screen today were the twin embryos, lapped on each other like two bubbles sharing the same air. Dr. Tom Bowman bobbed the transducer over the uterine horn, gauging the potential viability of each fetal sac based on its size, shape, and clarity. He was trying to pick the one to pinch. The young mare strained against his intruding arm, and the plug in the wall socket was not perfectly grounded, causing static on the screen. Five minutes passed, ten minutes. His left arm resting gently in the mare, his right hand on the contrast dial, Dr. Bowman waited. If he tried to rupture one while it touched the other, he might lose both. The mare's involuntary defense of the crushed fetus could overwhelm the survivor.

We have never had a day when all the ultrasound examinations revealed early pregnancies. Today, however, we were seven-for-seven when the twins arrived inside the eighth mare—*nine* conceptions, more babies than mares bred. The lighting program gets results. Still, two's a crowd. Doc Bowman persevered. Finally, eight maiden mares each carried a single fetus. Bowman is a marvel, the touch of concert pianist in the body of a fullback.

Wednesday, March 15. His forehead was sweaty and his speech halting, but today was the first day of the rest of this new boy's life; yesterday, our job applicant had emerged from an alcohol treatment center.

"What do you know about horses?" I asked him.

"Well, I've done a lot of three-day eventing, and I foxhunted with Green Spring Valley. I'm not too good at treating sick horses, though."

We gave our Valley boy a brand-new, thirty-seven-dollar, five-pronged pitchfork, and set him to mucking. By 9 A.M., he looked sicker than any horse he might encounter. I watched him shake, then said a prayer: Please, God, grant me the serenity to give this boy a chance before I pass judgment.

Thursday, March 16. HCG. The initials stand for *human chorionic gonadotropin*, a hormone injected into mares to hasten ovulation. Keeps covers to a minimum—saving the stallion from overuse, the mare from repeated breedings.

This afternoon, a mare shipped in from the Eastern Shore. In anticipation of a satisfactory cover, she had been given a shot of HCG *before* leaving home. This is common practice on breed-and-return mares. However, we don't administer HCG until *after* the mare has been bred. Much can happen between booking and breeding. The van breaks down, or never shows up. A bridge is closed. The ride shakes the follicle loose. The mare teases out in new surroundings. The stallion refuses to ejaculate. The stud man has his foot stomped and his backup has the flu. Anything can happen in the breeding game. Anything.

Friday, March 17. Can you believe it? We breed a mare named I'm Kelly on St. Patrick's Day. At seventeen hands, she could dropkick our little Northern Dancer stallion through the goalposts of life. We did not give her the opportunity. Michael correctly gauged the amount of stone dust necessary to raise Carnivalay's adjustable breeding ramp, then took I'm Kelly's left front leg and did not give it back to her until the little stallion had safely completed his mission. When there is a hand-and-a-half difference between the stallion and the mare, position is everything.

Saturday, March 18. Headlines and text from our local county paper:
Preservation Efforts Fall Into Disfavor—Established in the mid-1970s, Maryland's farmland preservation program is the prototype for other states...but the best price seen in recent years has been $1,400 an acre. "My land's worth $10,000 an acre," said the farmer. "Would you take $1,400 for it?"
Unrest Over The Land—"The time for worrying about preserving farmland is long gone away. For most of the younger generation, farming is dying in Harford County."
This is why I like the foaling season so much: It is such a beautiful distraction.

Sunday, March 19. Today, against the green pastures of an early Virginia spring, the black spokes of Ellen's Amish courting buggy spun the calendar back to an era when horses were regarded with a different esteem. On a path near here, Thomas Jefferson rode on horseback from Shadwell to read the law in Williamsburg. Later, at Monticello, he recorded the births of foals from the proud line of Fearnought. On the nearby banks of the James River, Sir Archy was foaled in 1805. The *American Turf Register* in 1829 recorded: "Most of the best stock in this country...are from the loins of Sir Archy." To the southeast, Stuart's Cavalry rode circles around McClellan's forces, providing Lee and Jackson with the position of Union troops set to ride against Richmond. Fifteen miles north a century after the Civil War, Secretariat was foaled at Meadow Stud.
On this brisk March Sunday in Virginia, the woods and fields of Goochland County echoed with the sounds of horses and horsemen who had passed this way before.

Monday, March 20. "This foal has an *undershot jaw*," observed Dr. Riddle upon his routine examination of last night's filly, "the opposite of a *parrot mouth*. Her upper jaw doesn't extend over the lower jaw. If she has no trouble nursing, don't worry about it. But by the time she is broken as a yearling, her teeth will have worn unevenly. Sharp enamel hooks will interfere with the bit, and require frequent rasping."
Dental care is on the checklist of horse husbandry. On a young horse, impaired mastication hinders growth. And we often see an older mare tilt her head as she eats, or waste grain, or "quid" her hay, as the English say, chewing but not swallowing it. I scratch my head and wonder why I can't put weight on her; meanwhile, Peter reminds Dr. Riddle to float the old girl's teeth.

Tuesday, March 21. Weather hits us over the phone before hitting us over the head. "Raining here in Kentucky," declares a Bluegrass client. "Snowing here in Virginia," says a Middleburg van driver. A one minute call to WE-6-

1212 provides a weather update, but our outdoor compatriots in states to our west are better gauges. This afternoon, under a clear blue sky, every four-footed hide on the farm marched into a deeply bedded stall. By suppertime, sleet cracked against the windows.

Wednesday, March 22. Richard Harris, our West Virginian who can out-work any four Marylanders, took a jab to the head this morning, courtesy of a colt who might be singing in the choir before he is a long-yearling. Granted, spring is in the air, but a studdish colt with a modest Maryland pedigree is a prime candidate for Dr. Riddle's scalpel.

With our power hitter out, the opposition threw fastballs all day. Five hundred bales of alfalfa pulled in as Rich pulled out. Alden's Ambition was only teasing when she acted as if she were not twenty-one days in foal. No veterinarian showed until four. The van for Kentucky came at five. The Buckpasser mare was heavily waxed at six. The old pickup overheated and didn't cool down until seven. The night girl's daughter has the flu, we learned at eight.

Looks like spaghetti again, maybe by nine.

Thursday, March 23. Two days into spring, our nutritionist arrives with ten pages of analysis on our pasture, hay, and grain. He's a math teacher piling on the homework when he visits in the busy days of spring, but he keeps it simple, stressing two basic guidelines—the "calcium-to-phosphorous" ratio, and the "forage-to-grain" ratio. Master these two building blocks, and the nutrition puzzle falls into place.

"The synergy between calcium and phosphorous is critical to bone development, because these two minerals comprise seventy percent of the minerals in the body," he explains. "Too much of one and not enough of the other throws off metabolism."

Today, we targeted suckling foals approaching three months of age in April. Inexpensive weight tapes girthed around young horses are surprisingly accurate. Our foals at ninety days should weigh three hundred and seventy-five pounds, and consume one and one-half pounds of alfalfa (about half a flake) per day, together with four and one-half pounds of pelleted feed (at eighteen percent protein)—fed either in the stall or in the creep feeder. Our Ca:P ratio is 1.6-1; the forage-to-grain ratio is twenty-five percent to seventy-five percent.

This is what our foals *should* be doing. I am grateful when they condescend to nurse vigorously, every day, rain or shine, oblivious to this complex dietary regimen.

Friday, March 24. Another equine suicide attempt was foiled today. Immediately after a routine cover, and impatient from being restrained, the mare backed up furiously, her momentum pitching her over onto her rump. Almost in slow motion, she rolled completely over, thumping her neck on the ground. She appeared unhurt, but walked stiffly out of the breeding area.

"She'll be sore for a few days," explained Peyton, as he palpated a sore area on her withers. "Since it wasn't too violent a fall, hopefully we're only dealing with soft-tissue damage rather than a fracture."

Treatment with anti-inflammatory medications and several days of stall rest were advised. A lightning bolt would not dare knock down the nickel mares, but the good stakes winners like this mare will roll over and die on you quick as they please.

Saturday, March 25. Less than two weeks ago, we pinched twins on a maiden mare at fifteen days past breeding. At twenty-three days, we found only a single surviving fetus. Today, twenty-eight days past breeding, on her *third* ultrasound exam, she was found to still be in foal with twins. The answer was that she conceived triplets, from multiple ovulations several days apart. Pinching would not work this time. The options available were either to leave the mare alone and let Mother Nature decide (a wrong outcome would be disastrous next spring), or to give Prostin to the mare, inducing abortion before the endometrial cups of pregnancy develop at thirty-some days past conception (mares that abort any later often will not cycle back in heat).

"The best mare on my farm just produced twins last week," Ted Carr said upon being presented with the alternatives for this Brookside mare. "We had to cut 'em out of her. It was a mess. But last spring, we pinched twins on her. Ultrasound exams afterwards confirmed it. The only explanation is that she had triplets in there, and we only got the one." A seasoned pro, he reflected on the traumatic delivery he had attended last week. "Better blow 'em out of there," he said. "How's everything else going?"

Sunday, March 26. All our equine Easter eggs were decorated in a flurry of Saturday afternoon breeding, leaving us to relax and enjoy the glorious seventy-degree weather on this holy Sunday. The stallions always benefit the most from a full day off. They spend their weekdays breeding—or worrying about it. Quiet Sundays, with no vans crackling on the gravel, allow their seminal reservoirs to replenish. Even Jazzman rested today, rolling in his sand pen, shaking off his winter coat. Various family members tended to private Easter rituals with in-laws or children. Horses were not a priority today.

Groucho Marx quipped that he liked his cigar, but took it out once in a while. That about explains how we feel today.

Monday, March 27. "Please give him some good news about his mares," Norah said in the office this morning. "His house burned down last week. He and his wife had to climb out the second-story window to escape, and their two dogs died in the fire. I was almost crying hearing him tell about it. All he wants to know is how his mares are doing, and if any have foaled yet."

Recently, Mr. Clagett's hay barn was arsoned, yet he called about the beautiful colt he had delivered the same night. Mr. Baylis suffered a stroke, but rallied at the prospects of his two-year-olds in training. Now Henry Rathbun loses his farmhouse, barely survives, then immediately seeks a pleasant diversion by inquiring about his expectant broodmares.

You might say these three fellows are close to their horses.

Tuesday, March 28. A stranger phoned today. He had bought a mare in foal to one of our stallions at auction. He said he had lost the foal at birth, asked if we would provide him a free return breeding, considering his misfortune. We explained that live-foal contracts extend only to the owner of the mare at the time she was bred, not to her subsequent purchasers. This custom is based on the premise that stallion managers consider the mare owner's horsemanship before entering into a live-foal contract. Goes back to the old days when you knew every fellow in the horse business. It is a personal, unassignable contract. Good thing for us, based on information learned today.

"You want to know why that mare lost her foal?" asked the mare's original owner, a gentle farm girl who thought her reserve at auction would weed out a bad horseman or the slaughterman. "You want to know why that foal died?" she repeated. "Because the impatient son-of-a-bitch *induced* the mare. He called me up and asked for her last breeding date. I told him. He said she was eleven months and five days. I said, 'She'll be all right. She always carries late.' She was only five days overdue.

"She hadn't had a foal in three years, and he induces her. My mare was in labor for almost two days, then the placenta came out first. The foal drowned inside the mare. Ignorant son-of-a-bitch. Doesn't know a damn thing about horses. I could kill him for what he did to my mare and foal."

Wednesday, March 29. Two foals were born within ten minutes of each other tonight, right after the evening news. Very nice of the mares to foal at such a polite hour. The mares dropped their afterbirth quickly, the stalls were cleaned and freshly bedded, and the foals were up and nursing in forty-five minutes. Sometimes foaling is as easy as it is hard.

Thursday, March 30. "I could no more wake up on a farm than you could punch in at my real estate office every day," coolly states the developer from Arizona. While he is speaking, though, his index finger nervously circles the rim of the water glass on the walnut table in the elaborate boardroom he built with profits from developing farmland. "Windfields Farm is for sale. Why don't you move down to Cecil County, expand your operation. How many acres is that? Eighteen hundred? And how much do they want for it? Fourteen million? That's not a bad price!"

He is reminded that he recently bought a farm himself, eighty-six acres to be exact. Paid less than a million dollars, then convinced a gullible Savings and Loan to lend him almost three million on the speculation that a development two miles from Bel Air, and twenty-five yards from the beautiful front fields of Maryland's oldest Thoroughbred farm, would be a cinch to appreciate in value.

"Well, I'll put on a pair of coveralls if you'll climb up on a pan and help me move some earth," he continued. "That's how different you and me are."

Schoolyard bullies draw lines in the dirt with a stick, then dare you to cross it. Some grow up and become callous developers, still drawing lines.

Friday, March 31. Tornado watch in effect today. These old bank barns are built of stone cleared from fields plowed two hundred years ago, of poplar beams hewn right here on the farm from stands of mid-Atlantic hardwood. Still, Mother Nature can exact whatever toll she pleases on the farmer, from drought to tornadoes, and everything in between. Maybe she's just gearing up for an April Fools' Day prank.

So March ends on the same note it began, with barns filled with horses out of the weather. We are one-third of the way through the unpredictable journey we take each spring.

READY FOR THE PUSH

Saturday, April 1. Jazzman cut a popular swath through the fields today. Every shaggy-coated ship-in mare wanted to meet our teaser. Problem mares three years barren felt faint conjugal stirrings. Maidens stepping off the vans threw off lily-white shipping bandages before showing heat. The powers that govern this sport moved the registration date for all Thoroughbreds to January 1st in 1833, a nanosecond ago compared to the evolution of the equine endocrine system. Today's in-heat mares are marching to the precise rhythm of the lengthening daylight, and April 1st is the mean date of first ovulation in the Thoroughbred breed.

God's lights are on. The breeding season begins again—this time naturally.

Sunday, April 2. Sundays won't be the same again until July 15th, when the breeding season deadline on our Stallion Service Contracts expires, and we become free agents again. Now there is no Charles Kuralt show, no lounging with the Sunday paper, no office hours eight-to-five Monday through Saturday. The farm at the moment is a hotel full of helpless, finicky, four-footed guests, reliant on us for three square meals, for clean stalls, for one-hole-at-a-time halter adjustments on foals growing three pounds a day, one hundred pounds a month. Take relaxation where it is found: A nap in a hammock, or under an old maple tree (a favorite of Dr. Bowman's after a long night with a foaling mare).

Mild April weather will bring weekend visits from proud parents of newborns. Sundays now promise the knotty feeling shared by restaurateurs for random visits from food critics. Yet I know that if we deliver good service in a clean environment, the patrons will keep coming back.

Monday, April 3. "Clemens checks the sign, reaches back, fires. *Strike One.* Ripken steps out of the box." The radio blares as the 3 P.M. breeding session gets underway. Anchoring first base, in his sixth season of holding mares,

31

is Richard Harris, who rises each morning at five to lift weights and do pushups. Mares sense his strength, few challenge him. "President Bush looks on as Ripken steps back in, on this glorious opening day of the baseball season in Baltimore."

Pitching for the Country Life team is Peter Richards, who would not think it cricket to divulge his secret for proper insertion of the stallion into the mare. He is in his fourth season since relieving Uncle John of a procedure vital to satisfactory covers. Switch-hitting utility man Michael takes the mare's left front leg as the stallion approaches, figure-eighting the leather strap for quick release once the stud is safely aboard. He came up from the farm system, has been with the organization all of his life. He tips rookie Steve Earl on where to set the buckets for Peter to rinse the stallion afterwards. "Ripken digs in. Clemens throws."

The stallion, held by another player from the farm system, checks the signs, nudges the mare's quarters, swings in behind her, and begins his delivery. Peter guides the stud, Michael releases the leg, Richard steadies the mare, Steve positions the buckets. "Fastball up high. Ripken swings, connects. A drive to deep center. It's outa' here. *Home Run*."

The team nods in unison as the stallion dismounts. High-fives for the home run.

Tuesday, April 4. Until this morning, the old gray mare was considered pregnant. On a routine walk through the heavy mares field, Jazzman sniffed out a minor disaster.

"It happens to everybody," consoled Dr. Bowman, as he palpated the mare, took a uterine swab, and treated her reproductive tract with a general antibiotic. "You check 'em until you figure they're safely in foal. Then one day, hey, they aren't pregnant any longer."

At some point between her last pregnancy check in December and today, she had lost her developing foal. You aren't likely to find a second-trimester fetus, the size of a basketball, in a thirty-acre field, in the middle of winter, near a woods where hungry foxes live. Gulping down the high nutrition of her in-foal companions, and thus fat as a house, her appearance was deceiving. Her status remained officially unchanged. (This mare also happens to carry late, usually three hundred and seventy days, causing annual concern for The Jockey Club Rule 4.A.2., requiring blood-typing and parentage verification if gestation is more than three hundred and eighty days.)

The mare's best foal is a stakes-placed colt named Fun Bunch, but an *un-fun* bunch of Country Lifers gathered round when Bowman confirmed Jazzman's diagnosis.

Wednesday, April 5. The mare's owner, upset that Assault Landing would not ejaculate after three solid attempts at covering his mare, blamed everything and everyone. He warned us the mare was difficult. Still, Assault Landing has bred fifty mares every year for five years. He's a tractable, willing stud. I sent him to the showers before he could get discouraged, or become sour on this game of breeding. On our home field, we make the calls. Right or wrong, decisions have to be made, on the spot. Perhaps today I made a mistake by putting the stallion away after three tries, inconveniencing the mare owner. Well, mistakes are part of the game. I know one thing for sure: The mare's coach didn't like my call.

Thursday, April 6. Our little Northern Dancer stallion had his *first* starter yesterday. Perhaps awed by the stately oaks in the paddock at Delaware Park, or by the echoes of champions which once graced this down-at-the-heels dowager of a racing plant, Stablemate somersaulted away from his handlers, then tried to run the race without a rider or competition. Lassoed and resaddled, the young colt finished second, beaten by a well-prepped Florida ship-in. The chart appeared in today's *Daily Racing Form*.

Thus, we began another vigil today, to end on New Year's Eve: Charting the wins and earnings of a freshman stallion's progeny. On such numbers depends the future of this little breeding farm.

Friday, April 7. "Three things in life are hard to do—climb a forward-leaning fence, kiss a backward-leaning woman, and say something clever when accepting a trophy. With that in mind, thank you very much for this wonderful award."

That was the quickest speech of the night, as our state breeders' awards dinner, held with the paint still wet on Pimlico's new Sports Palace, lasted hours. We worried about our inaccessibility to our night girl Sue, and when we arrived home at midnight to find Dr. Jacobson's truck at the foaling barn, a wave of guilt washed over us.

"Everything's okay," Sue assured us. "We had two, though."

"Not twins! No! Are they both alive? Is the mare all right?"

"Knucklehead!" Sue frowned. "*Two mares* foaled, and they're both all right. We just called Russ because one foal was coming funny, sort of sideways. We straightened him out. You need some sleep. Go to bed."

Saturday, April 8. Dr. Bowman was on a roll today. Four ultrasounds, four pregnancies. The good mare who flipped over after breeding is in foal, and she celebrated by cantering away for the first time since she banged her withers. Also in foal was a mare who resorbed *twice* last year, and Peter had a syringe of oral progesterone in her mouth before the twitch was off. She'll stay on progesterone until September or October, depending on her blood assay tests. She is the dam of a stakes winner named Bal Du Bois. French accents rolled clumsily off Marylanders' tongues as the cry went up: "Le dam of *Bal de Bois* is in foal."

The crew worked very hard on Bal de Bois. They deserve to act foolish, so long as she acts foal-ish.

Sunday, April 9. After dinner, the photographer fell fast asleep in the guest-room. At midnight, Sue called with word that the mare was foaling. I think the photographer slept in his clothes. He was out the door and down the road as if his feet were on fire. His motor drive was in overdrive as he captured the scene, the huge colt on the white straw, the mare nonchalant about being caught in a compromising position.

Now we have a colt and a filly to chronicle from the cradle to the starting gate. The perseverance shown by our photographer makes me wonder if he knows something I don't. Maybe we'll get a Preakness starter out of this project after all.

Monday, April 10. *Typical field-teasing talk:* "Jazzy, hold your ground. Don't let 'I'm Kelly' scare you, even if she does weigh close to a ton. Whew! What big feet she has, flying past us. There's Native Summer, showing in on her foal heat. Skip her, let her uterus bounce back. Hmmm. Kay Band sees you. She's four days off Prostin. Watchin' you. That's more than she usually does. There she goes. Vet list for her today.

"Look who's showing now! Your girlfriend Scold with the Amazon Queen at her side. Don't worry, bred her yesterday. Gave her HCG. Okay, eat some grass, Jazzy. Let the girls know you're here. All right, nothing else cooking. Let's take a stroll through the babies' field, let those new moms know there's still a man in their life. Come on, Jazz, we got miles to walk before you eat."

Tuesday, April 11. With a deft touch, attorney Thomas A. Davis lays bare the skeleton that supports the flesh of the horse business in the American Horse Council's *Horse Owners and Breeders Tax Manual*, the annual supplement to which arrived in this week's mail.

The AHC's annual three-day convention in June is always worthwhile, but it comes at our busiest time. If we can't get to the seminars, we'll stay on the farm, flipping through the dog-eared pages of the tax manual with particular attention to the chapter with the seductive title, "What Is a Profit Year and How to Plan Profit Years."

Wednesday, April 12. For two hours now, she had felt a tiny fleck of hay scraping her eye. She could not help rubbing it, but her hands were dirty from mucking stalls, and to ask one of the men to get close enough to look in her eye would have made her even more uncomfortable. She found a quiet stall and sat down in the straw. She thought about her two children at home with her husband, and she prayed that his idea for running a printing business from their basement would work out better than had his other ideas. She was certain she could not work on a horse farm much longer. The work was so hard: "Look at my hands! My back hurts from carrying bales. The dust from these stalls never leaves my nose. The mares step on my feet, even the foals do—with their sharp little hooves. I'll pick up the classifieds at lunch. I'll give them my notice, finish out the week. It's not their fault."

Crying had flushed the hay from her eye. After lunch, she pulled me aside. A few weeks later, Ellen told me the girl had taken a job as an asbestos-removal inspector. I was sorry she had never gotten past the mud and manure of March to witness the glorious April weather, to revel in spring in full bloom on the farm. I picture her now, a surgical mask keeping asbestos out of her lungs, a quiet Tennessee country girl keeping her family alive. She was good with the foals. I wish I could have made her job here easier, but there are no easy jobs on a working farm, and the pay never matches the effort.

Thursday, April 13. The red and blue van was forty years old, and its grillwork was straight out of a cartoon, as if it might become animated and begin smiling and talking.

"Why, I've been coming to this farm since before you were born," it would boast proudly. "The mare I've brought today, why, I carried her mother and grandmother *both* up here for Mr. Hugh O'Donovan to breed to Country Life stallions."

The inside of the van was a step back in time. Varnished oak purlins supported the wooden roof. The stall dividers had heavy lead panels glued to the wood—protection from decades of the laxative effect vans have on horses. The ramp slid smoothly, the wings were in good repair, and the mare and foal walked off as if pampered travelers on a pullman car.

The engine had the deep purr of good compression as Mr. O'Donovan pulled away from the loading ramp. I might have been watching a vintage Chris-Craft boat cruise across the lake at sunset—the sounds were so similar. I enjoyed standing on the grass near the stud barn, watching the van depart, knowing that my father, and his father before him, had felt the same mixture of commerce and camaraderie towards the driver.

Friday, April 14. When everything is quiet in the middle of the day, look out: It's about to get crazy. Right after lunch, a lady from Virginia pulled her trailer up to the office steps, passing three driveways clearly directing van traffic to the unloading area. "I don't get out of Upperville often," she said breezily. "Can't I just unload here and turn around?" Behind her arrived a trainer from Laurel who could not understand why his maiden mare had not cycled yet. He asked exactly what was the purpose of putting her on Regu-Mate for ten days. We explained that often mares remain "transitional" unless their reproductive tracts have a chance to quiet down, to get regular, to quit "horsing." The progesterone regimen sends them out of their halfway heat.

"So when are you gonna breed her?" he asked anxiously, as he calculated his board bill in his head.

"Soon," was the reply. "Soon as she quits kicking the teaser."

Joining the traffic jam was Dr. Fred Peterson, a specialist in the periosteal stripping procedure to correct weaknesses in young foals' legs. Then Dr. Riddle arrived, and the two colleagues conferred about Rollaids' chances for improvement through such an operation.

Finally, with horses in the barn and rain in the forecast, we had one quick chance to order a fertilizer buggy from Bel Air Farm Supply before *this* year's drought might begin. The tractor's headlights lit the way as Steve Brown, our farm's jack-of-all-trades, spread three hundred pounds to the acre of a mixture of nitrogen, phosphorus, and potash—a boost for the grass in the crunch of the boarding season. He finished at nine-thirty, as the first drops of an easy, steady rain began to fall through the darkness.

Saturday, April 15. It is so quiet tonight you can feel the earth spinning for morning, carrying a farm full of bedded-down horses towards tomorrow, with a promise of clear weather after the day of rain we gratefully endured. You can feel the farm sleeping, rows and rows of foals nestled in hay corners, their dams chewing in hypnotic rhythm, as if sleeping on a train. You don't know and you don't care what Saturday night in the city is like.

Sunday, April 16. "I shouldn't have eaten all that hay, but I did, and it's not moving through me, and I haven't made manure all day, and if somebody doesn't do something, I think I'll simply *explode*. Maybe I'll stare at my side, be real still. They'll notice me if I don't eat. Oh good, here they come. Uh, oh! Looks like the vet chute for me. Not another rectal. I'm in foal, what more do they want? *'Impacted?'* Well, that's a first. *'New Bolton?'* A great hospital, but really, I feel a little better. Check my gums—good refill. My pulse? Can't be over fifty. See! I'm not ready for New Bolton. But do something here! Now!

"Looks like they're getting out the rubber tube. I don't need worming! I need Ex-Lax. They're mixing mineral oil and Epsom salts in warm water. Wow, ought to taste great, a gallon of this stuff. What's the Ivory Liquid for? An enema? I can't hold three gallons of soapy water. I'm full. Don't they know what this feels like? Yeah, get that stethoscope out. Tap your finger on my back, you'll hear it. I feel like I ate a basketball. Just give me something for this dull pain. That'd be fine, some *dipyrone*—sounds great, sounds soothing.

"By the way, how long does this oil take to work, anyway?"

Monday, April 17. Peyton arrived at dawn to examine last night's patient. We had turned her out in a small field with good grass for laxative purposes and for the exercise of walking. The floodlights from the barn lit the field. She did not pass manure all night. Double-checking, we walked the paddock for signs of oil-coated manure, but found none.

"Not surprising," said Dr. Jones. "The oil sometimes takes twenty-four hours to have any effect. The Epsom salts pull in water from the lumen of the colon, and the oil lubricates the tissue. You only want to use the salts once, though, because they're a little harsh. Patience. As long as her vital signs are good, we need to be patient. The question I am concerned with is how long before the impaction begins damaging blood flow to her tissues. I would like to run some fluids in her, to be safe, to help her out a bit."

Several liters of electrolytes were slowly administered intravenously. Watching the drip, drip, drip, of the fluid through the I.V. tube was as relaxing to me as Chinese water torture. Patience? I wandered out through the fields, stumbled upon Rollaids chewing his tongue absentmindedly. How do *you* spell relief? I asked him.

Tuesday, April 18. At 2 A.M., the little filly started making manure, noted Sue on the Dry-Erase Board. Any parent whose baby has ever become constipated knows our relief. By 8 A.M., the stall was awash in glistening, oil-coated manure. I am singing the spiritual, "Oh, Happy Day!" as I make the morning rounds, seeing how the rest of the farm passed the night.

Wednesday, April 19. Tension tightens down the breeding area as a maiden mare gives us fits. The platoon is in a crowded foxhole, under heavy fire. Our Englishman rallied the troops: "She's got us jammed against this *bloody* wall. Let's get this business over with." And so we made one final assault, landing safely on terra firma upon dismount.

Bombing missions over hostile loins take a mental as well as physical toll on the breeding crew, but the exhilaration of success is immediate, and will be palpable in fifteen days.

Thursday, April 20. "Welcome to the world," Dr. Bowman said, wrapping his big arms around the colt born this morning in the field of broodmares, an hour after we had turned out. "This is the best you'll ever have it."

Under a cloudless sky on a Technicolor morning in April, the veterinarian displayed great affection for his calling. With a pen light bearing the physician's caduceus (serpents wrapped around a staff), Bowman peered into the eyes of the three-hour-old foal.

"First entropic eyelids I've seen this spring," he declared. "Lids are inverted. Do you have any ophthalmic ointment? Use it twice a day, keep the lids from abrading the corneal surface. Ask Dr. Riddle to check him tomorrow. Might want to suture the lids to their proper position."

He checked the foal's gums: "Beautiful color. *Textbook* color." Then he passed his hands along the foal's neck, across the shoulders, stopping to check for cracked ribs, for trauma suffered in passage through the birth canal. Next, he reached underneath the foal's belly to palpate the umbilical area and the external genitals. He stepped back, hitched up his blue jeans, and ordered us to take a good hold of the colt.

"I'll see two or three of these a year. This colt hasn't urinated yet. It's because of an adhesion on the prepuce, keeping the penis from descending. He's straining. Could rupture his bladder." Then, very carefully, he uncoiled the little colt. The results were immediate. We *all* felt relieved. Once again, I admired such thorough veterinary skill.

Friday, April 21. Our breeder from upstate New York, who presented us with prototype arrival information on his maiden mare in February, is picking up his now-pregnant mare on Sunday. He instructed us to administer ten cc's of injectable progesterone to the mare prior to his arrival.

"Studies from the University of Colorado indicate that progesterone levels *decrease* during shipping due to stress," Jack Conner explained. "I'll give her another ten cc's when she first gets home, then back off to five cc's for the next few days. Maybe it's overkill, maybe not. But I'll take any prudent measure to make sure I don't drive the New Jersey Turnpike again this spring."

Saturday, April 22. We could only take a few minutes off, so when the bell at the main house rang one two three four five times to signal the start of the Wood Memorial on television, farm help dropped hay bales and pitchforks and water hoses and threw dollar bills in the pool while Dad put a match to the fire and Mom made everybody have a seat and be quiet.

When Easy Goer won by three lengths, Dad said the chestnut colt looked like his grandsire, Raise a Native, whose mother, Raise You, was bred by Country Life in 1946. I thought about the letter from her owner, Corty Wetherill, to Granddad about the daughter of Case Ace: "I named her Raise You, which I don't think is very good, but The Jockey Club rejected everything I liked that had to do with poker."

When he said he had "high hopes" for her, I'm sure he never thought she'd be the grandmother of the likes of Mr. Prospector or Alydar, so we rooted for Alydar's son Easy Goer, then finished up the chores as I told myself: "You just never know if this foal you are caring for will *raise you* from long Saturday afternoons in the trenches of the breeding business."

It's a sweet drink, this game we play, but it can be such a long time between sips.

Sunday, April 23. Don't tell me. Did I miss another family member's birthday? No, it's tomorrow. Anyway, he and my sister Norah live here on the farm. He is to Country Life what *The Far Side* comic strip is to the Sunday funnies. He makes custom birthday cards by cutting your face out of your worst photo and gluing it into some glamorous but unlikely magazine photo: Shorty's face pasted onto the bronze body of a beach boy riding a surf board, Mom as Aida the Ethiopian princess in Leontyne Price's opera garb, somebody parachuting, somebody on the podium with Adolf Hitler.

Gary Larson in *The Far Side* drew the Headless Horseman returning from a day's work to a headless family: Headless dog leaping into his headless master's arms, headless wife, headless kids rushing to papa, headless fish in a bowl. Sick stuff, but funny.

After Mike Newberg ("New Mike" we call him, though his speech therapy students call him Mr. Newbug because they can't pronounce their r's) gets through with your photo, you'll feel right at home with the Headless Horsefamily of *The Far Side*.

Monday, April 24. Whenever a mare arrives without proper vaccination records, I remind myself: "It's the gun you think is unloaded that kills you." For instance, apparently healthy horses arriving from a farm where an outbreak of strangles (Streptococcus Equi) is occurring can shed the bacteria for months. Strangles is more prevalent in Standardbreds than in Thoroughbreds, but many van companies transport both breeds—from sale barns to race tracks to breeding farms. The current *S. Equi* vaccines can cause reactions and provide less than absolute protection. But unvaccinated carrier animals stepping off vans can jeopardize the resident herd at a commercial breeding operation, including income-producing stallions, and senselessly damage a farm's reputation for health care. Maryland's Windfields Farm weathered a strangles outbreak in the late nineteen-seventies, then began a vaccination program and never had another case. Yet for years thereafter, Kentucky breeders shipping to Northern Dancer patronized mom-and-pop farms in avoidance of the big farm.

Today, two mares arrived with *absolutely* no paperwork. Peter gave Norah a chance to phone the mares' home farms. Whatever vaccinations could not be confirmed were blasted into the mares at our isolation barn. On this Monday, when twelve vans either dropped 'em off or picked 'em up at the loading ramp, Peter didn't assume any gun was unloaded. Still, bacteria and viruses are like hoodlum gangs, happy to mow down innocent folks in drive-by shootings. A few big commercial operations never even saw the gun that killed the breeding season, or tarnished the farm.

Tuesday, April 25. Satellite antennas rose thirty feet skyward from the vans of three Baltimore news crews covering the county council hearings on the Master Development Plan. Citizens carried signs against all three of the million-square-foot malls proposed for Harford County: "We'll Remember in November." The seven elected officials quieted the standing-room-only crowd in the high school auditorium: "Tonight will be an exercise in democracy. Everyone will have their turn to speak and to be heard. Let's begin."

The hearing finally adjourned at 1 A.M. I remember the president's admonishing words to a zealous advocate of an upscale mall: "It is *not*, as you say, inevitable that such a mall will be built in Harford County."

Still, sometimes I feel like the Indian, taking the word of the white man. I see my people trapped on the reservation, surrounded by five-lane boulevards, besieged by settlers who live to shop. We pick their beer bottles up out of our fields, and watch the phosphates from the car wash hit the rapids and whirlpools of Winters Run, as oil residues from parking lot runoff glistens in the polluted stream. And I don't know how to keep us from ending up like Indians on the reservation, dreaming of the land of our fathers.

Wednesday, April 26. Alone in the bowels of the bank barn in the dead of night, Sue pulled the keystone bale from the straw mow, and a dozen fifty-pound bales cascaded down upon her. In the hour she was unconscious, the mare delivered a perfectly healthy foal. Sue was madder than hell that she missed the foaling, and left this cryptic message on the Dry-Erase Board:

"Who stacked straw? Almost killed myself. Mare foaled 1 A.M. Nursed 3 A.M. If not here tonight, expect pay."

Of course, she'll receive hazardous duty pay. She's tough. I thank God she's a country girl.

Thursday, April 27. Our five yearling colts practice equine wrestling techniques upon being turned loose. This daily ritual of takedowns and one-upmanship among the colts is crucial to their competitive spirit. They grapple for teeth-holds on withers, feint like boxers, dipping left, now right, lunging for the opening to bite the legs out from under the sparring partner. Then they race off, kicking in earnest at the chin of the pursuer, who throws his head back, cleanly avoids the flying foot, and bears down harder on his prey. The tables turn quickly, no bully dominates, and afterwards all five splash fresh water out of their eighty-gallon watering tub.

The colts are now more than one year of age. They each weigh close to eight hundred pounds. They consume nine pounds of alfalfa hay and seven pounds of fourteen percent pelleted feed per day, and are shedded out and blooming on the fresh bluegrass in their pasture. The calcium-to-phosphorus ratio of their diet is 2.1-1; the forage-to-grain ratio is fifty-seven percent to forty-three percent. They have nicks on their forearms and bites on their butts, but at the annual Maryland Horse Breeders Association Yearling Show on June 25th, we have a feeling that this year's judge, Shug McGaughey, will notice our athletes amidst the Wonder Blue Shampooed and Show Sheened contestants. No hot-house tomatoes are grown on this vine.

Friday, April 28. Each morning, Peter prepares our Daily Veterinary Report, based on his impeccable record keeping, and on my rhyme-or-reason teasing results. Peter recites idiosyncrasies of ovulation as part of the mare's history when Dr. Riddle rolls up his sleeves and begins rectal exams.

"This mare was a breed-and-return in March," Peter recounts about a mare who shipped in from Pennsylvania supposedly back in heat. "She had a good cover and ovulated after returning home. We teased her upon arrival. She was fussy, but she *did* show a little heat."

Dr. Riddle comments that the mare's uterine tone is good. Could she be in foal?

"Their vet didn't think so. That's why she's back here," Peter explained. Dr. Riddle pulled out his ultrasound machine, scanned the mare, found a thirty-day-old developing fetus, and froze the image on the screen. This happens two or three times each season. A mare is sent back here for a second breeding. She won't tease well on arrival. We'll double-check her with Bowman or Riddle, and whammo, she's in foal after all. These unexpected embryos bring the same excitement as snow days from school did while we were growing up on this horse farm, just something you feel like you got for being good.

Saturday, April 29. Oak rails thicker than a man's thigh rip through supporting posts as iron-legged horses smack their cannon bones on the twenty-two devastating fences in the toughest timber race in the world, the Maryland Hunt Cup.

Sometimes the rails bring these heavy beasts to the ground. Sometimes the rails bounce ten feet away, without exacting a casualty. Seventeen-year-old Patrick Worrall, astride the favorite, Von Csadek, lasted four fences, bruised his body, but bolstered his resolve to win this race someday. Meanwhile, forty-seven-year-old wizard Paddy Neilson safely piloted Uncle Merlin to victory over the four-mile course. In a happy coincidence, Uncle Merlin's sire, Easy Gallop, stood his stud career at Worthington Farm, over whose beautiful soft valley the timber horses lift their riders to the awesome heights of the Hunt Cup fences.

Sunday, April 30. From the stud barn, classical music resounds from the radio bought from Goodwill for the stallions' relaxation, and a familiar client startles upon hearing Mom's voice send out a special lyrical birthday sonata to Norah. I tell him Mom spins compact discs at the Community College on Sunday mornings. Our very classy stallions love the show.

From the broodmare barn, the Sunday crew of 4-H'ers and Bel Air wrestlers muck stalls as a pink boombox plays a tape of television's Greatest Hits. I am about to crack the whip on them as *Rawhide* blazes forth, but the teen captain, Jason, assures me: "There's No Need to Fear! *Underdog* Is Here!" and the manure flies down the aisleway. From the office, the sound of silence is music to my ears. The girls at the answering service say we're all clear. They have learned that frantic folks who call and say they need a breeding *today* are not speaking for themselves, but for their mares, and a sense of urgency is duly imparted. But not this day. From the foaling barn, the twenty-eighth baby of the season nurses with vigor. The score at the end of April is fourteen fillies, fourteen colts—but we lost one, the placenta preview. That colt's dam grazes with the single mares now, a sixteen-day-old fetus growing inside her as nature and time bury the misfortune of her spring. Ten mares still to foal.

Yesterday's rain has moved on. This evening, we head for May under clear skies, ready for the push that will come with the onset of hot weather.

A SPLENDID LITTLE WAR

Monday, May 1. "In the mid-Atlantic states, look for rain, heavy at times, beginning this afternoon," the weatherman advised. "Farmers from Maryland to Maine need this rain."

A promise made during the ninety-day droughts of the past three years to never, *ever* complain about rain again, was repledged while trundling foals into the bank barn, as a thunderstorm straight out of the Old Testament rocked the farm.

Tuesday, May 2. Several law school classmates developed ulcers during finals. They switched from black coffee to pink Pepto-Bismol very quickly. The same dispatch helps foals with symptoms of ulcers—foals who grind their teeth, lay on their backs, salivate, or have diarrhea. Ulcers are most prevalent in suckling foals under four months of age, a parameter neatly encompassing every baby on this farm: the twenty-seven dropped here, and the twenty who shipped in with their dams.

A mare named Bitte Schon, who has had only two foals to live in six years, foaled here last week. I think her name means "Thank You" in German, but I call her Sally, Six-Year Sally. I call her filly Thanks A Lot, because I'm not sure I will be grateful later. Sally's arrival papers included the pathologist's comments on her 1985 suckling: "Post-mortem examination revealed lesions of severe gastric erosions and ulceration. There was a perforation of the stomach. Virologic tests were positive for rotavirus."

The barn medicine cabinet contains a stockpile of Zantac, an ulcer drug for people. It is not available in an easy-to-give paste form, but rather, comes in little yellow pills that you crush and dilute, then syringe down a poor foal's throat. If Thanks A Lot so much as chews funny, she goes on Zantac. And she'll think cod liver oil is mother's milk, this Zantac tastes so bad.

Wednesday, May 3. Doomed now to become a highway linking Baltimore-Washington International Airport with the model city of Columbia, the University of Maryland Equine Research Farm is disposing of its herd of donated horses—tax-deducted and forgotten in this unforgiving market. There is a sister facility to the farm, where the detritus, only recently appraised for accountants' purposes, are euthanized by electrocution.

A mare in foal to Corridor Key survived the shock of donation, but was scheduled to be let out for bid with the remaining herd. We intercepted her today, and tonight she is in the foaling barn, dripping milk. The foal gallops inside a watery haven, punching the dam's belly, causing her head to lift in surprise. Tonight it feels so good to have undone a donation.

Thursday, May 4. She is on our mind through the afternoon tasks. The first foal bred by a young client, produced from one of Stuart Janney's mares, sired by a home stallion, raised, broken, fired, and blistered here, today's first-time starter is a sentimental favorite. Dad phones. Marva answers: "She won by five!"

From his perch on the tractor as the spreader kicks manure, Richard punches the air with his fist upon seeing the thumbs up sign. Ellen is liming the stalls down, ready to bed, and lime pops like flour off her hands as she claps with joy for the owner. Peter recounts with pride the precise procedure he followed to overcome the ankle problems the filly encountered as a two-year-old, then frowns at the lost opportunity to wager a few bob when he discovered she paid twenty dollars to win. Rookie Steve grins from ear-to-ear seeing everyone so happy, before asking: "Five lengths is a lot, isn't it?"

The filly is evocatively named Landing At Dawn, but her sterling debut is sure to keep morale aloft for many days after sunrise tomorrow.

Friday, May 5. The *Maryland Horse* office contains videotapes depicting various aspects of raising horses. Our new employees benefit greatly from these visual aids, and our old employees enjoy the refresher course in fundamentals. With the imminent foaling of the farm's new mare, however, the opportunity was at hand to produce a video detailing our particular foaling procedures, with our own heroine, so aptly named Follies Star.

At 8 P.M., milk dripped from Follie. At nine, Sue punched in for the night. By ten, we were more uncomfortable than the mare and fought off fatigue. No film at eleven, either, as Follies Star dawdled. Sue took over the director's chair while we crawled off to the casting couch, alone and discouraged. At 1 A.M., an hour into her exact due date (using three hundred and forty days), Follies Star delivered an exquisite filly. So I'm told. I missed it.

Saturday, May 6. Everyone's talking about the Kentucky Derby today, except Uncle John, who is busy giggling about the check-out girls at the pharmacy who rung up his acquisition of breeding shed and foaling barn supplies: A dozen tubes of lubricating jelly, five hundred disposable gloves, ten bottles of baby oil, and six enemas.

"The girls were afraid to ask why I needed so much of everything," Uncle

John laughed, preening for his captive audience in the farm office. "I told them I was going to a Derby party this afternoon."

He watches the Derby with a cadre of cronies who have been breeding mares here since the nineteen-thirties. I see the scene in my head—Sunday Silence upsetting Easy Goer, Uncle John smiling for Arthur Hancock's success, then regaling the old boys, tastefully of course, about his morning at the pharmacy.

Sunday, May 7. Divine comedy is the new filly at play, rearing off her front feet, springing off her hind legs, forwards and backwards, over and over, burning off energy provided by Follies Star's bountiful bosom. Insidious drama is the knowledge that the foal's blood test, taken once at fifteen hours, again at twenty-four hours, clearly showed failure of passive transfer of the mare's antibodies, due no doubt to that same bosom leaking vital colostrum in the two days prior to foaling. No use crying over spilt milk. What now?

"Well, the small intestine will not absorb colostral antibodies after eighteen hours," explained Dr. Bowman. "The use of plasma to lift the foal's IGG count is your only choice now. Purified plasma is commercially available for about two hundred dollars, including the vet's time, but since she's *your* foal, and no liability for an adverse reaction to a transfusion, let's collect plasma directly from the mare."

Our freezer is full of frozen colostrum, a dozen precisely identified Nasco Whirl-Paks with twist-tied tops—courtesy of various mares stripped through the early part of this foaling season. But the time has passed for a colostrum-bank withdrawal. Now, it is either plasma therapy, or risk any willy-nilly virus blowing in the breeze.

Our *own* foal is helpless against infection. Well, Mom always said it is the shoemaker's children who go barefoot. We better go shopping for plasma first thing tomorrow morning.

Monday, May 8. To harvest plasma from Follies Star, Peyton bled three liters of whole blood from the mare at 9 A.M. He returned in his jeep at 3 P.M., brandishing a liter of plasma—slightly more than one quart of viscous, protein-laden, antibody-rich, life-supporting amber liquid. Most but not all of the red blood cells had settled and were removed from the container. (The commercial product is virtually free of *all* red blood cells, further minimizing risk of adverse effects.) We hope to raise the foal's IGG reading from two hundred to four hundred milligrams per deciliter with the plasma, administered intravenously very slowly. The same time-consuming process will be repeated in several days, hopefully lifting her to six hundred mg/dl, and assuring her of playing on the same level field as dozens of her schoolyard chums.

We could have saved ourselves this effort by weighing Follies Star's colostrum at birth in a simple colostrometer, the specific gravity of a mare's first milk bearing a direct correlation to its quality. But that is hindsight brought about by lack of foresight.

Tuesday, May 9. Our little foal born in the field on April 20th, with entropic eyelids and an undescended penis, is a walking anomaly. It seems the only correct thing about him *is* his IGG level; everything else keeps acting up. His jugular pulse is visible, perhaps the result of a heart murmur generally present in newborn foals, but which persists in him at almost three weeks of age. And despite daily handling, he is a wild deer when approached for leading. He cranes his stiff neck to see who is coming, as if he suffers from a pinched nerve. Today Ellen tried to figure out this unfortunate fellow, with the help of Dr. Bowman.

"Better have Dr. Riddle X-ray this foal tomorrow," Bowman said. "I can't find anything broken, but even a tiny fracture in the area of the neck would be extremely painful."

At 10 P.M., Sue phoned about another foal. Seems Six-Year Sally just got another year tacked on to her sentence, as she has apparently stepped on Thanks A Lot's hock, and tonight it is swollen the size of a softball. I pull out my *Manual of Equine Neonatal Medicine*, Dr. John Madigan's terse outline on foal problems. I read the six clinical signs of joint ill twice, because Thanks A Lot's injury puzzles me, and because I don't grasp vet lingo on the first go-round. Symptom No. 5 stops me: "Sudden onset of lameness with owner history of trauma to foal by mare (stepped on by mare) i.e. any lameness in a foal less than 45 days of age *should be proven to NOT be* related to an infection."

Hard to sleep when Thanks A Lot has either a broken leg or joint ill.

Wednesday, May 10. It's a splendid little war, these few months of anxiety known as the breeding season. Today, as the dull thumping of a trio of helicopters from the nearby Aberdeen Proving Grounds drowned all other sound, in a moment's respite from consecutive days of monsoon rains, our farm lay under siege.

Dr. Riddle raised the tailgate on his veterinary truck as three vans sat stalled at the loading area, waiting for the first van to get its throttle unstuck. The stallions paced, the yearlings galloped, and this morning's foal had not nursed after six hours and now needed tubing with colostrum. A maiden mare was found to have resorbed at sixty days. X rays were taken of the two sick foals—a filly who can't use her hind leg, and a colt wrapped in the shape of an S from the pain in his neck. The DMSO applied to the injured foals soaked through Ellen's plastic gloves and made her feel nauseous.

Messages piled up on my desk from playing phone tag with mare owners. A teaser limped from an abscess in his hoof. Tractors laden with manure passed pickups replenishing hay and straw to the isolation barn. The crew in soaking wet baseball caps and waterlogged leather boots trudged to bring in fifty foals from ten fields. Meanwhile, the thump/thump/thump/thump/thump of the choppers' rotors whirled the whole wet mess into a frothing brew. Country Life was a M.A.S.H. unit in full cry.

Maybe we are simply like soldiers who keep reenlisting. Maybe we'd go crazy without the war. Maybe after the rain stops, after the foals mend, after the breeding's done and the horses stay out all night, maybe then, we can reflect on why educated grown people stand in mud and manure for months at a time, raising a crop that invites heartache. Maybe it's for the leisure of living on the farm in the off-season, or because we don't have to commute to work.

Tonight, Ellen made hot turkey sandwiches, and garnished them with fresh mushrooms picked this morning by our manure man. I smiled with irony thinking about the tasteful by-product of all the stalls we clean. Tomorrow, we go to battle all over again.

Thursday, May 11. "I love my yearling by your stud," the horseman said today. "Only trouble is, he cut his coronary band as a weanling, caused some nerve damage and a quarter crack I haven't been able to close. Won't hurt me, though. He's got the pedigree for the select sale here in September. I'll fill the crack with putty, bond it all up. A buyer will never see it."

No, not quite never. Soon as a rider gets on the colt's back during breaking, the crack will surface. The colt will be sore, and the buyer will be miffed. He might avoid yearlings by the stallion in the future, thinking he throws bad-footed horses. He might get wise to the trick played on him at the sales, and sour on the whole game. Whatever the outcome, the horse business is its own worst enemy at times.

Friday, May 12. Doctors work thirty-six-hour shifts, and mothers hold down jobs and come home to feed screaming kids. I think about similarly situated folks in different callings when the job of working on a horse farm exhausts me. It's the twenty-four-hour days of May, the phone that rings just as sleep finally comes, the muddy boots and the scent of foal scours on blue jeans stacking up at the washing machine until there's no clean pants because who has time for laundry when mares are stepping on foals and ulcers are grinding teeth and follicles aren't ovulating, and I can't breed this mare on her third heat-cycle! What is she doing to me?

Family sees stress and together lift it. They know that foals who stand and nurse sometimes fall and die. They know that tired minds and one thousand-pound animals form a lethal combination. "Pull over and let us drive," they said today. So I walked home across the dike on the pond, at the blue heron's feeding hour. He stopped swallowing tadpoles and gawked with unusual boldness. We moved in unison, one step forward for me, one step backwards on his stilts for him. He lifted off the dike, landed on the board running atop the wire-mesh fence, rested as the sun streamed through the clouds for the first time in five days, then flew off towards the house by the creek—as if part of the family conspiracy, showing the way home.

Saturday, May 13. "Looked out my window and saw the mare's head down, with the foal on the ground in front of her," Ted Carr told me this morning, recounting yet another equine suicide attempt. "They just didn't look right, so I went out to the field. I saw the filly's leg stuck in her mom's halter. That mare had been shaking that foal like a dog. Broke her shoulder. Couldn't even operate. Couldn't even *touch* that filly for days. She healed, but she'll never race. So I sent her to you to breed."

We remove nylon halters from ship-in mares or foals and replace them with leather halters at the isolation barn at check-in time. Nylon doesn't break. Sometimes leather doesn't either, but generally, it will rip at the stitching before a horse's neck breaks. Thinking about this morning's lesson, I took a stroll through the fields. I used my best boot-camp voice to caution this year's recruits: "I don't want to see *no* foals shaking like *no* dogs in *my* fields this spring. Understood?"

Sunday, May 14. A well-intended Good Samaritan left a pile of mowed grass in Jazzman's sand pen this morning. But lawn clippings can cause colic if no other forage is available. Dr. Riddle grimly jokes that lawn mowing season brings plenty of revenue to veterinarians from emergency colic calls. Jazzman was vexed when he saw his pile of dried grass hit the manure pit. But on this Sunday, I told him, Father knows best.

Monday, May 15. It's Preakness week. Local news, Dr. Alex Harthill on TV, Sunday Silence, bruised foot, maybe no Preakness. Front page, Donald Trump in Gov. Schaefer's ear, photo ops at the Pimlico Special. Talk on the farm is of racing, and how to get grass mowed before Thursday's Preakness Party at the big house.

Four new mares with foals arrive. Four pregnant mares depart. Rain for thirteenth time in May. Thunderstorms. Slick fields. Hard walking. Muddy gates. Wet clothes. One more mare to foal. Finished at five. Feel great.

Tuesday, May 16. It is a *Chooseday*, as Peter pronounces this day in his *veddy* proper English. Dr. Bowman is laughing as he performs six consecutive ultrasound exams confirming fifteen-day pregnancies. The mares' names on Peter's vet list were unnecessarily foreboding: Big Scare, Headache Hill, Bad Penney.

Big Scare is Rollaids' mom. She had been reluctant to consider a sibling to follow in the splayed footsteps of this year's Poster Child. But two weeks ago, she condescended to conceive. It is no surprise when mares with difficult deliveries prove tough to get back in foal. Contracted-tendon foals like Rollaids often traumatize the mare's reproductive tract. A fallow year after such pains often is nature's mandate. Hoping to lose, I wagered crabcakes for the farm crew that Big Scare had not conceived. Bowman pointed to the phone the moment the black sphere of pregnancy appeared on the ultrasound screen.

"A pretty good *Chooseday*, wouldn't you say?" Bowman needled. "Call in the crabcakes, thank you."

Wednesday, May 17. This afternoon, seven May foals see a bright day for the first time in their lives, and an alert disc jockey sends out a special request to the newborns: "Little darlings, it seems like years since it's been here. Here comes the sun."

Thursday, May 18. Mom started this party in 1961, Carry Back's Triple Crown year, to accommodate the restless members of the Fourth Estate. She called it the Press Party, and Joe Nichols and Red Smith and Whitney Tower and anyone else who could write but not sing would swing and sway to Mr. Nichols' piano as *Peg O' My Heart* bounced across the lawn to the budding dogwoods and back.

Mom throws this party for the folks who keep this farm going through good times and bad. Tonight, the Good Lord turned off the heavenly sprinklers and we had a thanksgiving of sorts for the good fortune to still be kicking.

Friday, May 19. It is 6 P.M. when Dr. Riddle finishes the Friday vet list. He says one mare definitely needs a cover this evening. She is only twenty-six days past foaling, but showed heat several days in advance of the usual twenty-eight-day post-foaling heat.

"Find out if her uterus had been flushed with antibiotics after her foal heat," Dr. Riddle advises. "The fluids stimulate the release of prostaglandins, hastening the next cycle."

Peter checks her history. Dr. Riddle is correct. We breed her. Ellen tapes the local evening news devoted to pre-Preakness coverage. We watch the six o'clock news at nine o'clock and fall asleep wondering why it is so dark outside.

Saturday, May 20. Now only the vastness of Belmont Park stands between Sunday Silence and a sweep of the Triple Crown, after he beat Easy Goer by a nose today in a great Preakness. On the ride home from Pimlico, I thought about another game colt who stood where Sunday Silence now stands, and about this sport's wonderfully repetitive challenges.

Carry Back attempted his sweep twenty-eight years ago. He was bred, owned, and trained by a fellow named Jack Price, who brought his mare here in 1957 to be bred to Saggy, a speed sire whose greatest triumph came when it rained on Citation's three-year-old debut in 1948 at nearby Havre de Grace Race Track. For four hundred dollars, Jack Price purchased the stud fee that changed his life, then he drove off to Florida with Carry Back's mom, Joppy, the mare pregnant from Saggy's cover.

Across the field from the seven foals of May is an old farm, dissected now by a subdivision. It used to be owned by a horse trainer. When Havre de Grace was leveled in 1950, the trainer dismantled one of the barns and reconstructed it fifteen yards across Old Joppa Road from our farm. The barn still stands, overgrown and neglected. I point out to the May foals that Man o' War might have been stabled in that barn. Or Citation. Maybe Saggy slept there, listening to the rain on the roof the night before he stuck Citation with the only loss of his Triple Crown season.

A foal is a dream. Nobody knows where the runners are. They might be sleeping in the shadow of the barn from Havre de Grace, dreaming of the glories of the Turf. My grandfather was August Belmont's agent in the sale of Man o' War. My father stood a stallion who beat Citation. My earliest memories are of being told to be quiet while the family gathered at the TV screen to watch Carry Back in the Kentucky Derby. My job today is to keep the dreams alive, to fulfill the glorious uncertainty of raising Thoroughbreds.

Sunday, May 21. People who don't even like horses are talking about yesterday's race. And to have happened at Pimlico, where a few years ago the Preakness looked like a commodity that might be moved out on Mayflower vans, like the once-proud Baltimore Colts football team. Sunday Silence and Easy Goer brought the house down with excitement yesterday.

Monday, May 22. Dad points out that the jockeys who finished first, second, and third in the Preakness are each "in the program." Anniversary Number Nine comes this Thursday for Dad. Preakness week for him meant lining up headline acts for his backstretch meetings.

"Did you see Valenzuela on TV?" Dad asked me. "McKay couldn't give him but a second, but you could hear the jock telling kids: 'Stay off drugs. Listen to your parents.' "

This Preakness had it all for Dad.

Tuesday, May 23. Rich Rolapp spoke at the Maryland Horse Breeders annual meeting this morning:

"Is it tough to sell horses today? Well, yes and no. At the race track? No.

More horses are being claimed than ever before. At the two-year-old sales? The select two-year-old sale here over the weekend increased twenty-seven percent in average. What does this mean to the breeder? It means you have to take your product closer to the finished stage. An example? Sunday Silence—not accepted in the yearling market, nor in the two-year-old market. The breeder who will prosper in today's environment must practice superior husbandry in producing his crop, market it intelligently, and get a little lucky along the way."

Rich gives a great pep talk. He's right. For the umpteenth time in May, we spend the afternoon dodging raindrops while leading foals, but there is a lightness to our step thanks to Mr. Rolapp.

Wednesday, May 24. "I was at Delaware Park one day, watching him forge a shoe," our blacksmith Tom Rowles recounted this morning about a fellow farrier. "Prettiest shoe you'd ever want to see. Couldn't tell it from a Victory Racing Plate. Then he went to put it on the horse's foot. The colt pulled away, and this fellow got mad and smacked that young horse with the rasp. I said to myself, 'I'll always have plenty of work so long as he keeps beating horses,' though damned if he wasn't the best man in the fire I've ever seen."

Thursday, May 25. The term "Satellite Farm" is a misnomer, at least for the local folks who board mares booked to our stallions. Satellite implies a dependency on a larger body. However, Sue Quick's St. Omer's Farm, home for a dozen breed-and-return mares, is twice the acreage of Country Life. And the matter of dependency is flip-flopped: We rely on Sue for presentation of valuable mares for breeding, for keeping covers to a minimum, for good service to mare owners.

"I'm out the door to take my daughter shopping," Sue said at lunchtime today. "I've haven't had a day off in three weeks and I'm due back here at four to finish the afternoon chores."

She is twelve-for-twelve on conceptions of mares bred off the van. She is taking four hours off and thinks she is on holiday. The truth be known: Country Life is the satellite, revolving around the dedication of people like Sue, who enable our little breeding farm to compete against the much larger commercial facilities in this region.

Friday, May 26. "Looks much better," Dr. Riddle declared, shining his pen light into the cloudy left eye of the mare. Ten days ago, tears streamed down her cheek, the result of trauma to her corneal surface—maybe a scratch from a piece of hay, or from a thick-stemmed weed in her pasture. Who knows? But a watery discharge from an ulcerous eye can result in blindness if not treated. Dr. Riddle gave us two ophthalmic ointments to administer daily: Mycitracin, a triple-antibiotic to fight infection; and atropine sulfate, which causes the pupil to constrict, smoothing passage of the eyelids.

"Keep her in a dark stall during the day, turn her out at night," he also advised. "The pupil can't constrict in sunlight because it is held open by the atropine. The sun would only exacerbate blepharospasms—the constant blinking of the eyelids over the irritation."

Dr. Riddle is a walking veterinary dictionary. He graduated second in his class from the University of Pennsylvania. He is a board-certified surgeon, but that does not impress us as Memorial Day weekend begins in the face of a lengthy Friday afternoon vet list. We make him spell "blepharospasm" twice as we twitch up the last mare, trying to recall what it was like to be a student, on the ritual drive down to the ocean on this first real weekend of summer.

Saturday, May 27. The breeding crew is a five-piece band, jamming through the afternoon session, until we play Six-Year Sally. She is dynamite. Richard and Michael try to twitch her in the teasing chute, but she flies backward. It is painfully clear we are not the first to experience Sally's temper. Peter suggests putting the chain shank over her gums. We do this, and the restraint enables us to twitch her. She is thirteen hundred pounds of meanness, and mighty spiteful about her soft follicle. She corkscrews in the chute like a bronco waiting for the gate to spring. Rich backs her out because she will not walk forward with a twitch on. Reluctantly, she lets the teaser jump her.

Now for the real McCoy, and the stud wastes no time. Even with her left front leg in a strap, she turns a half-circle, forcing the stallion to thrust slightly uphill, and her high pelvis causes him discomfort. He does *not* ejaculate before he pulls out. Sally thinks we should be done, begins to pitch her front feet, furious at the restraint. Somebody is going to get hurt if we persist. There is the feeling as we put the stallion away that sometimes the compensation of our job is not worth the risk. I reason that as long as the boys in the band are safe, we will be able to try Sally again, when she might be more receptive. No sense spending this holiday in the hospital.

Sunday, May 28. Right on schedule at 8 A.M., the goose family heads to Sunday School—six flying churchgoers, low and beautiful against the blue sky above the fields. The white band of feathers around their necks is so brilliant in the early sun that at first I think maybe they have bread in their mouths, it is so white.

Always on a horse farm there is the contrast of vibrant life brushing against possible death. In the barns below, our two sick foals need prayers. The foal with the injured neck sleeps fitfully, moaning in his dreams. Thanks A Lot's temperature shows certain infection: 103.2, two degrees higher in the past forty-eight hours. I cannot shake the beauty of the geese in flight, thank God, as the decision is made to keep the foals here at the farm, with our nursing care, rather than send them to an equine hospital. The staffs at hospitals serving our area are wonderfully dedicated, but this is a major holiday, no getting around it. *Our* nurses have a lot more riding on these foals.

Monday, May 29. The drive to Ellen's family farm in Virginia requires four hours, and the travel day away from the farm during breeding season is like being stopped on top of the ferris wheel—the panoramic view of the world, other people alive in the distance, mowing their yards, cooking out, a river over the treetops. Round bales—huge, green African huts—cast spectacular shadows across the Virginia fields in the setting sun. The bales looked surreal, dotting the landscape like so many Salvador Dali watches. I disciplined myself to think about anything other than the breeding season, knowing full well that tomorrow, the ferris wheel will drop me back on the ground, with the carnival still going.

Tuesday, May 30. The crew got Sally bred in my absence this morning.

"Ah, go on," Rich said upon being congratulated. "We've bred tougher mares than her."

No sooner had Rich spoken when another powder keg showed up at the afternoon session. Couldn't twitch this mare, either. Had to lip her, then jump her, then move her into position for the stallion, just like Sally. I looked at Rich taking his time with this difficult mare, talking to her, keeping her relaxed while keeping us safe, and I thought: "Any fellow who questions the courage of this crew will have to play in our Thanksgiving Day football game." I wouldn't mention that Rich played for the Bluefield Beavers in West Virginia, that his hero is Hall of Famer Sam Huff; that Michael gained a thousand yards his senior year in high school; that Peter played the vital position of hooker in the rugby scrums in his native England. No, I wouldn't mention a thing before that poor fellow lined up against us.

Wednesday, May 31. Looking back over May, I don't see the rain, or the war scenes, or the foal with the hurt neck we put down this morning: Left a bare, empty stall, milked the mare out and turned her to run with the single girls. She hardly hollered. It was as if the colt was never here, except if you walked into the broodmare barn after Dr. Jacobson left you could see the new girls trying to keep from crying, and the seasoned help trying to focus on what's next.

No, what I recall is a scene from early May, a bluebird afternoon. When I drove in the lane after lunch, I saw Mom standing over a chestnut mare, keeping the other mares away while a chestnut colt was being born. Mare and foal deep in the green grass, Mom in a pretty skirt standing sentry, Michael coming over the crest of the hill with the post-op team in blue farm hats, me standing at the fence, knowing everything's all right, grateful to Mom for being so observant. That pretty scene will stay with me through the flies and the follicles and the fierce heat of June.

HOLDING ONTO ROCKS

Thursday, June 1. Last foal of the season—a colt who tied the final score at nineteen of each sex—was born at one o'clock this morning. He is owned by the same man whose foal we put down yesterday. There is a nice symmetry to nature sometimes.

It is as hot as Dante's *Inferno* in the hayloft of the bank barn as alfalfa is stacked, while the city bakes under a new record of ninety-nine degrees. The heat is on, in the air and on the phone lines. Mare owners are antsy. Not many chances left. One more heat cycle after this one, maybe.

Friday, June 2. Laughing with relief after five months of night watching, Sue visited the farm office in daylight.

"I need another T-shirt," she demanded triumphantly. "I want one for each day of the week."

She gets her wish, for a job well done. We used to wear rented blue uniforms, like gas station attendants wear, but the youngish crew and the teenage turnover mandated the summer camp look of T-shirts and baseball caps imprinted with the farm logo. Brothers Michael and Andrew still tease about the day Assault Landing retired to stud—in the era of the gas station uniforms. Assault Landing was a black panther in white bandages stepping off the green Locust Hill van. We stood for a photo, looking like Manny, Moe, and Jack, the tune-up trio, as if we were ready to fix Mr. Janney's Mercedes instead of take care of his stallion. Oh, well, just a memory now. Must congratulate camp counselor Sue for guiding this year's foal crop through the birth canal.

Saturday, June 3. It is nine o'clock on a Saturday night. My shoulder is wet from my wife's tears. She has gone to the barn to say goodbye to Roger, her old riding horse and everyone's favorite farm character. He is not a pretty sight right now. Blue Vetrap holds a catheter in his neck; he exhales DMSO, the smell of another world all over him as stored courage leaks out. I phoned Dr.

Jacobson: "Let's put this colt down," I said. Then I remembered thinking he is not a colt, but a gelding, a grand old Confederate Flag, bought for two hundred dollars when Ellen was fifteen. Came with pompons in his mane and no implied warranty for a particular purpose. He could do anything—marched in the Fourth of July parade down Monument Avenue in Richmond, past the statue of Lee on Traveler, past Stonewall Jackson and J.E.B. Stuart, past these great cavalrymen who would have welcomed his kind.

Roger gave us all a ride. Liability insurance was not a concern when Ellen lifted visiting friends up onto his back. He never stepped in a hole, never spooked at shadows, never came home riderless. He is twenty-four years old— give or take some leeway on the part of the consignor fifteen years ago—and Father Time needs a fresh mount. Some hidden tumor has burst its capsule, sending Roger's heart rate above one hundred and ten. Doc thinks maybe it is Horner's Syndrome. No one wants Roger to be a pincushion for vet needles at the clinics. We looked for pleuritis in his lungs with the ultrasound, but saw only his heart, beating a mile a minute.

Ellen has gone now to put her old boy in an end stall, so the Valley Protein truck can get him out on Monday. Such a clinical thought. Such a loss of a good friend. When Stonewall Jackson was mistakenly shot by his own sentries, he said it was time to cross the river and go into the trees. That is Roger tonight.

Sunday, June 4. Finney wrote that "this is the time for frayed dispositions on the stallions' part." I watched for signs of slackening interest as we bred the three mares on this morning's books, but the studs performed beautifully. No waiting for inspiration.

Monday, June 5. Ten mares leave for Kentucky tomorrow, and Dr. Riddle tubed mineral oil into each mare this afternoon. The van ride will take twelve hours, with stops every four hours to water off. Mares often will not drink the chlorinated water served at truck stops. The stress of maintaining their balance for such a long period also may dehydrate them. Chance of impaction is significantly lessened with a bellyful of mineral oil.

We spent the day pulling manes and hand-rubbing coats. These girls come in during the hot sun and are turned back out for the night. They glistened this evening, but I know the forecast is for showers, and that we will have muddy mares again by morning. Yet experience tells me that horses confined to stalls prior to a long trip often stock-up in their legs. I would rather hose them off tomorrow than blow them up tonight.

Tuesday, June 6. Ellen clipped fetlocks, Tracey wrapped halters in flannel, Jennifer hit the nameplates with Brasso, Peter reviewed ten sets of departure papers prepared by Norah, while all other hands pushed brushes over the flanks of the soon departed. Last night's mud had been hosed off. Tom the blacksmith rasped around the edges this morning. Dad sprung for crabcakes for the good folks in Kentucky who sent us these mares. Michael sprung for the cooler.

The van was scheduled to arrive at 5 P.M. It came at three, right ahead of two breed-and-return vans. Richard asked the Kentucky boys to sit on a bale for a few minutes. They obliged. One half of our crew bred the mares, the other half put finishing touches on the term paper we had been working on all semester. At three-thirty, ten handsome young mares, each more than sixty days in foal, lined up for boarding. Not one mare balked. The Sallee Van attendants backed them into brightly bedded stalls. I was not at ease until the van slipped away from the loading area, down the road past the oaks, headed south for I-95, then west on I-64, bound to be in Lexington by dawn tomorrow.

I looked back to thank the crew for such a presentation, but they had gone, like shadows, back down the hill, to turn mares out for the night.

Wednesday, June 7. In the cool of this evening, I walked out into the "hot" field, where the last dozen mares to be bred are cycling. Conditioned to seeing me in their field every day—albeit with the teaser Jazzman at the end of a shank—three mares stood off in the distance. I am *certain* they were disappointed by Jazzman's absence. Dr. Bowman has told me that when he interned at New Bolton Center, he used a mere tape recording of a stallion's voice to cause mares to show heat. Horses are patterned animals, creatures of habit, accustomed to the discipline of our field-teasing them *every* day, rain or shine. The girls know our habits, but it helps when the longest day of the year is fast approaching. The sunlight of June is nature's most powerful aphrodisiac.

Thursday, June 8. The blood work on the joint ill foal Thanks A Lot showed a much lower white blood cell count, a sign of abating infection, after weeks of Ellen's twice-a-day medical care. The stallions bred like it was opening day. Taps was a postcard sunset: Mares and foals framed against a backdrop of red roses, blooming now in the rain forest climate.

Friday, June 9. It rains. It stops. It rains. It stops. All day long. The crew was soaked, but Dr. Riddle rolled through another lengthy Friday vet list: Ten rectal exams, one ultrasound, one saline uterine flush, eight vials of blood drawn for Coggins tests for the June 25th yearling show, three mares' teeth floated, one stallion insurance exam, and one very important hoof examination revealing a not-so-serious abscess on a stallion with six more mares to breed.

The stallion had been walking like Chester on *Gunsmoke* for two days. Yet when Dr. Riddle pared away the overlaying hard horn, opening a shallow hole the diameter of a dime, infection drained out. The pressure was off. He then applied iodine, tightly packed the hole with cotton, and instructed us to repeat this procedure twice a day.

"Keep it clean, turn him out for brief periods on soft ground."

It is a scary sight to find a horse absolutely dead lame, and wonder what bones must be broken. It is a great relief, to both man and beast, to discover simply an abscess.

Saturday, June 10. Three inches of rain fell last night. Fish in the pond can't see their noses for the stirred-up topsoil that found its way down the hillsides. Back roads on the farm are washboards. The last white blooms on the "snowball" bushes are scattered in the front yard. But the sun was shining at dawn, and Dr. Bowman was due at ten, and teasing and breeding and feeding went on without interruption.

"I can't understand why her progesterone levels were so high," Dr. Bowman said quizzically when Peter led in the maiden mare. She had already resorbed once, and tagged "problem mare" upon rebreeding, she immediately went on the artificial oral progesterone product Regu-Mate, which, unlike the injectable progesterone, does not affect a blood assay test. For today's twenty-two-day ultrasound test, Bowman's curiosity was piqued.

"There it is," he said, spotting a second black circle on the screen. "A twin, from an ovulation three days after the first one. She's keeping two embryos alive. That explains the high progesterone reading."

He carefully pinched the twin, then reultrasounded it to confirm the white lines of a crushed vesicle interlaced in the black circle. Today, Bowman's intuition from a blood test captured a free radical twin, floating like a mine, alive and ticking, set to detonate eleven months from now. The whole scene went almost unnoticed as we hurried through to catch the telecast of the Belmont Stakes.

Sunday, June 11. "To be more aggressive, that was our plan," said Shug McGaughey after Easy Goer, the colt he trains for Ogden Phipps, won yesterday's Belmont, ending Sunday Silence's Triple Crown bid. We weighed the merits of being more aggressive ourselves this morning as we prepared to breed a two-year-old filly, owned by the estate of Mr. Phipps' brother-in-law Stuart Janney. Hand-delivered here, a February foal two springs ago, this filly was sired by Carnivalay and produced from Dress Ship. Ellen suggested the name Carnival Cruise to Mr. Janney, and he thought it perfect. Her racing career ended when six feet of intestine was removed in a colic case, but she is physically mature enough to be bred. Yesterday, Dr. Bowman gave the green light, but the filly gave us a two-way stop sign, using both hind feet to repel Jazzman's advances.

This morning, she was indifferent to Popeye. Her follicle was confirmed by Peyton, and I phoned for the A-Team. A careless kick on a Sunday morning in June can undo months of careful breeding practices. Some were dressed for church, some for rest, but the A-Team put Popeye on the young mare's back once more, then covered her for real with the stallion. My report to my superior at Sunday breakfast was succinct: Thanks A Lot's temperature a cool 100.8, Rollaids straightening up like nobody's business, Teen Land abuzz with weed eaters, answering service all clear. Oh! And by the way, we got Carnival Cruise bred. Nothing to it.

Monday, June 12. The mare who lost her foal the last day of May did not get back in foal. This was no surprise. Confined to a stall because of the foal's neck injury, deprived of the exercise beneficial to her uterine tone, and stressed by weaning, her stars simply were not in line for the miracle of conception. We have stripped her of the milk produced in the foal's absence several times. Yet her bag refuses to dry up, and today we again pulled milk from her.

We've had one case of mastitis this spring. A mare's swollen bag extended to the mammarian veins in the center of her belly. We spent two weeks giving her daily injections of penicillin, and manually milking out the curdled fluid before rubbing DMSO onto her bag—not much fun for any of the concerned, especially painful to the mare. Yet we tugged out the infection. When mares lose foals, mastitis can be an unhappy reminder of misfortune.

Tuesday, June 13. After twenty-four years with the Maryland Horse Breeders Association, general manager Mary Thomas stepped down last month. She was sunshine to every member of the MHBA. Among the books given to me from Maj. Stryker's library is a copy of Benjamin Franklin's *Poor Richard's Almanack*. Mary had inscribed it with "Happy Birthday Major." I flipped through the pages, thinking of the wealth of knowledge Mary carries of this game she loves, and Mr. Franklin's admonition seemed to speak to Mary's career:

> *"Hide not our Talents, they for Use were made:*
> *What's a Sun-Dial in the Shade?"*

Wednesday, June 14. Near Washington, D. C., this morning, moderator Tad Davis asked the audience at the American Horse Council's Tax and

Business Workshop for a show of hands: "How many accountants here?" Half of those in the room raised their hands. "How many lawyers?" The other half raised their hands.

Mr. Davis did not ask how many folks actually *owned* horses. Dr. Bob Lawrence did, though, and only a smattering of hands appeared. What does this tell me, I asked myself as I put my hand back down. That smart people don't own horses nowadays? Then Sidney J. Baxendale, DBA, CPA, CMA, demonstrated the mechanics of a software program for evaluating the purchase of a broodmare. He used model numbers to achieve an eighteen percent annual rate of return. He bought a mare for forty thousand dollars, sold her five years later for forty-six, and in between sold one hundred and seventy thousand dollars worth of yearlings from her.

"After expenses, it comes out to eighteen percent return on your money— better than a T-Bill," he declared, tongue-in-cheek. The savvy crowd repudiated the make-believe numbers. "Put up real figures! Put up current sales prices!" The murmur rose to shouted suggestions. Sidney plugged in the suggestions, then watched his computer light up and the model dip down.

"Well," he admitted with a sheepish smile, "this would be as silly as walking into a bank with fifty thousand dollars, telling them to hold the money *for free* for seven years. And oh, for their trouble, you just give them seventy-five hundred dollars, tell them: 'Keep it. Just keep it!' "

Washington is being called the murder capital of the world. Today's horse business can murder some capital pretty quick, too.

Thursday, June 15. In bunches, mares and foals loaded onto vans bound for Virginia, New Jersey, Pennsylvania. We are down from fifty-five foals three weeks ago to thirty-five this afternoon. The horses leave just when the summer help shows up, and we have too many boys in the band when all of a sudden we lose our drummer.

"Watch her striking with her front feet," I warned Steve Earl, who held the mare in the chute just as the stallion lunged to tease "up high," around her head. Some stallions prefer head-to-head contact—something wildly stimulating about asserting themselves face-to-face. (And stallions need to be assertive to settle the tough mares of June.) However, the mares often get scared, and strike with the suddenness of a gunshot. Today's mare rocked Steve's wrist off the twitch. Michael went for the mare, Steve went for ice, the stud went for procreating. It was my fault for letting the stallion tease the mare's face. That does not make Steve's wrist feel any better tonight. T. S. Eliot said April is the cruelest month. He never worked on a breeding farm in June.

Friday, June 16. Ordinarily, we are filled with trepidation when visited by yearling inspectors for the select sales. However, in the Maryland market, where the median price at the select sale over the past three years is only eight thousand dollars, and where thirty-two percent of last year's offerings were listed as "Not Sold," Timonium does not plan on installing an extra digit on the Bid Board, as Keeneland did in the boom years. Unconfirmed but undeniable is the notion that local breeders are keeping the real eye-catchers in their run-in sheds, preparing to take their *best* youngsters to the track, where purses might meet the cost of production.

Today, moments before the inspectors arrived, a hardworking farm girl brought her mare here for a breed-and-return. Last week, her one-horse stable, a homebred product of a two thousand dollar stud fee and a modest mare, was claimed for thirty-five thousand dollars in his fifth start. She took home the claim price, the winning purse, the breeder and owner bonuses, and promptly called a builder to erect the barn of her dreams on her little Eastern Shore farm.

Now, *that's* how to sell a horse these days.

Saturday, June 17. "I don't want her covered past June 25th," the mare owner said. I wondered why June 25th was his arbitrary cutoff date. It is almost a cliche to mention that Northern Dancer was foaled on May 27th, so Natalma must have been bred in late June. Or that Northern Dancer himself once covered a mare as late as August 1st, when a no-guarantee season threatened to live up to its terms. The resulting foal won a stakes. Nope, when a mare owner says, "That's it!" well, that's it.

Sunday, June 18. A friend with a swimming pool invited the family to celebrate Father's Day. Everyone was barefoot except the two-year-old colt whose shoes were heard galloping up the driveway as voices from the neighboring farm shouted: "Loose horse! Loose horse!" We tried cornering him in a plowed field, but he found his action in the soft going. Imperiling his would-be captors, he dodged adroitly, lunging past us in a frothy lather, showering clods of dirt in his wake. He went back down the driveway faster than he had come up, his shoes ringing over the countryside as if a Currier and Ives print had come alive. Dad did not have much faith in the crew who chased the colt in a golf cart. He took off in his car, Shoeless Joe on patrol, happy as a clam to be following the ponies, in one form or another, on this day reserved for him.

Monday, June 19. The old gray mare, she ain't what she used to be. Last week, Dr. Riddle removed one of her ovaries.

"She's doing fine," he reported today. "You'll get her back in foal next year, don't worry."

Her prolonged heat cycles were a tipoff to reproductive problems. Jazzman jumped the old gal in the field every day for two weeks. He became enamored of her, calling to her from several fields away. Three times we paraded her to the stallion barn for breeding. Wasted effort. Her follicle would not ovulate. Meanwhile, Jazzy fell in love. It broke his heart when Dr. Riddle diagnosed a granulosa cell tumor on her right ovary, after her left ovary had shrunk to a firm, inactive state—the result of steroid production suppressing other hormones.

Jazzman does not have many candidates for heat these days, and does not seem to relish his daily rounds as in earlier months. I hope he gets over his fling with the gray mare by next season.

Tuesday, June 20. Today, a foal out of a mare named Step Lightly was certainly stepping lightly, trying to escape the searing heat in his feet at the onset of the calamity known as laminitis.

"A *foal* foundering?" I asked Dr. Bowman incredulously as he felt the increased digital pulse in each of the baby's feet. Neither of us had ever seen a foal with laminitis.

"The clinical signs all point to it," he said. "He's a big foal, turned out on lush pasture, with access to the feed creep. He's overloaded his system with carbohydrates. It's a circulatory problem. The blood flow to his hooves is interrupted, and without oxygen to those capillaries, it's more painful than frostbite. Let's put him on Butazolidin immediately. Reducing the pain will get blood back into the hoof, keeping damage to the sensitive laminae tissue at a minimum. Put him on penicillin for infection, and since a correlation exists between the use of Bute and ulcers, start him on Zantac as well. Now go stand him in cold water. He'll love it, believe me."

Bowman left words of encouragement: "Foals are like cats—hard to kill. One day they're dying, the next day you wouldn't know anything was ever wrong with them."

Ellen dug a hole in the mud behind the equipment shed, and stuck the hose and the foal's feet in it. I know I will find them still wading in Lake Lightly hours from now.

Wednesday, June 21. "With satellite TV, I'll see ninety Orioles games this year," said Jack Sadler this morning in Aiken, South Carolina, five hundred miles from his hometown of Baltimore. "That's more than Dad'll see, and he lives fifteen minutes from Memorial Stadium."

Jack is the assistant trainer for Dogwood Stable, which winters its youngsters, and summers its lay-ups, in Aiken.

"Have you ever been at Pimlico in August, all that heat rising off the asphalt? Or Delaware Park in July? The humidity will drive you nuts. Aiken's no hotter than those places in summer. Can't use the main track here, though. They plant it in millet so it doesn't wash out in the summer rains. We use the five-eighths track to keep a bottom on these horses."

Aiken is red clay roads and magnolia trees taller than a house and oaks in rows and rows and rows, and it is fun to be here out of season. The ballpark is almost empty, but a few players peer out over their webbing, nursing the chips, the bows, the splints, the breaks of the game. When the millet reaches its seed-head in ninety days, most of these players will be back at the "show" up north, back in the "bigs," in the thick of the race.

I can almost see Jack on Labor Day, watching TV with his Orioles cap on, grazing through the channels for replays of races from Belmont, Arlington, Laurel—a Maryland horseman taking pride in the quiet restoration work he did in the Aiken heat of summer.

Thursday, June 22. All the foals over sixty days of age received their first series of vaccinations this morning. I expect to see some heads hanging as the killed viruses marshal the forces of the foals' developing immune systems. They will be sleeping on their bellies like cattle before a storm when Jazzy and I visit their moms tomorrow morning.

Friday, June 23. "You can have her. I give up," said an Eastern Shore farm manager, unloading a mare bred-and-returned here three times this season. When unable to settle a mare off the van, it is often helpful to ship her here early in her heat period, let Jazzman get to know her. We will draw fresh cultures, and treat her uterus with antibiotics, then breed her at the optimum moment.

The studs are not busy now. Soft follicles wait for no one. The longest days of the year are here, for farm managers as well as for the lights in the mares' eyes.

Saturday, June 24. Signs shout from places most likely to be visited by the general public—fencelines by the road, stud paddocks, fields across from neighbors' houses: "DANGER: ALL HORSES BITE."

Actually, some horses are more likely to bite than others, but to state that "THIS HORSE BITES" betrays a knowledge that the particular animal has a vicious habit, a fact crucial to a plaintiff hoping to establish negligence after he disregarded the signs and got himself bit. Me? I don't have a clue how appetizing *any* horse might find a child's plump little hand, so I warn: "ALL HORSES BITE." And yet, time and again, I have cautioned a visitor, then watched in amazement as the urge to pet a stallion on his nose overcomes all prior warnings.

Yesterday, a Mounted Police officer in Baltimore found out just how capricious even a trusted old police horse can be. On patrol among the tourists in the bustling Inner Harbor, the officer's eleven-year-old gelding Trinity was confronted by a baby stroller. Trinity reached down as if to be petted, then suddenly bit off the infant's ring finger. Two passersby put everything on ice and rushed the baby to a nearby hospital, where all is well today. The child's parents live near us, on a street named Hitching Post Drive, for the days of Bel Air's past when horses were not regarded as novelties to be petted.

Sunday, June 25. The judges over the fifty-five years of the MHBA Yearling Show are legendary Turf figures. From 1932 to 1942, James "Sunny Jim" Fitzsimmons spent an afternoon each summer pinning ribbons. Hitler and Hirohito were the only characters who could stop Sunny Jim. Judges in the post-War years included Preston M. Burch, Abram S. Hewitt, and Max Hirsch. (In 1948, Mr. Hirsch pinned a blue ribbon on a Country Life product, Loraine, a subsequent stakes winner who as a broodmare memorized her stall in the courtyard; she walked unescorted to her field and back).

These judges loved any yearling by Discovery. Five of six years running a Discovery was grand champion. Discovery was to Marylanders what Randolph Scott was to Westerns.

In the nineteen-fifties, Charlie Kenney, William du Pont, Jr., and Charles

A. Asbury did the honors. Humphrey Finney, Horatio Luro, Syl Veitch, Buddy Raines, and Elliott Burch stood in the hot sun in the nineteen-sixties looking for potential young athletes. Mack Miller, Sid Watters, LeRoy Jolley, and Woody Stephens left beautiful Belmont Park for torrid Timonium in the nineteen-seventies, and today, Shug McGaughey closed out the nineteen-eighties' judgeships in the footsteps of John Williams, Jack Van Berg, and Penny Chenery. A partial list leaves out worthy figures, but we are grateful to *all* the fine judges, even if we have not won a ribbon since Loraine.

Monday, June 26. Heavy fog burned off by 9 A.M. Thereafter, horses wilted, whether in stalls or in fields, as oppressive Maryland humidity turned itself up a notch. We put Fly Wipe on, but they sweated it off. Only the cool of the bank barn's stone stalls provided any relief. Gray horses seem particularly thin-skinned, as flies draw bull's-eyes and practice their fine art of aggravation with particular savagery. What little breeding is left this season had best be conducted in the cool of a morning, before the day fulfills its promise.

Tuesday, June 27. More Sign Talk (knock on wood): FIRE LANE.

Blue arrows on four knee-high signs along the driveway point the Bel Air Volunteer Fire Department towards the pond, where four hundred dollars worth of ten-inch plastic PVC pipe forms a hydrant extending five feet below the water's surface and fifteen feet out past the dock pilings. The hydrant was ordered from the Fire Department catalogue, and is listed on their hydrant maps. It is capable of handling one thousand gallons per minute, and the hose truck can drop its nozzle at the houses or barns and run a thousand feet of hose to the pond in five minutes—under optimum conditions.

"We might not save the barn that's on fire, but we'll sure water down the building next to it," explained Capt. Woodward. Tonight, two quiet mares moved under the shower falling only in their little pasture as volunteer trainees lifted clear cool water three hundred feet skywards. A wet "run-through" with the trainees once a year eases my mind, and cools down a few lucky horses as well.

Wednesday, June 28. Historic but extinct Windfields Farm Maryland, which closed its doors last August, is about to be parceled into manageable-sized lots, any one of which is bigger than this farm. Who knows whether proposals will be accepted, but it is intriguing to envision the idea of Northern Dancer Drive as the address for breeders' activities, or the training track serving as a launching pad for Maryland two-year-olds, or flocks of geese luffing into winds of change yet landing in familiar fields.

Thursday, June 29. At quitting time, those who commute to work sit in the quiet lobby of the broodmare barn, swapping stories, kids around a campfire. They do not want to go home, because no place is as pretty as the farm tonight.

It is all in color on this unseasonably cool day—blue skies, white clover, red gates, bay mares, brown foals, black fences, grass as green as Ireland's. We work to earn such evenings.

Friday, June 30. "I don't care how many angels can dance on the head of a pin," our county council president told me this morning. He winced as a ream of computer paper listing the one hundred and fifty members of the Winters Run Preservation Association—all registered voters, I reminded—fell from my hands and gathered momentum, like a Slinky, to topple down upon itself on the floor. "We don't rule by plebiscite. But frankly, I don't see a mall working there. I'd say the odds are one-in-ten for approval of that project when we vote on July 11th."

This was good news, on a summer day so crisp and delicious it was hard to imagine any outside force threatening the soft hills and stream valleys that roll away in all directions from our farm, blanketed now by lush growth from the rainy spring. At day's end, buoyed by Dad's call that an Assault Landing two-year-old made his debut a winning one, I walked through the pastures to Michael's house, to deliver a Junior Orioles uniform to a nephew whose first steps coincided with his first birthday.

For a moment, I dreamed I was a young boy again. I saw water rippling at eye level as we played commandoes in the stream, Huck Finn river-boys keeping our heads above water, pulling ourselves forward against the current by holding onto rocks on the stream bottom. I recalled the long view over the backs of broodmares grazing atop the hills, the mountains of my youth.

Years later, Richard Stone Reeves photographed the view from our hills as a possible backdrop for his portrait of David A. Werblin's stallion Travelling Music. Mr. Reeves observed that, "Spectacular terraces of land stretch for miles towards the north." I told him that was not our land, and gently suggested he photograph Trav against a hillside that might be with us longer. I regret that now. I would love to see Trav silhouetted against the proposed mall site, a Reeves landscape depicting the "Before" scene, a strip mall showing the "After" scene. The local papers would eat it up.

The sun flared orange in nature's own portrait this evening. In metaphor, we move into July as we did as children, keeping our heads above water, holding onto rocks. Now, though, we play King of the Hill in earnest—for the sake of children not yet born, and for the one-year-old who squeezed snugly into his Orioles uniform, a power hitter coming up through the farm system.

A ROUND OF PERPLEXING CARES

Saturday, July 1. The annual almost-the-end-of-breeding-season volley-ball game between Bonita Farm and Country Life is great for morale because there are no losers. Dr. Bowman sees to that. He supplies bushels of steamed crabs from his Eastern Shore home for the post-game picnic. Dr. Riddle plays for the Bonita team in a matter of seniority, Dr. Bowman plays for us, yet neither vet escapes the combined team effort that hoists them aloft like winning Super Bowl coaches, to be deposited ceremoniously in the deep end of the Bonita Lake.

Sunday, July 2. It ain't over till it's over. The last mare in Assault Landing's book was bred at lunchtime. Just as the stallion dismounted, I noticed a wide-eyed delivery girl at the gate, her pizza getting cold. I think folks who have never seen such large animals in the act of breeding regard it as a terrifyingly primal display. The TV critic who reviewed the recent program on the state's Thoroughbred industry enjoyed the show, except for the stallions captured in the act: "A bit too graphic. Could have done without the breeding scenes."

I think the pizza girl felt the same way. As a matter of fact, I'll be relieved, too, when we can do without the breeding scenes—until next year.

Monday, July 3. "The doctors took out a section of my colon *this* long," my brother-in-law Jack said, holding up his hands from his hospital bed as if indicating the size of a fish he had caught.

"That's nothing," I told him. "They take *six feet* of intestine out of horses sometimes. 'Colic' cases, we call them. A lot of horses are walking around today with less than God gave them, and they're doing just fine. You'll be better in no time."

Seems if you can equate human medical procedures with their equine counterparts, you'll never feel lost for words when visiting friends or family in the hospital.

Tuesday, July 4. The Fourth of July was a carriage ride across the country-side of Virginia, through the fields where purple buds of lespedeza competed with tight seedheads of timothy, white clover, and orchard grass. The baler will soon gather indiscriminately, but for the moment, it was quite a sight to see such rich fields of different grasses reaching above the steps of the carriage, and rising skyward through the rains of summer.

Wednesday, July 5. "No one really knows what makes a stallion a good sire," observed a respected friend who has managed stallions, both fantastic and dismal, for fifteen years. "Tesio said, 'Don't own a stallion. You might be tempted to breed to him.' If *he* didn't know, what chance do we have?"

He reflected on a colt he managed in the nineteen-eighties: "He was warmly received and bred quality mares. Then we waited and waited, and made excuse after excuse why he wasn't siring runners. The plain truth of the matter? He was a terrible sire. Terrible. And he wasn't going to get any better if he lived to be a hundred.

"What are the odds, really, of any stallion *hitting*? One in twenty? More likely, one in fifty. But lightning will strike, if you keep at it, and learn from your mistakes. The breeding business—unlike the business of racing—is not always to the swift, but to those who persevere."

Thursday, July 6. By profession a real estate attorney, Dick Abrams is by avocation a horse lover—and a collector of racing partnership documents.

"The common thread running through these offerings," he said today, "is that very few ever make any money." He noted that often a four hundred thousand dollar yearling can be found running for a twenty thousand dollar claiming tag. "The attraction is to own one part of five horses rather than all parts of one horse."

He looked out of his office window, off into the woods in the direction of Delaware Park, as if he were listening for the bang of the starting gate, the announcer's voice, the cheers from boxes. He said that almost all owners, alone or in partnership, lose money over the long haul. Then why was he smiling?

He told me he owns one-sixth of two maiden three-year-olds who are training forwardly and today are being vanned to Saratoga. A morning spent watching your own horses gallop on the Oklahoma training track at the Spa, the jumpers popping through the hedges in the infield, Hall of Fame trainers on horseback leading fresh sets across the grounds behind you, well, that is worth smiling about.

Friday, July 7. Rain on the Fourth of July forced the town of Bel Air to hold its fireworks display tonight instead, and as a fingernail moon began its ascent into the humid night sky, we strolled to the top of the big hill to gaze two miles towards the high school football field, the launching pad for the rockets.

Mares and their foals were startled by us at first, but then, like schools of fish, they began floating past. Foals pushed in play, grabbing pals, tormenting mothers—much like human youngsters in the moments before the rockets red glare stopped all nonsense. Then mares stood erect, staring straight ahead, their pricked ears silhouetted against the pink afterglow. Only the mosquitoes seemed oblivious to the lavish bombardment. The sounds of the brilliant finale were muffled by the heavy air, and it was soon peaceful again over the tree-covered watershed—just us and the horses and the traditions whose ritual observance mark the seasons.

Saturday, July 8. At a family picnic in an undeveloped valley of the county, the bluegrass band was picking and grinning and singing, "All I got left is a two-dollar bill, a two-dollar bill." Young Billy Boniface and myself admired our friend's fields of corn, head-high in July.

"What other crops besides horses could you raise on your farm?" I asked him.

"Hmm. I don't know," he said, their beautiful Bonita Farm purchased after his father trained a homebred, Deputed Testamony, to win the Preakness Stakes. "Maybe cattle. Maybe corn. I don't like thinking about it. D.T. has moved up the national sire list. And we've got some nice two-year-olds Dad is training. Who knows? We just try to keep the expenses down, to keep plugging till this thing turns around."

There were farmers' young in diapers in every lap. I think a bunch of these toddlers will grow up to be farmers because their fathers stayed confident through adversity.

Sunday, July 9. There is a sacred quality to animals sometimes, especially in the quiet of a Sunday morning, when the world seems all their own. Indians revered that quality. Cavemen became Rembrandts to record it.

Certainly the Great Blue Heron is the high priest on this farm. Perfectly still on the shore of the pond this morning, he reflected, preparing to ascend the altar of air he reaches with three strong flaps of his immense wings. Perhaps his parishioners anticipated his flyover. Foals obediently followed their dams to the top of the hills, where the exposure caught the morning breeze. Stallions atoned for their lifestyle and disregarded the siren calls of maiden mares stabled nearby. Dogs sat on command. Crows perched without arguing. Only the incorrigible teaser Dew Burns seemed unrepentant, pacing in frustration over his calling.

At eight o'clock, the heron lifted off, uttering a guttural sound that horses heard when they were only eleven inches tall and had toes: *Rackkkk, Rackkkk, Rackkkk.* The heron's voice passed over the land, and we all felt blessed by a higher power.

Monday, July 10. We yanked the shoes off the studs today. Stomping flies all day is tough on shod feet, and too frequent shoeing results in empty

66

nail holes in the hoof, cracks in the hoof wall. The studs will go barefoot for awhile. By late fall, when prospective breeders come to inspect the stallions for next season, the boys will be modeling fine footwear again.

Tuesday, July 11. The county council let us blow in the wind for another week in our battle against the mall across Route 1, but they voted approval for a mall closer to I-95. The councilwoman from another district opened the meeting with a fervent prayer: "Dear Almighty God, please give us the strength to make the proper decisions, for this tiny planet of yours, whose population keeps increasing but whose land does not."

It was a prayer she hoped would shift the burden of responsibility to God. It didn't. When roll was called, she puffed up against her microphone and tried to feather the blow: "A mall at this site, the entire acreage under single management, would best serve the citizens of this county." As she said "Aye," she dropped her head into her neck like a bird in a stiff breeze.

Hurt cries of "No! No!" burst from citizens who moved here to get away from malls and congestion, to see farms and sheep and horses. Reporters rushed to file their stories, and the council president ordered a five-minute break in the confusion. A trim man in a pinstripe suit remained in the first row, satisfied that his flight from his company's Texas headquarters for tonight's vote had been worthwhile: Five-to-two for the mall near I-95. Next week comes our turn.

Wednesday, July 12. Yellow fields, shaved closer than a Marine's haircut, rolled away from the baler kicking straw onto a trailer, as the farmer hurried to beat the wet forecast. Second-cutting alfalfa grew in a low, emerald swath towards fields of corn in a summer so soft and splendid that fears of global warming and drought seem remote. Mother Nature is a wet nurse to farmers—some say too wet. Me? I would rather be with Noah on the Ark than with Moses in the desert.

Thursday, July 13. Five folks sat down to improve the *Maryland Horse*, recent winner of the American Horse Publications magazine award.

"We should report on the entire mid-Atlantic region, not just Maryland," remarked editor Rich Wilcke. "Our stallion register pages peaked three years ago, with almost two hundred stallions. Now, we have a hundred and fifty stallion pages—a twenty-five percent decrease. We're losing farms and stallions in this state to development and economic pressures. Yet if we expand our coverage to the breeders in Virginia, Pennsylvania, New Jersey, and West Virginia who do business here, who support our excellent racing, we'd tap into a broad market for advertising."

Peter Jay, editor of a weekly newspaper in Havre de Grace, cautioned: "Don't lose sight of our special feeling—the steeplechasing in the spring, the fellow who breeds cast-off Thoroughbred mares to his Belgian stallion, the flavor of family operations in the context of a larger industry."

A little tinkering tonight, and a good magazine got a little better.

Friday, July 14. With cool weather in the forecast, Peter suggested field-weaning Rollaids and another foal more than five months of age. So this morning, we led Rollaids' mom and the other mare away quickly, leaving eight mares and foals to baby-sit. Rollaids and his pal tried in vain to alarm the eight unweaned foals. Nothing doing. Misery had no company. By dusk, the hollering had quieted. Sucklings had become weanlings.

Saturday, July 15. Final ultrasound results reward the perseverance of the worn-out rangers who, from the office to the stallion barn, put forth tremendous effort this spring. There is no champagne-popping celebration for the end of the breeding season. Rather, a calm like a gentle rain descends over the farm.

Most of us are too numb to envision doing this same job again next spring. We all suffer some degree of burnout right now, and I know from experience that some rangers will return, some will not. Turnover is a part of the process, but I will always recall with particular fondness the folks who roamed the range of Country Life beside me this spring.

Sunday, July 16. At times it was so dark today in the on-again, off-again rain that I would not have been surprised to see a Great Horned Owl sitting on the fence at high noon, as I saw in March during a partial eclipse of the sun. That day was scary-dark, Apocalypse-dark, and I suppose the owl figured he would see why nighttime was coming during daytime. He sat on the fence and called in a sonorous voice: "Hoo. Hoo-Hoo. HOO-HOO."

This horse farm is birdland, anyway. A Belted Kingfisher, concealed on the banks of the pond, startled and spooked into the trees when I drove across the dike this morning, calling a loud warning in the rattle of his penetrating voice. If I could only get stunning photos of the kingfisher, of the heron saying Mass, of the Canada geese with their yellow goslings, of the owl on the fence at noon, I bet *National Geographic* would help us keep the malls away. Maybe I've been going about this thing all wrong.

Monday, July 17. Craig Colflesh is a trainer based at Penn National. He's honest, works hard, shoes his own runners—even trailers them all the way to Atlantic City some evenings, where with luck he scores with a betting overlay, Penn form being mighty hard to handicap. At our farm today dropping off an equine *hors de combat*, Craig never moaned about the travails of cheap claimers and night racing. Instead, he carried on about the longhorn cattle he bought this spring from Walmac Farm near Lexington.

"The government saved the longhorns by establishing them on the Wichita Wildlife Reserve," said Craig with infectious excitement. "These were the cows they drove to Montana in *Lonesome Dove*. Hardy cattle. Easy keepers. The market for lean beef has driven their price up. They're a beautiful sight to behold. They eat pasture a horse won't touch. And they're *American*."

Just the thought of his longhorns grazing at sunset on his fifty-acre farm, the mountains of central Pennsylvania as a backdrop, seemed to make Craig forget all about the gelding with the pulled suspensory he had just led from the gooseneck trailer. People in any echelon of the horse business have wonderful coping mechanisms. Many horse farmers are rediscovering the benefits of cattle.

Tuesday, July 18. No windy prayers led off the county council meeting tonight, which I took as a good omen. When roll was called for adoption of the Land Use Plan, leaving the farm across the street zoned agricultural—not commercial—we breathed in relief: Round One to the citizens of the Winters Run watershed.

The council president lives in a house overlooking a one thousand-acre cattle farm near Darlington, a little bit of Ireland in the northern part of this county. He shook hands all around after the seven-to-nothing vote against development, and suddenly I no longer felt like the Indian on the reservation. For tonight, at least, I felt that citizens *could* make a difference. I will enjoy looking at that hillside again for a while. Then this fleeting satisfaction will fuel the energy to rejoin the battle.

Wednesday, July 19. One of the oldest manuals in my modest Turf library is Grandfather's copy of Capt. M. H. Hayes' *Veterinary Notes for Horse Owners*. I am certain Capt. Hayes' treatise also rested on the spinning lazy susan bookshelf that traveled with Finney from Maryland to his last home, Burnage, on the Pisgah Pike near Versailles, Kentucky. Finney often referred to *Veterinary Notes* in his *Stud Farm Diary*.

In his ninth edition in 1924, Capt. Hayes wrote laconically of such ailments as umbilical hernias in foals, which "can scarcely escape the observation of the most inexperienced person. It may vary in size from that of a hen's egg to that of an ostrich's egg."

We have a five-month-old foal with a hernia. Dr. Riddle usually will not suture a hernia closed until the foal is a long weanling, when the stomach wall has thickened. Other vets prefer to reduce the hernia and keep it in place with elastic wrap. Our foal has glaring conformational faults as well, which could "scarcely escape the observation of the most inexperienced person." I'm not sure this foal is worth the vet bill. It added salt to the wound to read Capt. Hayes' remark, "The existence of the hernia naturally depreciates the value of the animal suffering from it."

Thursday, July 20. "Too much rain for good straw this year," said Bill Moore, who usually puts five thousand bales of straw in our hayloft each summer. "I shredded my straw—cost me twenty thousand bales. It was just a brown, ugly mess. I sell my straw for a buck fifty a bale. Of that, it costs me thirty cents to pay somebody to bale it, another thirty cents to move it to your farm, so it hurts a little when it rains a lot."

Hay and straw will be dear next spring, but for the moment, with lush grass shin-high, foals thriving in the overcast weather reminiscent of Peter's homeland of England—their mums dappled out like the broodmare class at the Devon Horse Show—singing *Who'll Stop the Rain* is not popular around here.

Friday, July 21. "Last week, I put down an old gelding the farmer wanted to donate to the Hunt Club," related a visiting Virginia vet, Dr. Tom Newton— over lunch, of course. "If the animal is going to be fed to the hounds, I can't use barbiturates. The drugs will knock the hounds for a loop."

I hesitated before putting a fork in the crabcake. If vets can't use the drug, then I know the alternative is the gun. The macabre subject of shooting horses immediately brought forth war stories. Peter, for instance, was assistant trainer in a Newmarket yard when a legendary English trainer culled the racing stable.

"I held six horses in a row for the vet, who shot them all. Drew an X on their foreheads, between their eyes and their ears, and shot them all," Peter declared matter-of-factly one day this spring. The rookies holding mares in the vet chute swallowed their chewing gum.

"I hate guns and don't carry one," Dr. Newton continued through lunch today. "The farmer shrugged. He didn't have a gun either. But his wife spoke up: 'I keep a pistol right next to our bed.' Her husband had no idea, and they've been married *thirty* years."

I poked absently at my crabcake as the stories unfolded.

Saturday, July 22. For equine ailments seventy-five years ago, Capt. Hayes extolled the pain-killing properties of cocaine: For eye examinations, for relief of lameness, even for painful cracked heels—when "it is well to use a solution of five grains of hydrocholorate of cocaine in twenty drops of oil of cloves."

We had a girl working for us who has refrained from this addictive pain-killer for over a year now. Her anniversary party on July 5th swelled to one hundred Narcotics Anonymous members. Warned in February of the migrant-worker seasonality of a breeding farm, she watched in July as mares and foals boarded vans for year-round homes, and knew her job was driving away as well. She was laid off this week.

"I couldn't have asked for a better introduction back into society than you all gave me," she said yesterday, taking life one day at a time. "I feel so confident now."

I am going to give her Capt. Hayes' manual, so she can see how this whole cocaine business got started. When it is her turn to chair an N. A. meeting, she can wow 'em with a little history. I hope she comes back in January, straight and strong and reaching for a fifth—the fifth of July, Anniversary Number Two.

Sunday, July 23. Trees, rivers, the Bay, blue crabs, and Henry Clark are among Maryland's great natural resources. The Hall of Fame trainer names his homebreds for small towns in the state, and he is taking his sweet time with a two-year-old colt by Carnivalay named Emittsburg.

"He's sixteen hands tall and too fat," he said today at Laurel. "Must get that from Obeah—she was good sized." Mr. Clark trained Obeah, Carnivalay's dam, to win the Delaware Handicap twice in the halcyon days of that track. "I told the girls not to feed him so much. He'll be along in time."

I was glad Mr. Clark was in the air-conditioning, out of the searing heat. He toiled through many a summer at Delaware Park. Andrew worked for him there for several years. On summer evenings, we would watch the boxing matches between grooms in the ring set up under the backstretch trees. Mr. Clark did not push his two-year-olds then, neither does he now. This is bad news for syndicate managers of freshmen stallions, but good news for young colts maturing into strong oaks.

Monday, July 24. The mare hates needles, hates vets, hates being tube-wormed—but she loves getting a cold bath on a hot day. We stand her in the chute, splashing water over her, rubbing her, talking to her. She is busy trying to drink from the hose as we pinch a fold of skin on her neck. No fright response. She is cool in the water. We insert the needle and administer a small dose of Rompum in the muscle of her neck.

Twenty minutes pass. She lifts her head sluggishly, allows us to twitch her, then worm her. She is the sixty-fifth horse Dr. Riddle has tube-wormed in his various rounds today. He treated her like she was his first. His patience kept us all from injury.

Tuesday, July 25. The Hagerstown *Almanack* reprinted a study from one hundred years ago on the "influence of occupations on the length of life." In 1889, one-eighth of all farmers lived to the advanced age of eighty years. No other occupation came close to such longevity, but the younger generation took no heed: "They start out to the neighboring city but soon find an anxiety of mind, a round of perplexing cares, is ever with them, and their life is very soon burned out."

One hundred years later, it is one hundred degrees on the farm, but the July burnout syndrome for this younger generation is abating. Family members return from vacations, refreshed. There is time to visit the Smithsonian sixty miles to our south, or to take day trips to the Amish country thirty miles to our north, or to speak unhurriedly to friends and family.

The *Almanack* also pigeonholed my Zodiac sign, Cancer: "Intuitive, but must fight against despondency." Thanks a lot. To spite the *Almanack*, tonight I cast aside my "round of perplexing cares" and look forward to doing this job into my eighties.

Wednesday, July 26. Dr. Peterson is preparing a seminar presentation on equine herd health. I hope he discusses vaccinating against botulism, a glaring omission on numerous health records I saw this spring. The organism is ubiquitous in nature, a spore-forming bacillus that when absorbed in the blood of foals causes a flaccid paralysis horsemen call "Shaker Foal Syndrome."

In the rapid onset of botulism poisoning, affected foals will walk with a stiff, stilted gait, the decreased limb flexion causing them to scuff their toes. Weaker foals stand for only a few minutes before muscle tremors (shaking) develop and they drop to the ground, recumbent, unable to sit up. The antitoxin available from hospitals is very costly. When we administered a single dose of antitoxin to a foal this spring, I felt I could have rolled up a one thousand dollar bill and burned it, for all the good it did. The label said it all: "Extreme Caution: Botulinum neurotoxin is the most potent biological toxin known to man." The disease does not allow time to run to the vet clinic. Many large farms keep the anti-toxin frozen as part of their medical inventory.

Drs. Peyton Jones, Tom Bowman, and Bill Riddle between them saw a half dozen local cases of botulism this spring, and Kentucky has a higher incidence than Maryland. Our unfortunate foal was the product of a mare shipped here from Kentucky just days before foaling. Mares should be vaccinated three times at monthly intervals during gestation, with the last dose administered two to three weeks prior to foaling. An annual booster protects adult horses thereafter. Foals may receive their first dose as early as one month of age if botulism has been reported.

After we lost that foal this spring, we disinfected his stall with bleach until our eyes teared, sent stall floor samples for cultures to Dr. Robert Whitlock, the botulism guru at New Bolton Center, and stepped up our vaccination program. Yet there was a senseless feeling of unnecessary loss afterwards, because the episode could have been prevented. If science has developed a vaccine against the deadliest toxin in nature, why doesn't everyone use it?

"Because they haven't been burned yet," lamented Dr. Whitlock.

Thursday, July 27. Splashed across today's Baltimore *Sun* were photos of dozens of swimming horses. Exciting shots, but simply another ritual of summer in Maryland—the pony swim from Assateague Island across the tidal channel to Chincoteague, Virginia. There the local fire department auctions off the ponies in an annual culling process, protecting the wildlife refuge from being over-horsed.

These ponies are not the gentle pets of Marguerite Henry's children's book, *Misty of Chincoteague*, and there is some dispute whether they actually swam ashore from the holds of shipwrecked Spanish galleons. What is certain is that Park Rangers share our sense of futility towards invitees who refuse to heed warning signs. "Please Do Not Feed The Ponies" shouts at every tourist on Assateague, but Ellen and I once watched as a man offered one apple to a stud pony, while concealing a whole bag of apples behind his back. The pony heard the cellophane rustle, wheeled abruptly and fired both hind feet to land squarely on the fleeing backside of the surprised tourist, who, as expected, dropped *all* the apples.

Ponies are to Thoroughbreds what mutts are to Irish Setters—smarter.

Friday, July 28. The sun rises at the old Bowie Race Course, now a training center supplying one thousand runners to Pimlico and Laurel. Starter Duck Nigh readies a set to break from the gate. Mosquito Coast, a patient this spring in Peter's R & R ward at Country Life, is back at the races, and she needs a fresh gate card after a six-month layoff. She stands calmly as Duck hollers down the track to riders whose mounts might startle when the gate clangs open. He turns to the riders in the gate: "Get tied on!"

Then the bells ring and the doors swing open, metal banging on metal, and Mosquito Coast springs crisply away, outbreaking four other horses. All I can see is her haunches, propelling her forward through her last long effort before her scheduled return to action on August 11th. Andrew does not want to wind her too tight. He knows that horses have only a finite number of races in their tender limbs, and he does not want her wasting herself for the benefit of morning railbirds with ownership interests. She gallops out the six furlongs in 1:16, returns to the barn hardly blowing. Andrew nods as he splashes water on her.

"Very nice," he says. "Very nice."

Saturday, July 29. Some of our weanlings are almost six months of age now. I can only guess that the forage-to-grain ratio of forty-to-sixty is being attained, because these babies eat all the lush grass they want. I am a little uncertain as well about the calcium-to-phosphorous ratio of 1.68-1, but we are at least close on this score because the pastures are in balance. Creep feeders are filled daily, totaling eight pounds of sixteen percent protein feed for each weanling. At six months, these foals weigh approximately five hundred and fifty pounds. Frankly, it is unnerving to think that Rollaids weighs more than Hulk Hogan.

Sunday, July 30. Today was the first day of my vacation, and the Reverend Converse Hunter preached: "Be ye not anxious." He explained that anxiety was unproductive, that Robert Frost said beautiful things, poems, for instance, were "created from concentration, not worried into being." I got the message, thank you, and blocked home and horses from my mind, even though I had brought along an anthology of wonderful poems, bound just like a hymnal, entitled *Song of Horses*.

In his column from Saratoga in 1957, Red Smith established a link between Daily Double winnings and the collection plate, noting the "happy affinity between horse playing and piety." So I sang along with the lively Irish hymn "Lord Of All Hopefulness," and I thought: How appropriate! We're singing the song of horses after all!

Monday, July 31. August is coming cool to New England this summer. Today, I stood on a street crowded with cars with bicycles on their backs. Later in the day I passed a horse farm carved out of a mountainside—mares and foals and run-in sheds and people working—just like home. I opened *Song of Horses* as evening fell, stopping to read S. E. Kiser's *The Passing of the Horse*. From the turn of the century to this last day of July, its words ring true:

> *When the bike craze first got started*
> *people told us right away,*
> *As you probably remember,*
> *that the horse had saw his day,*
> *People put away their buggies*
> *and went kitin' 'round on wheels;*
> *There were lots and lots of horses*
> *didn't even earn their meals.*
> *I used to stand and watch 'em*
> *with their bloomers as they'd flit,*
> *And I thought the horse was goin' but,*
> *he ain't went yit!*

ALL GOD'S CHILDREN GOT DREAMS

Tuesday, August 1. Prevailing westerly winds blow across the Earth's temperate zones. Easterly trade winds blow nearer the Equator. In between is a calm region sailors called the "horse latitudes," where the air rises straight up. Here, sailing ships would sit for weeks on end, using up provisions before being forced to jettison livestock. The waters of the region often were dotted with floating horses.

It's August. Breeding farms are idled. Feed is being used up. In Maryland, a hastily called mixed sale for early September drew one hundred and fifty entries almost overnight. In Kentucky, breeders decide what livestock to jettison as they fill out auction entry blanks before setting sail for the healing springs of Saratoga, the surf of Del Mar, the beaches of Monmouth.

Maybe a breeze is stirring. The Keeneland sale was up, Saratoga bodes well. Farms are being bought again in central Kentucky. Maryland Million Day is next month. With luck, the refreshing winds of a vacation will fill the sails, and we'll tack through these horse latitudes, back into the trade winds.

Wednesday, August 2. In his 1918 short story *I Want To Know Why*, Sherwood Anderson wrote about Saratoga through the eyes of a fifteen-year-old boy:

"They don't have paddocks under a shed, but saddle the horses right out in an open place under trees on a lawn and the horses are sweaty and nervous and shine and the men come out and smoke cigars and look at them and the trainers are there and the owners, and your heart thumps so you can hardly breathe. If you've never been crazy about Thoroughbreds it's because you've never been around where they are much and don't know any better. There isn't anything so lovely and clean and full of spunk and honest and everything as some race-horses."

Saratoga opened today, and the sports pages of the New England papers are chock-full of the racing action at the ancient Spa. We will follow it vicariously until next week, when we will be chomping at the bit to see that world Mr. Anderson first described for us.

Thursday, August 3. Near here is Thomas Choate's Willoughby Ridge Farm, a beautifully kept farm chiseled from twenty-seven acres of a granite mountain above a New Hampshire lake. White fences stand out against the evergreens. A trout stream flows down into a landscaped pond. A steep ascent from the state road leads to a dozen white farm buildings, where stalls on the second floor of an old bank barn have floors made of first-cutting hemlock.

The horses in the paddocks are smaller than Thoroughbreds and have straight shoulders and plain heads, but a foal is a foal—and no difference exists between the dreams of the men who raise Standardbreds and those of the men who raise Thoroughbreds.

"The Hambletonian is this Saturday," Mr. Choate said. "It has a purse of a million dollars. I'd do a lot of improvements to this old farm if I ever win the Hambletonian."

Willoughby Ridge sends a dozen youngsters a year to various meets. Some make it to the Meadowlands, some only to the Vermont State Fair in Rutland. One of the best-bred horses, Trafalgar Square (Mr. T., for short) developed wobbles, but through acupuncture, veterinarians poked him into being a carriage horse. On Sundays, Mr. T. pulls Mr. C. and his family in any number of different carriages—a Spider Phaeton, a Piano Box, a Queen Victoria—mothballed in 1908 by Mr. Choate's father-in-law when the automobile burst onto the scene. A family enamored of carriages chose the equine sport where a buggy is standard equipment, and, as the poem says, the horse "ain't went yit" from the hills of New Hampshire.

Friday, August 4. Eighty-five percent of New Hampshire is covered in forests. Yet only a century ago, almost the entire state was cleared and in farmland. Settlers used draft horses and loaded rocks onto stoneboats to be carried to farm boundaries that nowadays are stone walls leading straight up tree-covered mountainsides, past tumbled-in cellars, to summits where it seems only a goat could reach. The Civil War took the cream of the crop of New England farmers. The survivors returned to pack and move out West, and the textile mills along the rivers claimed the rest. The farmland was abandoned. Forest regenerated in the process called succession.

I cannot look at these mountains that were once farms without standing in awe of the hardy men and women who clawed a living from this difficult ground.

Saturday, August 5. The Harford County Farm Fair is happening today back home. Last year, planners expected fifteen thousand folks, and forty thousand showed up—a minor Woodstock, as countians new and old choked the tiny back roads for a glimpse of the pig races, the equestrian show, the draft horse pull.

The August issue of the *Maryland Horse* is out, and in the "Looking Back 50 Years Ago" department is a handsome photo of the old Bel Air race track on the season's opening day, eight thousand patrons admiring the new steeplechase course designed by William du Pont, Jr. An eight-race card drew two hundred and ninety-three entries, an average of thirty-six horses per race.

Perhaps racing fell behind because it did not seed itself for regeneration. The grandchildren of the racegoers from the 1939 Bel Air meet are at the Farm Fair today, gambling that the black pig will outrun the brown pig to the feed bowl. I think an impressionable, youthful market was lost forever to this sport when the little tracks fell to the wrecking ball.

Sunday, August 6. Finney's *Stud Farm Diary* from fifty-five summers ago carries the passage: "Stopped by the stud farm of a friend and saw some very fine foals by a son of dead Light Brigade which looks like carrying on in his sire's place. A good Fair Play scion and a Futurity winner of the fast Domino breed make up a well-balanced trio of sires."

Finney's book should be required summer reading between semesters in the horse business, and it is flattering to know he stopped at Country Life to inspect the three stallions—Crack Brigade, Ladkin, and High Strung—that we started with back in the nineteen-thirties.

Monday, August 7. News from home today concerned Travelling Music, a stallion whose biggest enemy always has been his feet. Nine years ago this month, he stood on top of the world: Winner of the prestigious Sapling Stakes, twice victor over champion Lord Avie, the pride of Sonny Werblin's Elberon Stable. His glory days were over after the Hopeful Stakes at Saratoga. A fractured coffin bone did him in, and we took him for stud service. Ancient history now. Trav's on his way to New Bolton Center to have a tumor removed from his right hind foot. X rays show a cyst on the coffin bone. Trav's threshold of pain is just slightly higher than the lion's in the *Wizard of Oz*. He is a massive stallion, not a good candidate for the surgery table. And though New Bolton is a tremendous facility, where impossible cases often are cured, a trip there is about the same as a trip to the principal's office for truant students. Some never come back.

Tuesday, August 8. More summer reading, this from James Herriot's *All Creatures Great and Small*:

"I am not, and never will be, a horseman—am convinced that horsemen are either born or made in earliest childhood. I have the knowledge of equine diseases, but that power of the real horseman to soothe and mentally dominate the animal is beyond my reach. If a cow feels like kicking you, she will kick you, she doesn't give a damn whether you are an expert or not. But horses know."

Herriot then describes getting kicked in the thigh by a stallion. For me, the enjoyment of this summer is heightened by having made it through the breeding season unscathed, with all my fac-, with all my facul-, with all my f-a-c-u-l-t-i-e-s intact.

Wednesday, August 9. "Back! Back! Don't go over there! *Joseph*! Whoa!" hollered Leon LeBlanc, tugging the reins with a surgeon's touch on the harness of his ten-year-old Belgian workhorse. They were pulling oak trees down a mountainside so steep my legs cramped coming up the hill. Workhorses are popular among New England landowners selectively timbering their mountains. Logging trucks and huge skidders tear up the pristine woods. Joseph can drag a thirty-foot oak log down a deer path.

Anchored by a whiffletree, Joseph at rest gives no indication of his power—until Leon hollers. Then the huge chestnut horse bows his neck and drives deep into his harness. A Swedish-made Husqvarna chain saw swings from the hanes of his collar. The muscles in his hindquarters bulge as he digs in. Chains clank as the log lurches, breaking the surface tension, bouncing along now behind Joseph, a timpanic pinging reverberating through the woods as the oak pole travels across rocky ground.

Leon and Joseph work alone, ten hours a day. When they stop to rest, Joseph will sit on his haunches like Francis the Talking Mule. As they work, Leon's fingers, hands, and feet are mere inches from being crushed should Joseph make a wrong move. I have never seen man and beast work so closely, so respectful of each other's power. As I left them in the New Hampshire woods, Leon hollered from above: "If you see Joseph returning to camp without me, you'll know I'm in trouble. If you don't see either of us by sunset, you'll know we're in *real* trouble."

Thursday, August 10. The bark-covered longhouses of the six Iroquois nations once dotted the wilderness now known as central New York state. The healing medicine springs of the Great Spirit of the Iroquois attracted the white settlers, and Saratoga, in fits and starts, was founded. I found it in 1970, and there appeared to be one Indian left: James W. Maloney. He had rich deep color and high cheekbones and a weathered face and he spoke in stutters, as if English was not his native tongue. I think he spoke horse. I groomed for him the summer Loud won the Travers Stakes, and Dutiful took the Adirondack Stakes. I lived in a longhouse on Max Hirsch Drive, near Barn Twelve, and was introduced at fifteen, just like Sherwood Anderson's narrator, to the world of Saratoga.

Mr. Maloney trained forty-two stakes winners, including champions Gamely and Lamb Chop. Today, five years after his death, he was inducted into the National Museum of Racing's Hall of Fame. He was a Great Spirit.

Friday, August 11. Alfred Vanderbilt leaned against the trainer's stand and pulled his porkpie hat down against the dawn. His eyes are not what they used to be. His memory compensates. He appeared in good health: "I promised my family I wouldn't drink until I was eighty years old, and I've got three years to go."

His mind jumped back almost sixty years as if it were yesterday: "You know Discovery raced in your grandfather's silks, until I bought the colt late in his two-year-old year. Tried to buy him earlier—brought my silks up here for the Hopeful Stakes, as a matter of fact.

"Walter Salmon had everything tied up in real estate when the Depression hit," Mr. Vanderbilt continued. "To keep his Mereworth Farm going, he sold his entire yearling crop to me one year for a thousand dollars a horse—I got five stakes winners from twenty horses—and he leased the racing qualities of Discovery to your grandfather, who was his advisor. I was scheduled to go to Africa, and I told Mr. Salmon my final offer, take it or leave it. I bought Discovery in 1933 for twenty-five thousand dollars. He never *could* beat Calvacade in the big races, but he beat everybody else, and carried great weights."

Discovery carried one hundred and thirty-seven pounds or more eight times. His highest impost was one hundred and forty-three pounds. Retired to Vanderbilt's Sagamore Farm, he sired the dams of champions Bold Ruler, Native Dancer, and Bed o' Roses. Grandfather had stipulated to Vanderbilt that Salmon would receive a breeding right for the first five seasons Discovery stood at stud, and from those matings descended much of Mereworth's later breeding success. Discovery is enshrined in Racing's Hall of Fame. I read his wall plaque today in the Museum, but I learned more from Mr. Vanderbilt's oral history lesson.

Saturday, August 12. Saratoga does not have a fancy circle where winners are photographed. So I suppose Angel Cordero, Jr., might have overlooked the little sphere of lime where jubilant folks in coats and ties rejoiced in the moments after a breathtaking Alabama Stakes. But Cordero intentionally galloped Open Mind past the sphere in the mud, down to the grandstand terraces where throngs of thrilled racegoers pushed against the rail to glimpse a riding legend aboard a rising legend. Cordero and the filly were breathless, but their sportsmanship carried them to the fans, then down the stretch again to the arc of lime for the historic recording of today's Alabama.

Sunday, August 13. In the darkness of a Sunday night, stable help walked home inside the hurricane fence surrounding the track. I thought about their living conditions. This morning, I saw a groom duck into a bathroom where only one sink out of five had a mirror for shaving, where the floor was flooded above his flip-flops, where half the commodes were out of service. Later in the day, horse breeder Jim Ryan of Maryland told me he had just resigned from The Jockey Club.

"I was out of sync, I was told. They said they're saving everything for a rainy day," he explained, extending his palm symbolically into the lightly falling rain. "I told them, 'It's pouring out there!' "

Dad runs weekly Alcoholics Anonymous meetings at Pimlico in conjunction with the backstretch aid program Mr. Ryan started. Dad brings home stories of troubles in horsemen's lives, stories that will make your skin crawl, stories that just *might* compel a fellow to resign from The Jockey Club, if he felt strongly about its apathy.

There is more to this problem than the rulers of this sport let on. Their Round Table discussion today was entitled "The Problem of Drugs in Thoroughbred Racing," but the problem they addressed was with horses—not horsepeople. As I thought about this matter, I listened to the grooms joking in Spanish, and I wondered about this difficult issue for which there is no easy answer.

Monday, August 14. Paddy Cleary hoses a Phipps' two-year-old and in a thick Scottish brogue talks all the while: "Learned about 'orses as a plowman at 'ome. Percherons come on the scene. Did the work of two 'orses. Learned about *this* game from a man named Neloy." A gap attendant walkie-talkies the times of working horses up to the clocker's stand: "I take the Pinkerton papers and memorize horses and trainers. The riders know I'll follow them back to the barn if I think they told me the wrong horse." A pickup delivers bushels of freshly cut red clover: "I'll sell eighty trash cans full of clover in a morning, three dollars a tub. Before fire regulations, we heated fifty-five-gallon drums for horses' baths. I sold sweet corn then. Put it in a sack, drop it right in the boiling water. The smell of corn cooking was wonderful."

The driver of the horse ambulance is parked along the rail, asleep at the wheel. Had a busy morning last Friday: Census, winner of the first Breeders' Cup Steeplechase, dropped dead during a workout. All quiet this morning. Jane

Lunger and Buddy Hirsch pose with a two-year-old colt by Carnivalay. The groom coaxes in a Spanish accent: "Stand still, Diablo. Look up."

Red awnings and pavilions and tailgaters and folks buying tomorrow's *Form* before the last race today. Bookstores and art. Bicycles on Union Avenue. Ironworks in windows. Impatiens in flower boxes. Oil paintings on porches. Herons sewn into lace curtains on Regent Street.

Saratoga is a series of unbroken gestures.

Tuesday, August 15. On this day in 1935 at Finney's Holly Beach Stud, he wrote: "We got the stable lawns all mowed and the edges of the gravel walks trimmed—it is constant care to keep the place up to scratch."

Constant care is what the crew performed in my absence, and despite the constant rains, the place is up to scratch. The fields are mowed pretty as a baseball diamond. Horses are thriving on grass tender in the wet weather. Silver Queen corn and garden tomatoes welcome hungry travelers, and the remarkable kindness of nature continues. My only lament is that the hay crops in our locale look so poor.

Wednesday, August 16. Martin Friedman, a Carnivalay shareholder, prepares himself for the misfortunes that accompany his small racing stable. I have never seen him disconsolate about bad luck. He claimed a colt for fifteen thousand dollars, discovered he was a ridgling, sunk another eight hundred dollars into him to have the troubling testicle removed, then received a phone call: "I've performed this surgery three hundred times," the vet said. "First time any horse died on the operating table."

So Marty spotted a filly who had won one hundred and sixty thousand dollars the hard way at New York tracks. He claimed her for twelve thousand. She came back roaring like a lion, her vocal cords paralyzed: "She can't breathe. I don't need another broodmare, but I won't punish a horse either."

Today, Marty phoned with great news. His homebred Carnivalay colt, Betcha Penny, provided our freshman stallion with his *first* winner. Marty's wife, Joanne, had selected Carnivalay as the mate for their mare, Bad Penney.

Tonight, Marty phoned again, crediting teamwork, his wife, his good fortune, the colt's heart, the great sport of racing—his mind jumping forward to November and his cherished Breeders' Cup tickets. Marty and thousands of small owners and breeders like him are the lifeblood of this sport, persevering because they love the game, and understand its peaks and valleys.

Thursday, August 17. Early fetal death in the broodmare is a direct cause of premature baldness among farm managers. I pull my hair out wondering which two of the twenty mares so plump and fat in the front field will fulfill the statistical certainty of ten percent early fetal death. Some farms put the entire broodmare band on Regu-Mate, at least for the first five months of pregnancy, after which the placenta assumes progesterone production from the ovaries. From a defensive position, this practice allows the manager to inform the mare owner: "Well, we had her rechecked and she's empty. Not our fault. Had her on Regu-Mate." But the additional two hundred dollar expense per month, *per mare*, sometimes is the straw that breaks the owner's back.

It's a guessing game. No category of mares—lactating, maiden, or barren—has proven especially susceptible. Only mares with a history of endometritis, or chronically low assay tests, are true red-flag candidates. Even rectal palpation can be inconclusive. Sometimes an interval exists between fetal death and collapse of the fetal sac, blurring the distinctive feel of a normal pregnancy. Stress, pain, weaning, disease, steroids, van rides, nutritional changes, frost on fescue—almost anything can contribute to early fetal death. Well, that day in late summer when the mares roll through the vet chute is just one more anxious milestone on the long road to a live foal.

Friday, August 18. I phoned Rollaids' owner today, told him I did not think his colt would make it as a racehorse. Rollaids' legs have straightened out passably, but I believe he suffered mild hypoxemia—a lack of oxygen— during his difficult birth. Mentally, he lags behind his classmates. When he whinnies, he sounds as if an old car is starting up. And his wonderfully placid personality is deceiving; the lights are on, but nobody's home.

So we took Rollaids off his owner's account today, and put him on ours. No doubt he could fill Roger's vacated role as the riding horse for the uninitiated, or Ellen might break him to the carriage. It's full circle, farm-style: Roger moves on, Rollaids moves in.

Saturday, August 19. Eight of our fifteen foals have been weaned. I look for warning signs of epiphysitis—the swelling around the growth plates of certain long bones. Nutritionists used to call it "Metabolic Bone Disease," but that was too easy to remember. So they renamed it "Developmental Orthopedic Disease." By any name, growth-related skeletal problems often begin at weaning time. Affected joints, such as the ankles and knees, display a boxy, puffy appearance.

If we find early signs of epiphysitis, we'll put that weanling on a lighter feeding regimen. I've found, however, that foals who enjoy great freedom of exercise are less likely to develop symptoms than those who are hothoused eight hours a day. The condition is exacerbated when the ground is dry and hard, but our fields have been buoyant with the blessed rains this summer has provided, and the weanlings look well.

Sunday, August 20. At 9 A.M., the fields erupted with the cacaphony of forty geese discussing seating arrangements as they plummeted from the sky to feast at the ground feeders. They picked and pecked—some standing sentry, some too hungry to care—until we approached. Then they waddled in an indignant wave up the hill, honking their displeasure, a lone snow goose standing out like an albino. In two months, the flow of geese headed south will be constant over the farm. For today, though, our feathered friends merely presaged the turning of the leaves.

Monday, August 21. Two sales companies presently serve the mid-Atlantic: Venerable Fasig-Tipton and upstart Equivest. Both battle the same problem: How to create an upbeat atmosphere at a rundown facility. The colorless Timonium sales pavilion resembles a minimum security prison. Buyers crick their heads at forty-five-degree angles to examine yearlings walking on pitched asphalt—horse, groom, and rainwater all headed down the same drain. Sellers numbly block out sale companies efforts to spruce things up: "What flowers? I didn't see any flowers. No buyers, either." Inside are plastic seats and cigar smoke and the drone of auctioneers culling out horses.

I dream of a sale under a red and white tent on a beautiful farm in historic Baltimore County, of inviting prominent East Coast trainers to a day's racing at Pimlico, to a catered sale of fifty of our best local yearlings, to stay over at farmhouses, not Holiday Inns. Whimsy? Maybe not. Bold Forbes stepped out of a tent in 1974 and two years later won the Kentucky Derby. There is precedent for such flights of fancy.

Tuesday, August 22. My niece Karianna Johnson worked as a groom at this month's Saratoga yearling sale. She wrote me a letter today describing *her* life at sixteen at the Spa:

"How, I ask myself, did 5 A.M. arrive so suddenly? It seems only minutes ago I laid down. I've got to get my stalls mucked before the truck comes at five-thirty. We stop at Dunkin Donuts. The only ones awake are other grooms or exercise riders, and some cats, a goat next to a nervous horse, and the birds. I catch my stalls, top off buckets, wash my brushes and sponges. The yearlings are bathed, the tack room cleaned. Time for breakfast. I never thought I could be hungry at six-thirty in the morning."

For almost everyone, nights are a blur at Saratoga. Memories of mornings are crystal clear.

Wednesday, August 23. Uncle George—unrelated except in the way all members of the horse family are related—runs a Bel Air seafood restaurant bearing his name. He used to be the chef at the Pimlico Hotel. He loves folks who supply him with information as fresh as the bushels of blue crabs clawing at the baskets in his kitchen.

"This fellow gives me a tip last week," he said tonight, his gray hair starched skyward by the steam from the crab kettles. "I bet *five big ones*. The horse wins. This guy tells me four more good ones. They *all* win. So I've made ten big ones, and the next day he calls and tells me how to bet. But I don't wheel the horse—just bet him to win. He gets beat a nose, it takes the stewards ten minutes to look at the photo. I lose everything. I hate the track somedays. I say to myself: 'Frank De Francis, I love you, baby, but I could kill you for this.' I was only *joking*. Really, I feel terrible I said that. My pal Frank died two days later."

Uncle George looked up to heaven, palms together, finishing his story. He *really* liked De Francis, who died last Friday from heart problems. I hate to imagine what Maryland racing would be like today had not De Francis revitalized the game. He made bettors, like Uncle George, feel welcome again at the tracks. He knew that the fan was the most important player in the game of Thoroughbred racing.

Thursday, August 24. After being turned back out from their morning feed, two old quiet mares and their May foals gather the eight weanlings at the top of the hill, in search of a breeze. Almost asleep on their feet, the mothers do not notice it is not their own foal at their bags, but a steady procession of hungry orphans taking turns at amenable milk machines.

Rollaids and his cohorts have perfected the art of stealing milk. It reminds me of Spanky and Our Gang peeking through the hole in the center field fence: "Easy, Buckwheat, one guy at a time," I can hear Rollaids admonish. I like the calming influence the old mares have on the weaned foals. But I must be careful not to let these rapscallions drag down their baby-sitters.

Friday, August 25. Ellen returned tonight after a day in the hot Pennsylvania sun, having outbid an Amish farmer for an old buckboard wagon at an annual auction of horse-and-buggy items. John King, our senior employee, rarely comments on anything other than the Orioles' pitching, but he shook the buckboard knowingly.

"Tell Edna the wheels are loose," John declared. He never gets anyone's name right. He calls Ellen "Edna," Assault Landing "Salt Lake Landing," Allen's Prospect "Ol' Alley Cat, by Mr. Proskecker." He mustered himself to his full five-foot height upon recollection of the buggies of his youth. "I'm not gonna ride in this thing. No."

Truth is, he would not ride in it even if the tires were cemented on. John was dragged by a team of plow horses when he was seventeen. He has been extraordinarily cautious about horses ever since. He is my little farm watchman, with an imagination for worrisome possibilities exceeded only by my own. Before Rollaids runs off with Edna, we will replace the leather bushings on this new old buckboard.

Saturday, August 26. The weanlings associate us with the pleasure of eating, and so have not developed unsociable habits. Some farms tie weanlings in stalls while they eat, a restraint technique I am uneasy about for such young horses. We do, however, rub them down and pick their feet up, which adds years to the life of Tom the blacksmith.

This season, every weanling is sired by a home stallion, out of mares which live here year-round. We are attached to these young babies in more ways than one: If they succeed, we succeed.

Sunday, August 27. The Maryland State Fair today drew fifty thousand folks. Part of the lure is the ten days of racing at Timonium. Imagine being a youngster atop the ferris wheel, gazing down the midway at speeding horses, then dragging your dad to the rail for a closer look at the rainbow of silks flying past, to hear for the first time the noise—the unforgetable hollow thunder of hooves, of jockeys shouting for position.

In Baltimore, parents give five-year-old kids lacrosse sticks instead of baseball bats, and the city is the cradle of lacrosse. Perhaps Timonium does the same thing for Maryland racing.

Monday, August 28. We've saved sales catalogues since the day the mortar dried on the new farm office six years ago. Today, we cleaned house. How many mares in these catalogues are now out of production? Where are the foals we sold, by sires we used to stand? How are the people who played this game just a few years ago? I recall so fondly Dr. Beegle from Charlottesville. Dr. Beegle raised Dalmatians, of course. He presented us with a spotted puppy we named Alien, who was hyper and ran into Route 1. By that time, Dr. Beegle had sold his mare and quit the Thoroughbred game.

This sport changed forever in 1986 with the new tax laws. It will never be the way it was. Probably for the better, I try to convince myself, rifling through back files of long-gone owners and farther-gone mares. There is always the feeling that as soon as you throw something away, you will need it. So I was glad when it started raining, soaking the discarded papers in the bed of the pickup. It seemed Mother Nature was saying: "Move on, son. Clean the shelves. Forget the good old days, even if it *was* such a short time ago." No ghosts were in the office late tonight when I shut the windows against the rain. Just a clean workplace, ready for the future.

Tuesday, August 29. The happiest racing scene on film occurs in the Marx Brothers' *A Day at the Races*: Tap dancing and piano playing and everybody singing *All God's Children Got Rhythm*, the camera drawing back to show the flimsy backstretch houses animatedly swaying with soul, Harpo on the flute and Groucho wagging his finger and Chico rolling his knuckles up and down the keyboard, singing wide-eyed: "Who Dat Man? Dat's Ga-bri-al."

Gabriel was at Monmouth Park today. Our guardian angel carried Groscar, a Carnivalay colt, across the finish line first in the Open Mind Stakes, and by this evening, all the houses on the backstretch at Country Life were swaying with the good news as we sang wide-eyed: "Who Dat Stud? Why, Dat's Car-ni-val!"

Wednesday, August 30. For better or for worse, for richer or for poorer, a little farm such as ours is married to the shareholders in the syndicated stallions. Five years ago, we set out together on the long road towards proving a stallion. Twenty quarterly statements ago. Four full breeding seasons ago. One hundred and twenty offspring in the interim. Maybe a dozen different advertisements— photos of a young stallion, photos of foals, photos of promise.

News too good for a mere memo was spread by phone today. Mrs. Jane Lunger, who bred the stallion, said it was the happiest news she'd ever heard. (Breeders in general are parched for glad tidings these days. And usually I have only news like: "Your foal just died," or, "Our rates went up.") At day's end, we made one last call, to Monmouth, to order the win photo, a little bit of promise fulfilled.

Thursday, August 31. Sometimes this business just gives Michael the fantods (as Mr. Anderson's young narrator would say): A state of high irritability. Groscar's owner phoned today to find out about the Maryland Million; the colt he loves is *not* eligible. It is not Michael's fault, but nevertheless he is quietly furious with himself. As the nomination deadline approaches every December 31st, he sends memos, plugs it in the farm newsletter, then grabs the Foal Reports and hits the phones—cajoling, coaxing, convincing folks to nominate their weanlings and yearlings. Thereafter they are eligible for life—or ineligible for life. No supplements. No horses of racing age can be made eligible. It is a program with teeth. Michael feels as if he got bit.

Sixty percent of all foals sired by the home stallions are eligible, but not Groscar. Michael will sit in the stands on September 10th and cheer for the winners because he has a high sense of sportsmanship, but he will be dreaming of the one that got away.

It was a spectacular day to end the summer—eighty-four degrees, not a cloud in the sky, a breeze moving us towards autumn. The familiar flock of forty geese landed on the hillside behind the office, the snow goose beautifully stark against the green grass and her gray friends. Tonight, we will sit outside on the porch, listen to the Orioles game under a sky filled with stars. The Birds were in last place last year, the laughingstock of baseball. They have been in first place most of this season. They are dreaming of the pennant.

What's the difference? If we couldn't dream, we would be out of this business in a minute. Might be better off for it, seeing as how so many days are nightmares. But even the name for the worst of dreams has a horse in it. Michael, didn't you say that colt was eligible for the Breeders' Cup? Well, go on, little brother, dream a little. All God's children got dreams.

AUTUMN SKY

Friday, September 1. An eccentric in-law gave us a pair of aerator shoes, green plastic gizmos with spikes on the soles. Ostensibly, the shoes would poke holes in the pastures, break up root-bound thatches of grass. I told her the best fertilizer for our farm would be the manager's footprints, and I would be easy to track in my strap-on shoes, thanks to her.

I have never worn these golf shoes for farmers, but I saw them in my closet today, and they reminded me to do a little pasture maintenance: The fall growing season is here.

Saturday, September 2. This morning, Hal Clagett phoned for a Stallion Service Certificate to register a foal. Mr. Clagett, who is at ease in the minutiae of The Jockey Club requirements, knows that we hold the formal certificate until the breeder requests it. We used to forward them as soon as The Jockey Club mailed them to us, but breeders often misplaced them in the long interval between pregnancy and registration of the resulting foal. (This certificate is also a means of last resort for collecting unpaid stud fees.)

"You recall we bred this mare to two different stallions during the same heat period last year," Mr. Clagett reminded me. "The bay stud, who didn't like her, then the gray stallion."

In the old days, before blood-typing and parentage verification, the names of both covering sires often were listed on the foal's pedigree, even if one stallion bred the mare in February and the other stud bred her in June. In the modern Rule 1 (E) of the *American Stud Book*, The Jockey Club assures it "will make every effort" to eliminate the incorrect stallion. Yet there are old-time mare owners who interpret "every effort" as not good enough, and exert unreasonable pressure on stallion managers to omit the first covering stallion on the Report of Mares Bred. Not Mr. Clagett. He knows the rules, and he understands the principle of two-coat color inheritance: A chestnut sire and a chestnut dam *must* produce a chestnut foal; a gray foal *must* have at least one gray parent.

Susan Bates of The Jockey Club told me only two of the fifty thousand foals born in the most recent fully registered foal crop will carry double parentage: "Any mare owner who is afraid their foal will carry the names of two sires just doesn't know the facts." Nevertheless, I prepared an affidavit setting forth the particulars of the mating of Mr. Clagett's mare.

"This handsome colt will not suffer for paternal ambiguity," Mr. Clagett said. The popular television show *Matlock* portrays a Southern lawyer who is country clever. I tease Mr. Clagett he should star in an equine version entitled *Fetlock*, because Hollywood cannot top the legal bard of Upper Marlboro, Maryland.

Sunday, September 3. The word canter supposedly derives from the easy, miles-consuming pace at which mounted pilgrims rode to Canterbury to pay homage to St. Thomas á Becket. I saw Rollaids canter today, and his gait bore no resemblance to that smooth, brisk pace. His hobbyhorse hustle was a pilgrimage to the creep feeder, where St. Thomas a Bucket awaited.

Monday, September 4. Dan Derr from Darlington is our soil consultant. He unloads his all-terrain vehicle from his pickup and bounces out across the fields, where he sticks a hollow pipe into the ground and withdraws four inches of soil. He calls back a week later:

"Your pH in the back field needs to get up above 6.4, so lime it this fall. Copper levels are good. Zinc, boron, manganese, iron—all good. Rotate your pastures like you've been doing. Dress up the front field with a little fertilizer. You're on target everywhere else."

Dan charges a modest fee for a service that is free from the Agricultural Extension Service, but I have not had much luck with the the fellow from the county, who tries to tell us how to keep our green acres green.

Tuesday, September 5. "Poke me with a broom. Twitch me if you want. I am *not* getting on that van. No yearling sale for me. Richard Harris might be the strongest man in the world. Big deal. I weigh nine hundred pounds. Richard is *not* going to push me onto that van. Not today. Not ever. I was on a van last week when my owner shipped me here. He drugged me all up. I said, 'Never again.' Haven't been here a week, and you bring another van. No way. No sir. No thanks. I like this farm fine. And I *don't* like that barn on wheels. I'll sit down if I have to. Hey, what's this? They're taking me back to the stall. I won! I won!"

(He lost. He lost. We gave him three cc's of Rompum, and he changed his stubborn mind.)

Wednesday, September 6. John Williams is a Marylander who moved to Kentucky, where with Lee Eaton he presents fine consignments at major auctions. He cut his teeth, however, on the rank-and-file yearlings marching through the Timonium sale ring. He has told me more than once: "Pay attention to the hind end. Most folks can't look at a hind end, but a good horseman can. Does the yearling break over in his hocks? Does he look strong and athletic? Does he look quick? If he doesn't, go on to the next one."

There were maybe a half dozen real athletes among the one hundred and fifty yearlings in the sale tonight, and the trainers spotted them as if picking the cat out of the drawing. The trainers, though, left before midnight, when Michael found the last cat in the sale. He bought a modest colt from the first crop of Allen's Prospect, with a hind end on him like a Quarter Horse. The colt is plenty strong enough, looks athletic, but only time will tell if he is quick.

Thursday, September 7. At the 6 P.M. starting time for the mixed sale, I did not see fifteen people in the pavilion. By the end of the evening, forty-three of the eighty-three horses sold brought seven hundred dollars or less. As Ross Peddicord wrote in his *Evening Sun* series, a lot of these horses are taking their last ride.

I left the sale and went to a crab feast at Pimlico, where Mary Thomas received the Humphrey S. Finney Award. It is a curious thing that just when this horse business has me totally frustrated, I am able to draw strength from the good people in it. I sat at the same table with Mary, and it felt great to crack crabs and listen as she humbly accepted congratulations from the breeders for whom she served so tirelessly.

Friday, September 8. The voice of Jim McKay was a wall of sound, bouncing off the Sports Palace chandeliers, his steady countenance on TV monitors lining the black walls of Pimlico's high-tech gambling emporium. He presented awards to Maryland horsepeople at the Breakfast of Champions. His kind words made every recipient feel like a gold medal Olympic winner. To *Unsung Hero Award* winner Danny Fitchett, a man of granite anchoring the starting gate crew: "The first time I went racing and saw men wrap their arms behind two-year-olds kicking and rearing at the gate, I told myself, 'That must be the most dangerous job on the track.' Shaking Danny's strong hand this morning, I understand why the horses decide to get in the gate."

To the *Outstanding Backstretch Employee Award* winner, McKay splashed colors on the palette of groom Robert Sterrett's life: "The backbone of successful trainers...taking care of top racehorses since 1957...rising before dawn for three decades." By the time Robert took the mike, he was so uplifted he declared: "It took me thirty years to get this award, and I want to thank everybody—including *myself.*" It was the moment of a lifetime for him. He cradled the heavy glass trophy, tried to find the inscripted side for the camera, while McKay patiently waited, nodding assuringly. The tables were turned when Duck Martin presented McKay with the *Outstanding Media Person Award*. This morning's standing ovation flustered McKay.

"I'm okay on the giving end of an award," he stammered, "but when I get in this position...the words just...stop."

Saturday, September 9. In a throng of five hundred people at the Maryland Million cocktail party, Lynda O'Dea looked lonely. She still grieves the loss of

her good friend Frank De Francis. Lynda was the inspiration for the Sports Palaces, for the renovations of Laurel and Pimlico, for the little touches of glamour that transformed racing plants into glistening casinos for the horse-player.

"He did everything for the fan, and in the process helped the breeders," she said. "We're gonna miss him terribly."

Sunday, September 10. In the paddock after Countus In earned the winner's share of a one hundred and fifty thousand dollar purse, Charlie Middleton clutched the victory blanket of flowers. I thought maybe we would have to water the flow-ers *and* Charlie to keep them both fresh enough for the folks back home in Louisville to see. They had laughed when he said he was taking his filly East for the Million. Her form certainly did not excite anyone, but Charlie likes Maryland, had paid the eligibility fees and, well, with Julie Krone riding, he took a shot.

"And I've got her half-brother in the sale at Keeneland tomorrow!" Charlie glowed.

Maryland-sired horses flew in from California and Canada, and Woody Stephens and Laz Barrera, Tommy Kelly and Phil Johnson, each had runners on the card. Gov. Schaefer greeted them all, and Maryland racing enjoyed a day that would have pleased Mr. De Francis, who probably lobbied the Lord to have the event simulcast to the big Turf Club in the sky.

Monday, September 11. At a party tonight, Dad mistakenly picked up a loaded screwdriver instead of his ginger ale and orange juice. This is the fear of anybody in the program—to be innocently, and instantly, overcome by proximity to the devil. A local alumni of horsemen, graduates of the same school of thought, take strength from each other, keeping depression at bay, stressing the positive. The devil cut down a classmate last month. The memorial service is this Friday. Dad put the screwdriver down. He thought about the cunning nature of this disease. He thought about the young horsemen he helps at the Pimlico A.A. meetings.

God bless Dad. God bless his alumni. God give me the strength to weather the same pressures Dad felt when he ran this farm.

Tuesday, September 12. "I'd rather go door-to-door with my grandsons selling *The Book of Knowledge* than hold hands with green investors in a racing partnership," said John A. Bell III, as one of three thousand yearlings being sold at Keeneland this week kicked up dust in the walking ring. "They all want to sit in the box at the same time. They drive a trainer crazy asking about bucked shins. And they won't like it in the end when they haven't made any money. When I did it, the tax laws made it attractive for newcomers. Most were in the seventy percent bracket. No fun now. It's going to be hard to get new owners in this game."

He reflected on a tougher issue: "I don't think this industry has a bright future. Can't run a farm at a profit. Expenses are too high. On the horizon is 'All Sports Betting,' and if we don't beat them to the punch and put All Sports Betting in at our racing plants, we're dead. If they build their own facilities? But I'm old and you're young so I'd say sure, give it a whirl, try and get some new folks in the game. Just don't let 'em think they're gonna make money racing horses, 'cause the odds aren't with ya'."

Wednesday, September 13. Endoscopic examination the morning after the colt was sold revealed trouble. The seller was not sympathetic: "Get a second opinion, get a third opinion, get Dr. Copelan to scope him. It won't make any difference. You bought him. No warranty for wind at this auction. The colt is yours now."

The buyer had acted without his agent, and now he had "buyer's remorse." He had seen a colt late in the sale that he felt was selling cheap, and he had bought him. When the colt did not pass the vet exam, he refused him: "Let them sue me. I don't care if it takes two years in court. I'm *not* paying for a colt with bad wind."

Racing secretaries might someday add a special condition race—for three-year-olds and up, whose ownership status was delayed by warranty issues as yearlings.

Thursday, September 14. The excuse to fly into or out of Lexington, particularly on a clear day, is one of the unheralded pleasures of the horse business. From the air, lime green fields of tobacco quilt the bluegrass pastures, and black angus appear no bigger than insects as they wander down prescribed footpaths. The ovals of training tracks, the thin treeline boundaries between farms, the turquoise swimming pools of an agrarian Beverly Hills, the white fences of Calumet: This is Lexington from two thousand feet.

Friday, September 15. Twice in my life, I have seen neighboring farms destroyed by fire. On a windy spring day in 1963, a grass fire spread to a forest of white pines, setting off incendiary explosions in pockets of sap-filled trees as the fire sucked all the oxygen from the air, then crowned, blowing sky high, almost completely extinguishing itself before—*poof*—starting up again. Houses sprouted up on the property the next year.

While out in the fields this afternoon, I saw small clouds of black smoke rising over the abandoned farm across Route 1. Too windy to burn leaves, I thought. *That smoke's coming right from the barn area.* The only car handy was the old yellow Volkswagen Thing—Dad's temperamental vehicle of last resort. My first thought was that I did not want to be stalled out across the wooden bridge to the old farm in Dad's Thing, a fire truck roaring up and an arson investigator slapping the cuffs on me right there on my arch-rival's property—the Good Samaritan Rule gone amok. So I drove the Thing to the small construction office at the foot of the property, and summoned the secretary outside. When we both saw the smoke coming from the barn area, she dialed the fire department. Sirens started up two miles away. I phoned the developer's office. I have not seen him since his zoning request was denied in July, but when a man's barn is on fire, he ought to know. He wasn't in. I left a message.

Neighbors gathered on our hill as a gray fog crawled across our farm. The mares we weaned this morning hollered for their foals. There was danger in the air, acrid with the smell of smoke. It was not the barn after all; it was the two hundred-year-old stone farmhouse on fire. It burned all after-noon, despite hoses trained on it, supplied with water from Winters Run.

This evening, we drove down the old lane, overgrown vines grabbing at the mirrors of the pickup. A policeman stopped us. Arson, he said, and a damn poor job of it. He drove past. Ellen and I were alone at the charred mansion. It looked as if it were a house under construction, the mansard roof not yet finished, walls of fieldstone and chimneys of brick, dormers of blue slate scalloped in the high style of a bygone era. In the quiet, I heard water dripping down through the floors onto the linoleum laid by the last itinerant tenant.

The developer had plans to make a restaurant out of the old house. The Harford County Historical Society had approached him about using the house as its headquarters. Well, no more. This evening, the sharp smell of water on burned wood blanketed the yard. Indians believed rocks were alive, and an Indian could have heard the stone barns crying over the death of the house. The spirit of the old farm took flight tonight, gone with the wind that had fed the flames.

Saturday, September 16. The weanlings change very quickly. One week, all fifteen appear to be fit as a fiddle. The next week, two or three look too fat, grass bellies stretching the skin over their ribs. Weanlings with conformation faults suffer the most from being overweight, with increased trauma on joints and growth plates. A checklist of conformation problems (upright pasterns, toed-in or toed-out, bench-kneed, base-wide or base-narrow in front, post-legged or sickle-hocked behind) provides a fresh examination of foals we see every day.

The pot-bellied look in weanlings is sometimes caused by a high percentage of lignin in the forage. Lignin is the stemmy part of the hay, the cellulose, which is not easily digestible. Tough pastures can be the culprit, too. Yet in some ways, weanlings are just like children—sometimes you just have to leave them alone and let them grow. After all, in college I lived on spaghetti and fish sticks six nights a week, and I'm worried about *their* diet?

Sunday, September 17. Wanton acts form the blind side of life. The arson across Route 1 is the latest example. Does the arsonist live nearby? What can we do to protect ourselves? We installed spotlights on the barns a few years ago. We drive into the farm at night from different lanes, always looking for the unusual. We store gasoline, paints, and thinners in an old corn shed off by itself; it has no electricity, so a vandal can't switch on a light and find fuel to burn.

This farm on a Sunday morning is so tranquil that danger seems remote, but we are no different than the deer in our woods, our ears pricked, instincts alert, surviving in nature, but aware of the hunter.

Monday, September 18. Some stallion contracts state that a stud fee which is due on September 1st, ordinarily refundable if a live foal is not produced, becomes non-refundable if a pregnant mare is simply *entered* in a breeding stock sale. She might be bid-in and returned to her owner. She might be scratched from the sale and never even leave her home farm. Non-refundable, nevertheless. The rationale is that sales preparation, or the journey to and from the sale grounds, causes stress to the mare, possibly contributing to an abortion that otherwise might not have occurred.

Our contracts might include this clause soon. A fellow has a mare in foal to Allen's Prospect. He will receive his Stallion Service Certificate by paying the stud fee. He will then be able to enter his mare in the November sale in Kentucky, though he told me he does not *really* want to sell her. She will be eight months pregnant when she travels twenty hours from New Jersey to Kentucky, to arrive at a farm she's never seen, to go to a sale where she won't reach her reserve. She'll be home for Thanksgiving—but she is cross-entered in the December mixed sale at Timonium, just in case the bidding in Kentucky is not spirited. She should be home for Christmas this time.

The mare is due to foal on Groundhog Day. If she doesn't produce a live foal for the man who shipped her to two sales half a country apart, then he gets his money back. I have been thinking about that no-refund clause for a few seasons now. I hope I don't learn the hard way that I should have put it in *last* year.

Tuesday, September 19. In late summer of 1853, Thoreau wrote: "Live in each season as it passes. Breathe the air, drink the drink, taste the fruit, and resign yourself to the influences of each."

Two days before autumn, we are in hurricane season, and while Hugo punishes Puerto Rico, he is benevolent to us. The pastures drink deep of the first soaking rain since mid-August, the mares taste the fruit of the second growing season, and the crew resigns to wet clothes. Ground feeders are emptied of rain, mineral blocks are replenished, and stalls are bedded for the weanlings. In the barns, cobwebs are swept and fresh coats of paint brighten the darkness. The rain pelleting the tin roof is nature's percussion, loud as a snare drum in some barns. The voice of Winters Run, swollen with the life of its tributaries, echoes up the watershed valley, evocative as any opera.

Hugo might batter us tomorrow. It is his season, and "Nature rests no longer at her culminating point than at any other. If you are not out at the right instant, the summer may go by and you not see it." For the next forty-eight hours, Mr. Thoreau, we will be watching closely.

Wednesday, September 20. Pat Trotter of the Thoroughbred Owners and Breeders Association was on the phone, discussing the New Owners Seminar scheduled for Laurel in October.

"Our research indicates that new people are intimidated by folks already in the game," she said. "We try to overcome this by keeping the seminars small, maybe twenty people maximum. We walk to the backside, introduce trainers, touch the horses. At Monmouth, a veteran owner received the bad news that his horse had a fracture and would be retired from racing. He carried the X rays around to show how things can happen. He admitted to new people that he was disappointed, but said he would regroup and buy another racehorse next year."

Current management at Maryland tracks is making it worthwhile for existing owners, while the TOBA plows new ground. In the future, perhaps these concurrent efforts will harvest more consumers for the product we make.

Thursday, September 21. The farm today felt as humid as a tropical plant conservatory, Hugo pushing Caribbean air north. We felt the excitement of an impending snowstorm, in reverse. The sun raced to set ahead of brooding purple clouds pouring in from the south, while the weanlings grazed voraciously—instincts forecasting heavy weather, a reservoir of energy being stored. An eerily warm rain fell as the evening news prepared the East Coast for a Category Four Hurricane.

Tonight, South Carolina will see the eye of Hugo. Tomorrow, we will gaze in awe at him, safe by two states, I hope.

Friday, September 22. "I want to buy a yearling for about fifteen thousand dollars at Timonium this Sunday," a friend said today, planning to spend twenty-five percent more than this select sale averaged last year. "I expect to keep him in training until the fall of his three-year-old year. What are my costs?"

It's thirty dollars a day for ninety days to break the yearling: Twenty-seven hundred dollars. Ship him south to a training center to miss the mid-Atlantic winter—forty a day for another ninety days: Three thousand dollars. Back here to a Maryland track from April 1st to December 31st, at the going rate of fifty a day: Nine thousand two hundred and fifty dollars. Nine hundred dollars per year for the veterinarian. Six hundred and sixty dollars for the blacksmith. One thousand dollars for vanning, another thousand for mortality insurance. Total expenses Year One: Nineteen thousand one hundred and ten dollars, conservatively.

"With my luck, I better have Year Two in the bank—another twenty grand."

What my friend stands to gain in purses is pure speculation. But his fifteen thousand dollar yearling will likely cost an additional forty thousand dollars within two years of tonight.

Saturday, September 23. Hugo missed us. Just high winds roaring through the hardwoods like a noisy subway train. A cold front scurried the rain clouds at dusk, and Ellen walked with Boxy to see the silhouettes of fifteen babies and one mom strung out like reindeer against the low orange sky. The sun fell quickly, the light catching in the moist air, and a spectacular, thick rainbow rose over the watershed. We watched the refraction of light grow in only a few moments—hurried up like a time-lapsed photo, a ribbon of color. There is no sky like the autumn sky.

The weanlings cantered by, kicking at the cold air blowing against them, Rollaids grunting as he tried to keep up with the cool guys. We turned to walk home when suddenly a red fox, out after the storm, startled himself *and* the dog. The centuries-old pursuit was on, the fox on the high ground, hoping his scent would be lost in the wind, the hound's nose on the ridge. They ran *right* past us, down the rain gulley in the middle of the field. Ellen called Boxy off, and the dog stopped. The fox ran on, casually looking back over his shoulder. After all, we live on his turf, a la-la land of Saturday night nature.

Sunday, September 24. The hero in the book *The World According to Garp* was ecstatic when an airplane flew into the house he wanted to buy: "It's predisastered. Nothing worse could happen. I'll take it."

A local breeder, perhaps hoping to predisaster his yearling before tonight's sale, registered the crass name "Why Buy the Cow" for his beautiful filly by Allen's Prospect. The catalogue misspelled it, called her a Crow instead, but it did not matter. She had the look of eagles—not of crows, nor of cows—and a lady from California paid twenty-three thousand dollars for her.

I will have a hard time falling asleep tonight. Sales do that to me. We bid on the Eagle, auctioneer Laddie Dance gravelly asking for more. The lady from California jumped the bids in increments of two thousand dollars. Adrenalin flowed. We were out at fifteen, but Laddie kept working. "I'm gonna *sell* her," he threatened, his face beet red. Go ahead. The Eagle can join the swallows in Capistrano, but in my dreams tonight, I will see her migrate east in the fall, like the geese sailing daily over the farm now, to take next year's Maryland Million Lassie.

Monday, September 25. The stallions love this cool weather. Corridor Key plants himself firmly and kicks with both hind feet upon hearing the chain shank rattle in the morning. He wants to get out and play so badly his energy reserves boil over. We cool him down with carrots, coax him out of his stall slowly so he does not hang a hip on the doorway. If he persists in his foolishness, we start all over, patiently schooling this giant child.

There is a line of respect between groom and stallion that stretches like a rope bridge across a canyon. You must walk that bridge together, every step of the way, or the bridge will start to sway. That's when you are headed for a fall. These studs will push you.

Tuesday, September 26. I have read the self-help book *The One-Minute Manager*, but it only took me thirty seconds to ruin an eight-hour day. Steve Brown, who can fix a submersible pump, solder copper plumbing pipes, spackle drywall, tune a tractor, set up scaffolding, or shingle a roof, was forty minutes late for work this morning. Baby-sitter trouble, I found out later.

We have been together eight years. I know from experience that Steve is a horse best ridden with a loose rein. But I forgot, chided him, regretted it. Said I was mad on the outside, not on the inside, so forget it. He understood. I came home feeling small. I yanked *The One-Minute Manager* down from the shelf, stopped at the introduction, when Confucius said: "The essence of knowledge is, having it, to use it."

Wednesday, September 27. Opponents of the county council's approval of a mall ten miles from us gathered four thousand signatures on a petition, thus forcing the entire Comprehensive Land Use Map to a referendum on next year's ballot. Is there such a thing as "no growth"?

For a year, maybe. In the meantime, Gov. Schaefer has promised to put punch in the farm preservation program. The only asset many horse farmers have is their land. It would be great if the state compensated farmers by offering them *real* dollars to withstand the temptation of selling to developers, but sometimes promises made in the city of Annapolis never really reach the countryside.

Thursday, September 28. "Think she's ready?" Andrew asked Peter, as they turned the yearling filly in her stall. She had been "bellied" by Andrew, but had never felt the full weight of a rider sitting on her back.

"Let me give her a few more turns," Peter replied. He led the filly in a tight circle in the stall, instilling his confidence in her. "Okay, she's ready."

Andrew gently lifted himself up into the saddle. The filly was quiet, as if she had been ridden all her life. The most important part of breaking yearlings is the ground work.

Friday, September 29. "Syndicates of doctors and lawyers are great when their horses are winning, but not so great when they're losing," one of Maryland's leading trainers said today. "The head guy for one of these syndicates told me to sell their five horses in a racing-age sale in October. I asked him: 'Do you want to get rid of these horses? Then let me run them where they can win and get claimed.' "

Almost on cue, the allowance winner came to the wire six lengths on top in twenty-five thousand dollar claiming company. The colt won sixty percent of an eighteen thousand dollar purse, *and* he was claimed.

"I just made thirty-five grand in cold cash for that owner," the trainer smiled, like a player with a hot hand at the blackjack table. "He can get caught up on his bills now."

The trainer created capital to fuel the parade of expenses generated by horses. It would trickle all the way down to the breeding farm, where September 1st stud fees will be a month overdue in two days, where board bills are mounting. Racing feeds this whole game.

Saturday, September 30. Rookie Steve Earl is toeing the line today, and has been all week—ever since last Sunday morning when he failed to show for work. Steve spent two years in the Military Police. He came to us last January in combat fatigues. I thought he was going to salute me when I ordered him not to throw so much clean straw out with the manure. Last Saturday night, Steve ventured into the big city of Baltimore, to the tough wharf area of Fells Point.

"I didn't do nothing really," he told me. "I was just standing at a bar when a cop came past with his billy club pulled. I told him I'd been an MP and he didn't need that club. He thought I was being smart, so he put the cuffs on me, instead of on the guys making all the noise. They gave me one phone call, but I was too ashamed to call you. They let me out at noon. I'm sorry. I've never been in trouble in my life. Never!"

The police report told a different story.

Steve works hard, knows all the horses by heart, loves living on this farm. He made a mistake. It scared him. On Friday, he gave me back his paycheck. He was still upset with himself. I deducted for the day he missed, gave him the rest. I told him everybody makes mistakes. I was thinking of the time I toured Tartan Farm in Florida with John Nerud. One of his men had broken the rules the night before. Mr. Nerud gently scolded the man, said he was disappointed, said he didn't expect to make this speech again. Then he praised the fellow for all the good work he had done.

"Leave the man his dignity," Mr. Nerud told me as we drove off. "A man is nothing without his dignity."

Between Nerud and Confucius, I am learning—slowly and often painfully—that managing a horse farm sometimes has very little to do with horses, and everything to do with people.

A LINE THROUGH TIME

Sunday, October 1. John King is seventy-something, wheezing with emphysema from decades of Chesterfields, his eyes going, his hearing gone in one ear. But he starts each day by pushing a half dozen cats off his bed at 5 A.M., so he can get to the stud barn to feed the stallions. In the pitch-dark, through the fog, he navigates his five-foot frame down the driveway, tiny steps at a time, leather soles dragging the ground. Along his route, an old walnut tree drops its bountiful crop, and John regards that twenty-foot stretch of pavement as if it were a minefield, his feet never rising above the height of a walnut. Twice a day, he sweeps away the walnuts, but by the next morning, another fifty will be back in his path. One of the biggest fears in John's life is to be tripped by a walnut.

Raised in an orphanage, an intelligent but uneducated man, John has given this farm his loyal services for longer than I can remember. Without his responsibilities, he would be lost, resigned to his house, to soap operas and talk shows and baseball on TV, to the company of his cats. I do not know how to tell John to take it easy. I do not know how to plan for his future. Will Medicare pay for all his doctor bills? Where will we bury him? He would be scared to death to leave this farm for an old folks home. I watched him kick through the walnuts this morning. I need to study on this situation a little more.

Monday, October 2. The most valuable weanling on the farm is also the most attractive. She is gray, with an irregular blaze and light spots mottled throughout her coat. Her sire is the gray Corridor Key, her dam the Sir Ivor mare Allegedly. The filly stands wide in front, her hind end bulges with muscle, and she has the temperament of her mother's irascible half-brother, Alleged, winner twice of the grueling Prix de l'Arc de Triomphe.

The filly's paternal ancestor, Mumtaz Mahal, was a spotted gray wonder bursting with speed. She won five of six races as a two-year-old, and was champion of her generation. She became the granddam of Nasrullah, Royal Charger, and Mahmoud (sire of Corridor Key's third dam, Mumtaz). Federico Tesio, who felt that the color gray represented a disease of premature senility in the horse's coat, wrote:

"The Army found little use for the gray—too visible a target in time of war. By 1815, the year of the battle of Waterloo, only twenty-eight gray broodmares were in the whole of England. Then in 1884, a gray horse appeared in France: Le Sancy. In 1911, a new gray wonder appeared in Ireland: The Tetrarch, grandson of Le Sancy. And so the disease of the gray, which was dying out in France, broke out with renewed virulence."

Mumtaz Mahal's sire was The Tetrarch. I have a photograph of Mumtaz Mahal. Her coat is mottled; she has an irregular blaze. Tesio could trace the ancestry of the weanling at our farm back to the Alcock Arabian, a gray born in 1722. I prefer to stop at Mumtaz Mahal, and let my mind take flight.

Tuesday, October 3. "Hold her head still," Peter gently asked Ellen, who was calming the impatient gray filly while Peter took Polaroids of the weanling's obscure facial markings that extend beneath her lower lip. "No, no, I'm afraid you're in the sun. There's quite a glare. I can't see the foal's face."

Ellen backed the filly under the courtyard overhang. Peter waited for the red light on the camera, but the filly did not, and moved her head just as the flash went off. I watched this ordeal safe in the knowledge that my days as the foal holder are over.

As an undergraduate in this business, I took the course Registration As An Art Form, with the weanling's name on the chalkboard in every photo, the right background, the foal standing perfectly, but Dad could never get the Polaroid to work, and the deadline was always tomorrow, and it never failed to be the hottest day of the summer. I remember photos of flies on foals' noses as comic evidence of our efforts. I pitied the poor folks at The Jockey Club who had to examine such down-on-the-farm photos. Today, I gave thanks for the professional assistance of Ellen and Peter.

Wednesday, October 4. "You cannot step twice into the same river," a Greek philosopher said, "for other waters are continually flowing on."

Every year I try to attend the fund-raising dinner for the local halfway house, a changing river of reformed souls, most bent on charting a new course, a few simply adrift. Some of Country Life's best talent has come from this whirling pool of recovering young men. Steve Brown, Richard Harris, Mike Price—boys as hard as their names—did not drift by. They stayed. Around these rocks, we built our fences, sheds, and houses. Given a second chance, these young men flourished.

As I watched the young men in the buffet line tonight, I tried to spot the look that Steve and Rich and Mike had in their eyes. I did not see it. But that does not mean that in three months, when the mares start shipping in and the foals hit the ground, I will not look again in these waters, which are continually flowing on.

Thursday, October 5. The week before Private Terms went postward favored for the Kentucky Derby, his breeder, Stuart S. Janney, Jr., visited his mares being bred to Assault Landing.

"Had some bad news this morning from Kentucky," Mr. Janney said. "Lost the yearling sister to Private Terms in a paddock accident at Claiborne. Must've gotten kicked. Died of internal bleeding. Really nothing anyone could do."

He delivered this news with such equanimity. In the annals of equine misfortune, certainly Mr. Janney's loss of the brilliant Ruffian ranks near the top, but Fate continued to exact her price from this gentle horseman. Assault Landing galloped in the Gotham Stakes, then limped off the track two days before the rich Wood Memorial Stakes, his classics campaign over. When Mr. Janney bred Private Terms' half-sisters to Assault Landing, Dress Ship conceived twins and had to be aborted, and Light of Foot resorbed in late summer. Last fall, a yearling by Assault Landing out of stakes winner Wedding Party died at Mr. Janney's Locust Hill Farm after injuring himself in play. The score at the end of three mares: Four conceptions, zero runners. This lost opportunity really hurts a young stallion, not to mention adding snakebit frustration to my job as stallion manager.

Such are the vagaries of the Turf. Today, young Stuart S. Janney III, who took over after his father's death last year, told me Light of Foot's two-year-old by Assault Landing is training well. Ah, there is hope yet for Mr. Janney's well-laid plans.

Friday, October 6. Three lawyers sat around the conference table, discussing potential new investors in the horse business.

"If he is simply a yuppie, driving a BMW one size too big for him, with a pretty wife who spends his money as fast as he earns it, and a mortgage on a big house, this is *not* a good investment for him," said the first lawyer. "Don't get gamblers, either. They'll play the ponies in earnest instead of for sport. One final type to avoid is the gal who's never worked a day in her life and has plenty of time on her hands. She'll second-guess every decision. Red-flag that type of person if possible. I've seen 'em all, and everything's fine until the horse deal goes south. Then those nice people can get awfully nasty."

The second lawyer studied me intently.

"Look," he said. "Just be careful. Use common sense. Any misgivings whatsoever about a person, don't sell him a piece of a racehorse."

The third lawyer, a tax specialist, paid no attention. He was mulling amortization aspects and capital gains changes. At the end of the skull session, I was not sure which will be the harder task: Racing the right horses, or raising the right owners.

Saturday, October 7. In beautiful Indian Summer weather, the boys are busy with chain saws on a part of the farm timbered last year, hauling firewood for the numerous woodstoves and fireplaces scattered among the farm's houses. However, this is "bee season," and a close look is wise before turning over any logs. Two weeks ago, Richard grabbed a boxwood for transplanting and did not see the yellow jackets. He wrapped his arms around their hidden nest. Forty-seven stings later, he stopped jumping long enough for me to apply baking soda, while Ellen ran to the pharmacy for a prescription of Benadryl. That episode would have killed some people. Richard just gritted it out, his face swollen. Rich is no weekend warrior. I was helping him with the boxwood brigade, and all I could think of while painting the baking soda on him was, "But for the grace of God, there I go."

Sunday, October 8. Spent the day in Virginia, a state whose natural beauty needs no makeup. Mother Nature is painting her in golden hues. Like jewelry are the leaves in autumn. The sweetgum trees and maples are brilliant, but the many oaks are browned and crisped, as Thoreau said, "like a loaf that is baked. The order has gone forth for the trees to rest. As each tree casts its leaves, it stands careless and free, like a horse freed from his harness, with concentrated strength and contentment."

Monday, October 9. I am fortunate to be surrounded by experienced horsemen. Peter, for instance, trained for Windfields for years, won the Canadian Oaks with Solometeor. Today, he pointed out that one of the yearlings has developed a curb. A curb is a thickening of the short ligament just below the point of the hock, appearing as a bump, a slight swelling, at the head of the cannon bone. This ligament helps stabilize the hock, and when strained, it bleeds internally, the inflammation subsiding only with rest. Once it has cooled out, Peter and Dr. Riddle will pin-fire the enlarged area over the ligament, then apply a blister. Peter is wonderful with the lay-ups sent here from the race track, and he has a keen eye for problems in young horses.

Tuesday, October 10. Joe Hamilton manages a farm two counties over. He is an excellent young horseman, but he has that race track belief that assumes everyone awakens at 5 A.M. rarin' to go. After coffee, he rifles through the sports pages. If a Country Life stallion sired a winner the previous day, Joe is on the phone immediately. He imitates Robin Williams' revelry from the movie *Good Morning, Vietnam,* singing out: "Gooood Mornin', Country Life!"

"Lucky Lady Lauren won by a head," Joe told Michael this morning as I walked into the office. "Looks like she was all over the track, but there's another winner for your little Northern Dancer stud."

Some mornings after tough foalings, I've wanted to strangle Joe for putting me on his early schedule, but he can call us about a winner for a freshman stallion at *any* hour.

Wednesday, October 11. The most unlikely stories can break the ice at life's difficult moments. Back at Johnny Sullivan's house today after the funeral of his father, with a roomful of old friends searching for the right words, the recent death of Secretariat came up, and a horseman in the group recounted the particulars of burying such unwieldy animals. Hardly fit talk for a wake, but the absurdity of the topic struck a funny bone of black humor.

"I read where they buried him in his entirety," said Kevin Kellar, the stud man for nearby Worthington Farm. "None of this 'head and heart' business. That's a big hole. We buried Lady Dean that way. I had to climb down and get the chains off. If one of those safety folks from construction sites had been there, we'd of been fined for sure—the backhoe leaning out over the hole, me ten feet down next to a dead horse. I swear, I loved Secretariat, but the stories in the papers of burying him made me remember my own experience, and it wasn't fancy with a bunch of reporters around and racing colors in the casket. You got any horses buried at your farm?"

Thursday, October 12. As Andrew led Mosquito Coast from the receiving barn to the paddock at Laurel today, four huge red tractors dragged the deep track into perfect corduroy rows, and grooms swung blinkers and bantered with pony girls. The awnings and the tents and the handsome round paddock came into view, sunlight so clearly catching fresh coats of paint, and railbirds quizzed us for tips.

I thought to myself: "If the TOBA could bottle this moment and sell it at their seminars, new owners would flock to this beautiful sport."

Friday, October 13. Three months after the crash that started the Great Depression, a farm worker here stuffed that day's newspaper between the poplar boards of the milking room of this former cattle farm. We found this wrinkled insulation recently when Steve Brown tore the wall apart in a renovation project. The sports headline in the Baltimore *News* of Saturday, January, 25, 1930, read: Blue Larkspur Gains Prestige For West.

"The West, which seldom produces a horse to lead every division, now has that distinction as a result of Col. E. R. Bradley's Blue Larkspur. The honor has long been held by the East, due to the reign of Man o' War, Crusader, Pompey, Sarazen, Zev, Pillory, and Morvich. The East in 1928 again attained precedence through the success of High Strung as a two-year-old."

I think an imaginary line runs through time, linking us inexorably to our ancestors, bearing us back into the past. I see a cold farmer in January of 1930, three years before Grandfather bought this farm, reading about Blue Larkspur and High Strung, then stuffing the newspaper into the wall to stop the cold wind from blowing in. Then he went on with his milking. Safely archived for sixty years, the paper became parchment, preserving an era when "West" meant any track to the left of New York on the map. It was the era of High Strung, which so coincidentally came to stud here as one of Grandfather's first stallions, and Man o' War, sold by Granddad as agent for August Belmont II.

Blue Larkspur, Zev, and Morvich, handicappers' picks at Havana, Miami, Agua Caliente, and New Orleans, even Irving Berlin's latest movie score, *Love in a Cottage*, all were preserved like cave drawings, or fossils in Olduvai Gorge, safe in the walls of the milking room, our own Steve Brown stumbling like Richard Leakey upon equine archeology. I cannot bring myself to discard the torn brown newspaper, because it is irrefutable evidence of what preceded us on this very land.

Saturday, October 14. An orange moon came over the treetops as if being tugged aloft by the setting sun, and tonight Ellen will not rest until she has walked the banks of Winters Run by the light of this harvest moon. The woods on such nights are Gothic—black trunks of trees reaching across the water, limbs on one bank touching limbs on the opposite bank, the stream like melted silver reflecting the moonlight.

The weanlings will call out to each other and graze easily through the night, child actors imitating behavior learned from forgotten mothers. The theater season has started in Baltimore. Restaurants fill with Saturday nighters. Mother Nature's stage tonight is a Globe Theatre of animals. Rollaids plays the comic Falstaff, the Allegedly filly is Juliet, the handsome colt out of Promenador is Prince Hal. Ellen and I can laugh and applaud seeing them in the moonlight, and take the cue from *Richard III:* "A horse! a horse! my kingdom for a horse!"

Quite a lot can be seen by the light of the moon, before the curtain falls on a full day.

Sunday, October 15. The deadline for nominating foals to the Breeders' Cup is tomorrow, and I am reminded of a phone call from Mr. Clagett on Friday.

"I've got twelve weanlings, but I'm only budgeting two thousand dollars for Breeders' Cup nominations," he said. "That's four foals at five hundred apiece. I don't hold much chance of winning a Breeders' Cup, but I've already bred a Maryland Million winner in Little Bold John, so I'll put all my foals in that program at two hundred and fifty apiece. My Maryland-bred registrations will each cost me seventy-five dollars. The Jockey Club registration and blood-typing will each cost another one hundred and sixty. Thus, four weanlings will each cost nine hundred and eighty-five dollars in various fees. The remaining eight weanlings will each be four hundred eighty-five. Total outlay in fees for all twelve foals: Seven thousand eight hundred and twenty dollars."

It hurts the wallet now, but in two years, when his dozen homebreds hit the races—with three of his four Breeders' Cup weanlings sired by our stallions—I hope and pray Mr. Clagett has the last laugh.

Monday, October 16. The adage that says "Only the good horses get hurt" is true. Believe me. I am not superstitious. It is simply true. We have fifteen broodmares in the twenty-acre front field. The gray filly's mom, Allegedly, has an appraised value greater than the fourteen other mares combined. Her stakes-winning son, Per Quod, is a leading handicap horse in Europe. If Allegedly could produce a stakes winner by dear departed Lyllos (our first and most memorable stallion failure), she must be a great broodmare.

Allegedly, however, has a mind of her own. In a squabble in the field this morning, the pecking order reached an impasse. Hind feet flew. Allegedly took a kick on the cannon bone of a hind leg. She went lame immediately. We wrapped the leg to give it support, and called Dr. Riddle.

"It's not broken, but it's bruised," Dr. Riddle said, taking X rays as a precaution. "Hose the leg in cold water three times a day. It's the best therapy—keeps the swelling down, and acts as a massage on the injured area."

True as well is another adage: "Good horses get hurt right before their owner visits." June McKnight lives in Vermont, but she's in town. She phoned while Dr. Riddle was here. No kidding. Michael gave her the preliminary diagnosis. She will be here tomorrow. Her mare will be dead lame. Mrs. McKnight is the only breeder of the winners of both the Arc, maybe the toughest international race in the world, and the Maryland Hunt Cup, certainly the toughest timber race. She is a pro. She will take it in stride. But still, can you believe the timing?

Tuesday, October 17. Through the fog at five-thirty this morning, the weanlings are spirits in the mist, startled by our presence. Ellen and I cannot see them, but the sound of hooves echoes on damp ground as seventeen babies canter in a pack. Ellen keeps up her easy talk, and the weanlings stop running. The waning moon is just a dim sphere over their backs. Rollaids comes up first, then one at a time they follow his trusting lead, until the herd surrounds us, a petting zoo of spirited ponies. We catch the weanling scheduled to leave on a 6 A.M. van to Ocala (the van company having thoughtfully phoned us at ten last night).

Ellen leads the weanling to the barn. I go to the office and wait. It is almost seven now. It is still dark. Trees are silhouetted against the gray mist. Dew drips from leaves to splash lightly on the roadways. Walnuts fall randomly on the tin roof of the lower barn, a sound as if single shots of gunfire were muffled by distance. Then the phone rings. The van is here.

Wednesday, October 18. The Budweiser International—still the Washington, D. C., International to me—is this Sunday. Two years ago, we met the trainer for the Polish entry. I thought about him tonight, about his country's rich heritage for *all* modern horses, not simply Thoroughbreds, when I read about Przevalski's Horse, the last genuine breed of wild horse, in the enlightening equine reference book *The Noble Horse,* by Hans Dossenbach. This book traces the evolution of the horse from Eohippus, a hare-sized creature who lived fifty million years ago, to Secretariat (as if evolution reached its pinnacle in the spectacular physique of the 1973 Triple Crown winner). In between, *The Noble Horse* depicts how the development of man paralleled that of the horse. Hunters of the Ice Age, the Persians, Alexander the Great, Mohammed, Napoleon, American Indians, all relied on horses for various needs.

I saw chilling photos of equine gas masks worn over the bridles of horses during World War I. During World War II, the surviving horsemen of the Polish Cavalry who had so bravely ridden against Hitler's oncoming tanks were conscripted to ride against the Russians. Hitler was short on horses by a million, but nevertheless sent two million of them into the battle against the Russian Army, which had three million horses. I can't even imagine that many animals, yet wars down through history have dispatched countless horses.

Cowboys to jackeroos to gauchos. Aintree to Sapporo to Longchamp. Carts to coaches to carriages. Belgians to Shires to Punches. This is a sit-by-the-fire-and-learn book. I would like to meet the author, Mr. Dossenbach. For his sequel, I want him to consider photographing Rollaids, a throwback to the sturdy wild horses of Poland.

Thursday, October 19. "I put an ad in the Newark paper," said the Standardbred entrepreneur, a local fellow who has assembled many racing partnerships. "It read something like, 'Are you interested in participating in Standardbred racing at the Meadowlands? Call this number.' I hired an answering service up there. But folks couldn't get satisfactory answers quick enough, and I lost potential investors. So I called the Newark phone company, arranged for a local line, with call-forwarding down here to my office. Now folks get *me* on the line instead of some service. It is very important in marketing that when you get a lead, you close on it."

Thoroughbreds, Standardbreds. What's the difference? We're both trying to get into the winner's circle. The journey starts with tiny steps, with a business plan evolved through thought, then bears fruit with leads that close. We are finishing our homework on plans for a large racing partnership.

Friday, October 20. Three days of cold rain finally moved out to sea this afternoon. It has been an unusual fall so far. Until last weekend, not a tree had turned, except the walnuts, which are the last to leaf in spring and the first to drop in fall. Then over the weekend with its cool weather, *zap*—poplars turned yellow, oaks browned out, dogwoods went purplish, ivy became a climbing orange banner to the lower limbs of green maples. I do not think I could live where the seasons never change. This farm provides a Skyline Drive panorama of nature on the move.

Saturday, October 21. In this morning's mail from Dr. Bob Lawrence of the University of Louisville was the brochure from the Alcohol and Drug Abuse Conference. For five days earlier this month, counselors at race tracks throughout the nation compared notes and shared experiences with tracks just getting programs underway.

"Actuarial tables indicate that eighteen percent of people with addiction problems will resort to dishonesty to feed their habit," Jim Ryan told me at the races this week. "The integrity of horse racing is at stake. Bob Babes, the counselor at Pimlico and Laurel, has been crying out alone about this problem. Now it looks like he's getting the industry's attention. We've got programs set up at forty-eight tracks now, including *three* in Nebraska. I didn't even think there were three tracks in Nebraska."

Mr. Ryan, Dr. Lawrence, and Bob Babes now have momentum, like principals at schools taking the classrooms back from the dealers, one day at a time.

Sunday, October 22. It was a day of "firsts" for me at Laurel:

The first time I had seen a horses-of-racing-age sale that was not a broken bones giveaway. Vetted and X-rayed, eighteen racehorses offered a new owner a shot at the winner's circle in the International this afternoon or in the Breeders' Cup on November 4th. Finney wanted this type of sale in the lovely round Laurel paddock forty years ago.

The first time I had ever seen an Arabian stakes race, with Jorge Velasquez, Angel Cordero, Jr., Jose Santos, and Randy Romero in the irons, no less. This experiment in Arabian racing—an outlet for hundreds of dish-faced horses bought by wealthy folks before their market crashed—is such a poor betting spectacle that its handle was only half that of the race for amateur riders. Nevertheless, these exquisite descendants of the desert ran proudly, albeit slowly, keeping their heads, and their tails, held high.

The first time a former claimer won the International. The first time the hot dog stand ran out of food. Quite a crowd jammed the joint for a great variety show, a circus of sorts.

Monday, October 23. White-tailed deer run like water between our property and the neighboring crop farm. Saturday night found a doe grazing near the driveway, indifferent to intruding headlights. This evening, two deer stood motionless in a clearing near Winters Run on the adjacent farm, brown coats napped out from the first few cold nights, tails straight up and white as paint against the woods. *Might as well be bull's-eyes,* I thought. Fifteen feet away were crude boards tacked to an old oak, leading up to what I used to think was a tree fort, but now know as a stand for deer hunters.

When the last leaf falls, deer season opens, and every year I ask myself:

Who owns these creatures? They lick our salt blocks and eat our hay in the fields in winter. We enjoy their fleeting company. I don't believe I have ever heard a hunter's gun fired on our land. Yet these deer also graze on our neighbor's winter wheat crop, and chew his corn stalks. Deer populations need to be thinned in some parts of Maryland. Hunters euphemistically call the killing "harvesting," and last year, they harvested thirty thousand deer in this tiny state. This year, the kill is expected to reach forty thousand.

Every November, we hear the gunshots. Some clumsy, once-a-year sharpshooter only wounds his prey. The deer runs off and dies in our woods, a safe place. A few days later, our dogs drag a deer leg up to the back porch, and we feel sad that a beautiful animal suffered. Yet I am a reluctant preacher, guilty of a double standard: I love the taste of venison, but I don't think *our* deer need harvesting.

Tuesday, October 24. This week in 1934, Finney greeted a new stallion to his *nom de* farm, Sleepy Hollow Stud, celebrated a birthday, took the train *George Washington* to Lexington, bought a mare at the mixed sale, visited Bluegrass farms, attended the Thoroughbred Club's annual dinner, arrived back home in Maryland, cared for a mare who had been kicked, presented his new stallion to fellow breeders, and finished the registration applications for his twenty-five weanlings.

Like the seasons, this game changes, but comfortably stays the same.

Wednesday, October 25. While the crabs steamed, Uncle George recounted a lifetime of betting:

"You hear things. You bet 'em. Years and years ago, my friend Ruthie told me, 'George! I just overheard a couple of jockeys talking about fixing a race! What should I do?' I told her, *'Go back, lean closer.'* I bet my lungs out on that race. My horse broke last. I was ten lengths out. I turned away, disgusted. Then I hear the stretch call. I'm on top. I look over to the track. It looks like The Three Stooges trying to ride. I win. I'll never forget it."

Without hesitation, he names the three jockeys who were banned from Maryland racing in the St. Valentine's Day fix.

"That couldn't happen today—I don't think. Lots of things used to happen that don't happen anymore. When I was a chef at the Pimlico Hotel, a lady come in from New Jersey with a suitcase, orders two Bloody Marys, opens up the suitcase, and I see two hundred grand. She bets it all on Marian Bender to show. Marian Bender was a really good filly around here fifteen years ago. She was the odds-on favorite, but she run bad, finishes fifth. Winsom Imp wins it. That lady comes back after the race, orders two more Bloody Marys, then drives off to New Jersey like nothing happened. Mob money, I'm sure.

"I was at Garden State, Thursday, April 14, 1977—the day the track burned down. I had the Daily Double. Smoke's everywhere. Tellers are running out of the grandstand. The place burned to the ground. I mailed my tickets in.

"One day I get a tip, but I can't leave the kitchen. We had a pregnant waitress working for us. I tell her, 'Run this bet over. You got four minutes to post.'

107

She didn't come back to work, and I'm worried sick. I call her husband. We call the hospital. She had a baby boy—eight pounds, four ounces. I find out later, she went into labor at the cashier's window, clutching her purse, screaming, 'This is George's money.' "

I don't bet much. But I know a sure thing, and Uncle George is it.

Thursday, October 26. Sometimes if I couldn't laugh, I'd go crazy. Spent the morning on the phone to see if mares bred this spring were still in foal. Instead, I merely confirmed my suspicion that the exodus of many small breeders from this game is continuing unabated. Sons and daughters of these breeders, young people my age, have no enthusiasm for their parents' pastime. Although this business historically has been cyclical, I am witnessing the protracted end of an era. I try to find some light in the comments I hear, but it is hollow humor, dark humor, like the joke, "How do you make a million in the horse business? Start with two million." I hate that joke, though sometimes it might be true.

The faces of these vanishing breeders are lost to me over the phone, but I know their eyes are lined with the crow's-feet of worry. Shake the hand of any gal who works on these farms, and you will feel the care they take of their horses. Yet their horses simply are no longer worth the cost of such care.

Oh, well, can't dwell on it. Must focus on the positive. Carnivalay's half-sister Go for Wand is in the Breeders' Cup. She just broke her maiden by eighteen lengths. Can she possibly be that good? And two of Carny's fillies, Valay Maid and Lucky Lady Lauren, are nominated for a $75,000 stakes at Laurel the same afternoon. Ten days from now, I might feel different about this business.

Friday, October 27. At the new owners' seminar at Laurel, Maryland trainers Katy Voss, Larry Murray, and John Hicks each took seven prospective investors to their training barns, explaining various misfortunes that afflict racehorses. The first three horses in Murray's barn revealed racing's whims: A runner which cost three hundred thousand dollars as a yearling is now three. He has earned only nine thousand dollars. Meanwhile, a modest homebred by Rollicking had earned four hundred thousand. A Damascus colt "looks like he wants to go long." In other words, he lacks speed, Larry said with a smile, and everyone understood that the colt was slow and probably not very talented. Down the shedrow we went, Murray's honest comments shedding light on a variety of questions.

I thought the trainers did an excellent job of presenting the risks, but a reporter from the *Wall Street Journal* saw it from a different angle. In "Race Horse Breeders Bet On The Hoi Polloi," she wrote, "If Bluebloods won't pay high prices anymore, who will? Breeders are betting on the common folk. Even while they move outside their traditional tony circle, racehorse owners still try to capitalize on the elan of the sport."

I saw new faces anxious to experience the roller coaster of racing. They did not feel that the owners and the trainers were patronizing them. They stepped across the puddles of water near the sore-legged horse being hosed in front of the manure pit. Russell Jones repeatedly stressed, "Don't check your brains at the door." There was no arm twisting, no hard sell, no hot-boxing potential investors. The folks who enter this game after these seminars have no excuse not to be prepared, for good fortune or bad.

Saturday, October 28. "I won't cry if someone claims her today," says King Leatherbury in the paddock at Laurel prior to a five thousand dollar claiming race. "She's a certified screwball." At that moment, the filly comes hump jumping out of her saddling stall, the groom a dangerous three steps behind her—in perfect line for a kick—but she misses her opportunity. She comes to a sudden stop, waiting for the next message from her brain telling her to explode. Leatherbury moves close, throws the saddlecloth over her back, then the saddle, but she blows when he cinches the overgirth, bucking in spasms as if electric shocks are hitting her. Leatherbury moves back in, reaches under her belly. He is quick. He gets the girth tight and in its keepers, then backs off—eyes on her.

Leatherbury greets Kent Desormeaux, and the state's leading trainer reminds the nation's winningest rider to take it one horse at a time: Don't get careless with this fruitcake, or Chris McCarron's record will be safe for another year. He lifts the kid into the saddle. The filly lunges out of control, Desormeaux' long legs squeezing for purchase. "Ride 'em, cowboy," a railbird shouts. The filly bounds out towards the race track, where a pony girl grabs her.

A few minutes later, Desormeaux is in the winner's circle, Leatherbury at the filly's head, hoping someone approaches with the halter of another trainer. These fellows each will receive ten percent of a meager purse, but they earned every penny of it.

Sunday, October 29. Joe Hamilton called: "Did you hear about Valay Maid's work? Went in 1:13 flat. Carlos only wanted her to go five furlongs, but the boy galloped her out six. Over the deep Laurel track, with all that sand on it, that time would've won most races. I'm tellin' you, he's sending her into that stakes off the win in her first start, but she's got a real shot."

I try to temper Joe's enthusiasm. After all, he and his wife, Betty, bred Valay Maid. But the conversation will sit like a leprechaun on my shoulder all week, whispering promises that might come true this Saturday. We will dance the Irish jig if Valay Maid carries the day.

Monday, October 30. In 1973, Finney introduced the second printing of his *Stud Farm Diary*: "More than forty years ago, it was the writer's pleasure to have served for better than a decade as Stud Manager for the Labrot family in Maryland. There the Holly Beach Farm spread over more than twenty-five hundred acres along the shores of the Chesapeake Bay. For obvious reasons, a *nom de plume* (Nothing Venture) was used. The location of the farm, its name and those of the horses thereon were likewise disguised."

As Halloween approached, we wanted to see through Finney's disguise. With the help of his daughter Marge, senior editor of the *Maryland Horse* magazine (which Finney founded in 1936 and for which he left Holly Beach late in

1937), we unmasked Nothing Venture's Sleepy Hollow Stud this afternoon.

In the late nineteen-fifties, the Chesapeake Bay Bridge was built, linking Annapolis on the western shore across the water to the eastern shore. The path of least distance was from Holly Beach Farm across to Kent Island. The farm's north end, containing the one-mile training track, was severed away by Route 50, the farmland given over to the state. Visitors to Sandy Point State Park must wonder what prosperous enterprise preceded them as they drive past the brooding hulk of a huge barn. For years after the Bay Bridge was completed, aerial photos revealed the sandy shadow of the oval training track.

The drive to the south end of the farm, some one thousand acres where the Labrot family still resides, has changed little over the years, except for the departure of the horses. Former pastures are sod fields, providing a vibrant green landing strip for the hundreds of Canada geese honking a spectacular greeting. Three-panel board fences erected in Finney's day still stand, in large part because the Labrot family was in the creosote business. A one-stall barn charged with housing the blind stallion Light Brigade stands sentry near the courtyard of the broodmare barn, whose slate roof and timbers remain intact.

I felt ghosts at Holly Beach today, as if by cupping my hands to my ears I could catch in the wind Finney's commands to the stable lads, as if by closing my eyes I could see the courtyard alive with the industry of raising horses. In the distance, the smoky gray columns of the Bay Bridge—leaping across water from one shore to the other—confirmed the power of imagination. It seemed facile to merely jump back in time. Tonight, I pulled the *Stud Farm Diary* down from the shelf to read again the introduction. About life on a horse farm, said the author: "There is little difference today from what made sense in 1935."

Tuesday, October 31. A warm rain fell overnight, and the mares all enjoyed a good roll in the mud this morning. We rotated the weanlings into a rested field, where they dove headfirst into the last grass of the year. Soon the frosts will turn the bluegrass brown, but for the moment, the beautiful weather continues.

Now November is calling, with Go for Wand in the Breeders' Cup, with Valay Maid and Lucky Lady Lauren pointing for a stakes, with the challenge of bringing new faces into the game while keeping smiles on those already playing.

And the seasons roll on.

TURNING THE CORNER

Wednesday, November 1. Clever Hans, the famous "counting horse," could tap his foot the correct number of times to solve simple arithmetic. His trainer would ask: "What is two times three?" Slowly, Hans would tap six times. The audience would applaud. Skeptics of the horse's intelligence, however, asked that the trainer withdraw so *they* could ask the questions. Hans continued to count correctly. Finally, they placed Hans behind a screen, where he could not see or sense his audience, and suddenly he was not so clever, tapping endlessly in response to the first question. Hans could only count correctly when in close contact with his audience.

While leading out the stallions today, I lost my temper, and asked the simple question: "What is a two-by-four? It's what I'll club you with if you don't quit biting me!" But I remembered Clever Hans. These big boys pick up on frustration or fright—the emotions of the audience—and take advantage of either. The key is to outthink them. Some days I am up to the task, and some days the task is up to me.

Thursday, November 2. The overnights are out for Saturday's What a Summer Stakes at Laurel, and both Lucky Lady Lauren and Valay Maid will go—one to serve as a rabbit, one to close hard on the tiring pacesetters. I tried to keep my mind from counting chickens before they hatch, and spent the evening reading through Grandfather's letters for bits of wisdom. In a 1934 note to a fellow horseman, he wrote: "Another good breeder has gone out of the business. Mr. George Herbert Walker has given me his mare Mistral and her filly foal. Last week we received word that Mistral's two-year-old in Kentucky is coughing again, and we let up on his training. Of course, Mr. Walker is a little impatient and wants to see his horses run quickly."

Grandfather never suspected that Mr. Walker would be the grandfather of our current president, George Herbert Walker Bush. Yet once again, his letters were a revelation to me. The repetitive nature of the horse business was clearly evident: The constant turnover in the breeding game, a two-year-old with a respiratory problem, a racehorse owner in a hurry to see a return on his investment. Grandfather's letter was written fifty-five years ago, but might easily have been written today—if President Bush owned a broodmare.

Friday, November 3. A leading veterinary firm is developing a test to determine the sex of the *in utero* foal. It may be possible someday to purchase a mare with the warranty that she is in foal with a filly. If she carried a daughter of Sir Ivor, that information would certainly add spirit to the bidding, Sir Ivor being such a good "filly sire."

To wit, the best mare on our farm: Allegedly.

Saturday, November 4. "I've been to the Super Bowl, the Stanley Cup, the World Series, and the Indy 500, and I've *never* had worse seats for a major sports event!" a racing fan yelled in my ear this afternoon as the Breeders' Cup got underway. Gulfstream Park produced bleacher tickets that looked remarkably similar to clubhouse tickets. Ellen and I were *way* over by the six-furlong chute—closer to the backstretch than to the frontside. But when Go for Wand wore down Stella Madrid to win the Breeders' Cup Juvenile Fillies, we were suddenly sitting on Cloud Nine. Carnivalay's half-sister! What an unexpected boost for a freshman sire, to have his little sister become a champion! Yet I recalled John Nerud's advice about the importance of the female line in selecting stallion prospects: "Don't follow an empty wagon. Nothing ever falls off them."

Obeah was sixteen years old in 1981 when she produced Carnivalay. She toted the wagon for six more years before Go for Wand fell out onto the grassy Pennsylvania turf of Russell Jones' farm. Today the wagon delivered fruit of the highest quality. I am not even curious tonight about what happened back home in the What a Summer—plenty of good fortune for one day. Can't be greedy.

Sunday, November 5. I hurried off the plane in Baltimore and dashed into the newsstand. Who won the stakes at Laurel? No luck. The paper was an early edition with no results. I felt the return of responsibility settle on my shoulders. "Too much to hope, to think we won the stakes *and* the Breeders' Cup. Back to work." Riding down the escalator, I saw the Maryland Horse Breeders airport display come into view—eighteen feet of photos of colorful horses, brilliantly backlit. My heart sank, though, because the white mare bounding towards us, followed by passels of foals, is Auntie Freeze, Country Life's "old gray mare," and I know she ain't what she used to be.

When I left for the Breeders' Cup, she was under Dr. Riddle's care, her blood chemistry showing kidney problems, her weight fallen off. I will have to put her down unless she shows marked improvement. When the photo was taken, she was eighteen. Now she's in her twenties, and gray horses simply do not live as long as non-grays—two years less, a study in England discovered.

Obeah was twenty-two when she produced Go for Wand. Auntie Freeze (one of Dad's best names, by Uncle Percy out of Heat Shield) will not see her twenty-third birthday, I am afraid.

Suddenly, Dad appeared at the foot of the escalator, smiling broadly, hands raised above his gray head: "Go For Broke!" he exclaimed, excited like a child at Christmas. "Ran one-two at Laurel!"

Monday, November 6. When Snowden Carter first retired as editor of the *Maryland Horse*, he filled his free time by riding Ellen's war-horse Roger through the autumn woods around Country Life. He was happy as a clam to rediscover the glorious bond between man, a horse, and the colors of November. He had written the stud news when Carnivalay first came to us. Snowden enjoyed seeing old friends Henry Clark and Jane Lunger during the interview for the piece. Now he writes the Maryland breeding column for *Daily Racing Form*. He called this morning. I think he was more excited than I was, and my feet hardly touched the ground all day. I felt happy that Snowden was happy. What a Monday! What a Summer! What a day was Saturday!

Tuesday, November 7. Mr. Momentum stuck around to help out Mosquito Coast, until today still a maiden after six tries. Andrew had wrapped her ankles in white support tape, highlighted against her black coat, against the orange blinkers on her face, against the orange and blue silks clinging to the jockey's ribs. When I noticed that the valet in the jocks' room had unearthed the correct cap, I went to the betting windows. She jumped out of the starting gate, almost fell down, but straightened up to find herself forcing the pace. She took off turning for home, winning by three lengths.

"I had begun to think I would never see this day," said partner Brian Keelty. Brian did not exactly have the luck of the Irish when we first met him. "I didn't think it was in the cards for me to *ever* own a racehorse who could break its maiden." He spoke about trials and tribulations, but said today made it all worthwhile.

Wednesday, November 8. We were deluged with requests to sign Stallion Service Certificates today, as the deadline nears for the fall mixed sale. I feel as if I am signing death warrants for mares I know will not bring a bid above the killer's price. I know these girls rather intimately: from handling Jazzman on the daily teasing rounds, from staying up nights to deliver their foals, from high-fiving their pregnant status on ultrasound days.

It is the *in utero* foals, though, I will miss the most. The percentage of "Registered Foals from Mares Bred" does not include the number of in-foal mares who vanish before producing those foals. Sales companies have raised the minimum opening bid—the upset price—from three hundred to five hundred dollars, causing the man with the cigar to get out of his seat and walk back to the barn to buy the "no-bids." Well, at least now the slaughterman has to work a little.

Thursday, November 9. "Mosquito Coast had a fever this morning, and she's coughing," Andrew said today. "Looks like we're out of business for a while."

Eohippus lived in Wyoming. I imagine he caught the flu trying to cross the Bering land bridge. The genus *Equus* died out in North America about ten thousand years ago, but relatives migrated from Alaska to spread out over Asia, Europe, and Africa. They evolved into modern-day zebras, donkeys, and wild horses and—after further refinement—into the Mosquito Coast's of today's racing world. All are subject to the contagious respiratory disease "equine influenza." Local trainers have battled the bug in their stables all fall. I am disappointed for Andrew's sake, as he had trained Mosquito Coast to the top of her game.

Friday, November 10. A suspenseful day as Dr. Riddle checked the sixteen "in-foal" mares for pregnancy prior to winter. This time of year, "in-foal" may be inaccurate, as Mother Nature quietly exacts her toll on mares' tender pregnancies. I was on the phone when Peter came into the office with the vet report, and he flashed the thumbs up sign. "Sixteen-for-sixteen," he said. I still don't know what the fellow on the phone was saying.

Saturday, November 11. "A race track barn is the perfect model for studying respiratory problems," said Peyton today. He is a clean-air activist, and he achieves wonderful results with a device called a transpirator, which allows sick horses to breathe fresh air.

"Once the polyethylene plastic goes up in the fall to keep the grooms warm, the problems multiply. The dust just rolls into the barn and stays there. Even good quality hay and straw are loaded with fungus spores. Set a cup of coffee on a windowsill in a closed-up shedrow. Within a half hour, a thick film of dust will have settled on the coffee. Imagine how much dust a horse breathes in twenty-three hours in that stall. When a flu virus is present, the air becomes the source of infection. Whole barns at a time are sidelined right now in Maryland."

Carnivalay's undefeated stakes winner, Valay Maid, is prepping for the rich Maryland Juvenile Filly Championship. She will be favored. No filly in Maryland appears capable of stopping her. I am holding my breath—and hoping she does too, at least while she's in the barn.

Sunday, November 12. What calm currents run through animals on Sundays! The weanlings snoozed in the warm sun as we put breakfast in their feeders. They woke slowly and stretched. Only the oldest foals hurried. Two fields away, their dams politely pushed their feeders as they established today's pecking order, paying no mind to the cacophony of Canada geese flying low overhead after feeding in the full moon in the neighbor's cornfields.

Sandhill cranes closed Charles Kuralt's show *Sunday Morning.* Our jealous heron must have sensed birds on the airwaves, because he made a low pass over the house, his shadow moving like a spirit across the walls. At noon, two red-tailed hawks screeched in the cool blue sky above Winters Run, never flapping their wings, but floating swiftly in circles on the convection of warm air rising from the land.

Mother Nature carries on every day, but I notice her so much more on a Sunday morning.

Monday, November 13. "Gooood Mornin', Country Life," Joe Hamilton sang into the receiver this brilliant Monday morning. "Have you seen the *Form?*"

The Maryland breeding column was full of Carnivalay's good fortune, while the Pennsylvania column featured Obeah and Go for Wand. The *Form* even carried a photo of Michael leading Carnivalay, the horse obscuring his handler save for the sneakers. "Tell Mike they got his best side," Joe laughed.

Today was Michael's birthday, Joe sang a telegram of good news, and this Monday was a winner: It broke on top and improved its position.

Tuesday, November 14. Peyton put down Auntie Freeze at dusk. She passed blood in her urine over the weekend and when I led her from the corner of her grassy field her joints popped like knuckles cracking. Near the woods behind the tractor shed he gave her a syringe of T-61 Euthanasia Solution and she gently rolled backwards into a bed of brown leaves.

Peter laid a blanket over her. He reminded us that all her foals were winners, and that her brother even won a stakes years ago. I wanted to be distracted and read the warning label on the T-61: "Avoid accidental self-inflicted injections." Well, of course. Peyton put his stethoscope on her skin behind her elbow and we were all very quiet. The light was fading and we walked out and looked at the weanlings and they were warm and alive and we put the deed behind us.

Wednesday, November 15. The barber once gave us lollipops to make us hold still, I recalled this morning when I gave Allen's Prospect a piece of sugar to bring his head down to a height where Ellen could pull his foretop. The success of Breeders' Cup weekend has folks inspecting the studs for bookings earlier than usual.

When Mr. and Mrs. P.F.N. Fanning arrived at nine, Allen's Prospect had that fresh look of a schoolboy stepping out of the barber chair. Mrs. Fanning said she wished his tail came out lower down his rump. She is a very astute horsewoman. She trains winners of the Hunt Cup with the same regularity as Woody Stephens trained Belmont Stakes winners. I never thought about his tail, but later I looked at his weanlings and noticed all their tails came out high too—an effect of muscular hindquarters. Now I will take it as a stamp of the sire. His first foals race next year. I hope jockeys get a good look at these tails.

Thursday, November 16. Over the stallion barn, the sky had become menacingly purple. Mom called from work in Baltimore. "Hold onto your hats," she said. "It's coming your way." The phone went dead just as the wind hit us. I stepped out of the office. Leaves struck my face, stinging like sleet. I looked towards the fields. The weanlings were scurrying about, trying to brace themselves.

"They're okay," I thought. "What about the stallions?" Out the driveway towards the stallion barn, the road was partially blocked by two downed cedar trees, their thick tops having trapped the sudden gale winds. A tree had fallen in a corner of Carnivalay's field, knocking a four-panel fence into two panels. Steve Earl spotted it from the stud barn and ran down the fenceline. Carnivalay, though, was too busy setting his rump against the strong wind.

Often in storms, it's safer to let the horses fend for themselves, but the stallions were all near their gates, so Rich, Steve, and myself triple-teamed each stud to his stall an instant before the heavy rains started. Electricity went out. At 10 A.M., the stalls were as dark as at dusk. Later, we heard radio reports of tornado sightings, but our barn roofs stayed on, the horses were unscathed, and the trees that fell missed the buildings. Tomorrow will be business as usual. Today we were blessed by not being in the direct path of a funnel cloud.

Friday, November 17. Set in the foothills of the Blue Ridge Mountains, the counties of northwest Virginia are home to a number of breeders who annually send broodmares to Maryland, and it pays to visit their turf periodically. The land around Upperville and Boyce and down Route 81 resembles the Bluegrass region: Rock walls of limestone cleared from the pastures, white columns of Georgian mansions standing tall.

The thought came to mind that, had it not been for the chaos caused by the

Civil War, followed quickly by an era of phenomenal success by a few great stallions in central Kentucky, Virginia might be the center of the Thoroughbred world today.

Saturday, November 18. A horse's ears are externally mobile, like those of a deer—sixteen muscles allowing the ear to rotate almost one hundred and eighty degrees. After the earthquake in San Francisco in October, seismologists remarked that horses detected the disturbance before it hit. I have seen this "sixth sense" before, as storms have approached or high winds were about to blow. Is it their hearing? Is it some innate ability to detect changes in the earth's magnetic field? Animal behaviorists feel that horses "tune in" to their environment, thus explaining why riderless horses often make it back to the campground, even over strange terrain. Maybe that is why horses in a fire run back to the barn—not for safety, but because the "magnetic map" of their home overrides even horse sense.

Daydreaming in the saddle today aboard a rented school horse, I almost came a cropper when my mount, Mickey Finn, darted sideways upon hearing a dog in the woods. My first thought concerned the prudence of the stable in having me sign a Hold Harmless Agreement. My second thought was how remarkably acute was the hearing of even quiet old Mickey Finn.

Sunday, November 19. Close your eyes. Listen to the wind, loud as the roar of a jet upon takeoff. Open your eyes. See fields of weanlings huddled on the ground, low, out of the wind. Inside, hear Mom's radio show. It is dramatic music, bowstrings of horsehair pulled rapidly against violin strings. Outside, watch the trees. They strip for the contest with winter. Brown leaves fly. The trees are black against the blue sky of a clear cold day.

November has come to the country.

Monday, November 20. "Joe Hamilton called," Michael told Dad at quitting time. "He said Valay Maid got the cough and wouldn't make the stakes this Saturday."

I felt Mr. Momentum rise and head for the office door.

"Sorry about the filly, son, but that's racing," Mr. Mo said, tapping his hat down onto his head. "See you around the grill."

Tuesday, November 21. Received a vet certificate for a mare in our consignment to the December sale at Timonium. "Mare aborted twins," said the report. One foal was born breathing, the other dead. The dead one would have killed the live one sooner or later. No one's fault. Ultrasound exams are not foolproof. Nevertheless, I checked to see if the mare had been examined here by our vets. No. A breed-and-return.

I had spoken to the mare's owners about a reserve prior to the sale. I told them that four hundred horses were catalogued, that there weren't that many buyers on the entire East Coast. At last Sunday's mixed sale, almost twenty-five percent of the mares failed to draw a successful bid. The sales company was too embarrassed to list an average price on the summary sheet.

However, the weekly newspaper *Lancaster Farming* reported averages for New Holland sales: Riding Horses $310-$750; Work Horses $615-$810; Lightweight Killers $235-$500; Heavyweight Killers $550-$700.

It appears that the heavyweight division is *very* strong this fall.

Wednesday, November 22. Michael has been working on increasing the awareness of the December 31st Maryland Million deadline. He is on the committee; he has to be a cheerleader. His motivation, though, stems in part from personal frustration. Groscar fell through the cracks, and thus Carnivalay was denied a golden opportunity in his first crop.

Luck favors the prepared, I tell Michael, trying to sound older brotherly. Yet I am talking to myself as well, reciting the recipe for success in this business.

Thursday, November 23. In the woods at dusk, we saw the force of last Thursday's wind. On a steep hillside on the banks of Winters Run, a beech tree had been lifted up by its rootball. However, before the weight of the tree could get behind its fall, two splaying branches—the main growth of the tree—were caught in another towering beech, thirty feet farther down the hillside. The effect was one of a giant slingshot leaning on a giant stake. We saw it in black and white. Four inches of snow fell last night, making this Thanksgiving feel like Christmas.

Friday, November 24. "My dream is to run a stallion operation and never board a single mare," the head of one of Maryland's most successful stud farms said today. "Mares shipping in to breed stay just long enough to cause a major inconvenience, but not long enough to make any money off them. Boarding these mares is a loser for us, but we do it to get them in foal to our stallions.

"Last year, we charged twenty-one dollars a day for a mare with a foal by her side, eighteen for a single mare, eighteen for a yearling, twenty-two for turn-outs and sales prep, and twenty-eight for breaking. But last year, alfalfa was a hundred and twenty a ton. This year, it's a hundred and seventy-five. Labor hasn't gotten any cheaper, either. We'll raise our rates next year, but still, no commercial stallion operation *really* makes any money boarding horses during the breeding season."

Saturday, November 25. Follies Star might prove a fortuitous acquisition. Her two-year-old, Follies Jumper, has won twice, and today ran fourth in the Maryland Juvenile Filly Championship. This is such a whimsical, random game at times. The first mare we ever bought cost us dearly. She colicked two weeks before she was due to foal. We had to induce labor, lost the mare. Her colt lived long enough for me to sign a no-guarantee contract at two thousand bucks for a nurse mare, then he died as soon as the ink dried.

This spring, we were practically *given* a mare by the University of

Maryland, and her two-year-old runs in the top Maryland-bred filly stakes of the season. I guess if you hang around the kitchen long enough, luck will find you.

Sunday, November 26. Sometimes it is incredible how little the nature of caring for horses has changed over the centuries. On a tour of George Washington's Mount Vernon plantation, I read the sign in front of the stable: "In addition to housing 10-to-25 horses belonging to the Washingtons, horses owned by guests were also stabled here. The tremendous number of visitors (423 in 1785) strained the general's finances and greatly increased the workload of the six slaves assigned to the stable."

You have to read between the lines: The horses almost broke him, and he had a help situation.

Monday, November 27. Tom Rowles, our blacksmith, watches the weanlings walk, notes where the toe breaks over, listens for the impact of the foot on the aisleway. He examines the stress points along the hoof wall. I am amazed at how light the foal's feet appear in Tom's hands. He bounces the hoof as he gets comfortable under the foal's weight. Sports medicine experts say the equine foot is lightweight to minimize energy during the "swing phase" of locomotion. They marvel at this miracle of bio-engineering. The hoof dissipates concussive forces by a slight outward expansion of the wall, especially at the heels and quarters, where the wall is thin. The pedal bone sinks and rotates downward, while the vascular network of the hoof acts as a hydraulic shock absorber.

Tom came to work two weeks ago with his knee swollen to the size of a pumpkin. He simply said: "Got kicked. Whaddya' think happened to me?" I theorized an equine pedal bone struck his vascular network and his knee became a hydraulic shock absorber.

"You better get back to pushin' papers, boy," Tom groused, smiling through his tough tone. "Leave a real workin' man alone."

Tuesday, November 28. George Washington named a dog "Sweet Lips," so I don't feel too bad about "Stingo." A traveling dog salesman came to see Travelling Music six years ago. We showed him a chestnut stallion; he showed us a chestnut Beagle. Now Trav is gone to another farm, but Stingo has never strayed far from the office woodstove. He seldom wanders even to Winters Run, where Ozzie, an immense Yellow Labrador, stands in the rapids and spears floating branches like a brown bear catching salmon. Another Lab, Austin, leaps from the dock on the pond like the famous Diving Horse at Atlantic City.

Nanny the beloved mutt looks both ways before crossing the driveway, but then sleeps under the cars of visiting clients, while Boxy the pointer gets behind the wheel of any vehicle with its windows left down. Lucy the Scotty immobilizes the UPS man. Any of these hounds can usually be found sleeping in the foaling barn.

A dog that chases horses is not welcome, but a horse farm without dogs is not quite a horse farm.

Wednesday, November 29. "I galloped horses for years at Belmont Park and never once got hurt. Last month, I gelded my two-year-old colt, and the first time back on him, he sends me sailing. I landed and broke both my arms."

Jill Gordon can still dial a phone, and she booked a mare today, but her conversation hung with me. My sister Norah is careful when she lifts a potted plant; as a teenager she broke her back riding. Andrew has broken his back, his foot, and his hand, but never his heart while training horses. Dad's wedding ring kept his finger on when he was bitten by the stallion Lochinvar. Peter almost had his eye put out when kicked in the face by a weanling two years ago.

I guess it's all relative. Dad went to a sports breakfast in Baltimore today. A cheer went up when quarterback Johnny Unitas, whose arthritic gait is legendary, and Bart Starr, who almost contracted frostbite in the 1967 title game against Dallas, were introduced. Ordell Braase, who played in the Greatest Game Ever Played, walked carefully on two artificial hip joints. Now he is on the Maryland Racing Commission. Maybe what you do with your life depends on what you are good at, and damn the risks.

Thursday, November 30. The barren and maiden mares go under lights tomorrow, and if the wind is right, you will hear the faint whistle of the train they call the breeding season, distant through the woods, but steaming steadily towards the farmland. Peter rotated the sixteen heavy mares into our largest field this morning. They spread out down through the hollow out of the wind and grazed contentedly in the warm sun. It was a beautifully serene sight, and I wanted to sit in the grass and let the dogs swarm over me and keep me warm, just to soak in the day, the blue sky, the mares, the sun. Then I heard the whistle blowing, and I made tracks back to the office, to prepare for December.

I am going to miss November. We turned a very important corner in the history of the farm this month, fueled by a spectacular Saturday three weeks ago. Miles still lay between us and the station, but we are all aboard, pulling in the same direction, the whole farm—Uncle John and Aunt Yvonne, Mom and Dad, all their children, and all the other good folks who live and work on this busy little enterprise. This month, we got a taste of what a little luck, and a lot of preparation, can accomplish.

THE CLEARING DAWN

Friday, December 1. Mentally handicapped people often become wonderful riders—something about the trust between horse and rider, the responsibility of one living thing for another. I thought Rollaids would be a good candidate to teach such children about horses, and called the county handicapped association to discuss this. I found a message on my desk this morning. It seems the head of the association owns a few broodmares. Perhaps he misconstrued my message. He said his wife books all his mares, and he was not interested in another horse at the moment.

The current market has caused an aversion to owning Thoroughbreds. I hadn't realized the effect had become quite so pervasive.

Saturday, December 2. Finney began his diary on this date in 1935. He described an apparent colic case: "However, investigation proved the horse had done a lot of running in yesterday's cold wind and then stood around. He stiffened up in the lumbar region. Haarlem oil for the kidneys, a brisk rub over the affected area, and a physic ball were his routine for the day."

A pregnant mare showed no interest at feeding time, and walked as if she were mildly cramped. I was not certain whether we had an impending colic patient or simply an in-foal mare suffering a particularly hormonal afternoon. I wondered if Finney would prescribe a "physic ball," so I consulted Capt. M. H. Hayes' 1924 treatise *Veterinary Notes for Horse Owners*.

"A physic ball acts as a purgative on the bowels. It is a bolus composed of five drachms Barbadoes Aloes, and two drachms of Ginger. The presence of Ginger diminishes the chance of griping. Treacle or lard is sufficient to make a ball."

I was getting nowhere fast. What is a "drachm"? (Sixty grains equals one drachm, and eight drachms equals one ounce.) What is "treacle"? (Molasses.) Finally, Ellen arrived, told me the mare was spooked by utility workers near her field today, and had been running. On cue, the mare passed manure and dove into her hay. I never found out what Haarlem Oil was, but I'm not griping.

Sunday, December 3. A gale blew as I stood looking at the Timonium sales grounds—clouds of sand lifting off the race track, barbed wire framing tin roofs on pole barns, the harsh orange of a December dusk showing no mercy. Sand crunched in your teeth if you greeted the five buyers who braved a visit to the barn. We were agent for eight mares, no reserves on any of them. We tried to bid each mare out of the meatman's range. We got stuck with one at eight hundred dollars. Michael sold her later for two hundred and fifty to a local farmer who promised a good home—her former owner's only request.

People sell broodmares in December because they don't want to feed them all winter. People don't buy broodmares in December for the same reason.

Monday, December 4. Another pregnant mare appeared uncomfortable at feeding time today, displaying the same symptoms as labor. This time, I called Peyton before bothering about physic balls.

"I hate to give medical advice over the phone," he cautioned, "but from what you've described, she might just be feeling the increasing weight of the foal in the last trimester of pregnancy. It could be a simple mechanical problem—pressure on a bowel, or stretching of the mesentery maintaining the intestines in position in the abdominal cavity. Take her temperature, keep an eye on her, and the moment she starts to go in the wrong direction, I'll be over."

Anyone who has ever labored to present a horse with a freshly bedded stall will attest to the laxative effect of clean straw on horses. Tonight was no exception. After a few minutes of pawing, the mare relieved herself and went to her hay. Peyton's diagnosis was on the money.

Tuesday, December 5. The attorney was a hard-boiled cigar smoker with a Corvette in the parking lot. He was married to a gal much younger. She worked in his law office. He squawked at her through the speaker phone. As he dialed her number, he whispered to me: "She can do anything. Sky dive. Drive a sports car. Anything." He smiled lewdly. I wanted to be somewhere else.

Sally here. "Yes, Sally, I have a gentleman here who wants to sell us part of a racing partnership. When did we buy our racehorse?" *Summer of eighty-eight.* "How much did we spend?" *Ten thousand to buy him.* "Did we have fun watching his foot mend?" *No.* "Did we like the people we met at the track?" *No, we didn't.* "Didn't the trainer say he'd make a great jumper?" *Yes, he did.* "And how much did we lose in the brief time we were involved?" *Twenty thousand dollars.* "Thank you, dear."

He rocked up and down in his swivel chair. I had to keep bobbing my head to maintain eye contact. He dismissed me: "Guess what? I'm *not* interested."

Wednesday, December 6. Richard jetted around the fields today, freezing on the tractor while chain harrowing before the wet weather hits. Parasitologist Dr. R. P. Herd of Ohio State wrote that stallions defecate an average of thirteen times in a twenty-four-hour period, mares seven times, and foals ten times daily. That's a lot of horse manure. Dr. Herd is an advocate of vacuum cleaners for pastures. The idea is sound, and in wide use in England, but if I asked Rich to vacuum the pastures this arctic afternoon, he might think I was some demon out of *The Far Side* comic strip. He looked too cold to take a joke.

Thursday, December 7. Guys with pure athletic ability can throw bull's-eyes in darts and sink basketballs at the state fair until their dates are loaded with stuffed animals. I have known such fellows, but the margin for error is not usually fatal in such games. It *is* in race-riding. Today, Kent Desormeaux came close to being killed in the second race. Turning for home, the kid urged his maiden mount through a hole on the rail. Surging to the lead, the horse suddenly shied, crossing over in front of the entire seven-horse field. One instant, Desormeaux is securely guiding a horse running at thirty-five miles an hour. The next instant, his right leg is flying wildly in the air on the near side of the horse. If he had let go, he would have been trampled. But like a gymnast on the balance beam, he swung his weight back over the horse, landed his right foot squarely in the stirrup, and resumed race riding—scrubbing and whipping to win by the slimmest of noses.

It was a circus act. It has been on all the TV stations tonight, replayed from various angles—sportscasters in dumbfounded amazement at the dangerous liaison formed when a man weighing a hundred pounds climbs aboard a horse weighing a thousand pounds.

Friday, December 8. It is snowing hard at 8 A.M., but the farm is abuzz. Dad rushes to attend a fund-raising breakfast for our Annapolis delegate. Peter prepares vet records for six new barren mares. Ellen beds down extra stalls. Steve Earl and Richard turn out the stallions. Steve Brown hooks the plow to the three-point hitch and wishes for a cab on the tractor. John King stays in his little house. (In October, tripping on walnuts is his biggest fear; in December, ice is.) Uncle John arrives with the mail. Alice copies breeding contracts. Michael contacts hay dealers for deliveries. Even the farm's communal grandchild, Philip, makes an appearance (baby-sitter snowed in). In the big house, Mom paints the dining room a rich golden color. Gone is the green of holiday dinners, yet I cannot wait for Christmas in the fresh surroundings.

Meanwhile, Mother Nature paints the russet earth white to the horizon, and it feels like the color is here to stay.

Saturday, December 9. The Amazon Queen was at her finest this morning, as though aware of the camera. Her personal photographer, Dave Harp, continues chronicling her life from the moment her mother entered labor until this leggy weanling reaches the starting gate. Dave arrived before a single horse had stepped onto the virgin blanket of snow. When the thirteen weanlings galloped away from the gate, Dave's motor drive captured the moment: The Amazon Queen in the lead, gliding past him in a cloud of snow. This handsome filly seems worthy of such a dedicated photographer.

Sunday, December 10. Received some statistics recently from Rich Wilcke, editor of the *Maryland Horse*. In 1962, the year the Maryland-Bred Fund Program was established, this state produced approximately five hundred foals, while Virginia produced eight hundred. By 1987, Maryland was producing two thousand foals, an increase of three hundred percent, while Virginia produced only a thousand foals, an increase of a mere twenty-five percent over twenty-five years.

The marriage between racing and politics has often been stormy, yet without the legislation that mandated the Maryland-Bred Fund, we might not have a job.

Monday, December 11. "Looks great, healing beautifully," Dr. Riddle declared as Ellen unwrapped the gauze bandage from the Slip of the Tongue's lower leg. On her third day at Country Life, the mare did not apply the brakes in time to avoid sliding into the four-panel fence. She tore a small flap of skin off the front of her cannon bone. But any time bone is close to a wound, utmost care must be taken to avoid infection. Dr. Riddle instructed Ellen to use a non-irritating dressing, such as Furacin, prior to wrapping the leg. Excessive granulation of the tissue often causes lasting scars and can be avoided by bandaging the leg and keeping it dressed.

I have not been able to reach Geri Hughes, the mare's owner. I don't know her very well. I'm hoping she's a good sport. Grandfather once had a client who simply could not understand that horses spend their lives trying to commit suicide. He finally referred the lady to Finney: "Our mutual friend Mr. Finney will tell you that accidents of more or less serious nature are liable to happen on any horse farm." I had Grandfather's conciliatory letter in mind as I dialed again.

Tuesday, December 12. Enchanting day on the farm, as snow covered the evergreen trees, beaded perfect white lines down the top board of the fences, and painted the rust spots on the barn roof. A visitor arrived from Camden, South Carolina, and we showed her the stallions, walked her through the fields to see their get—whistling for Rollaids to leave his pile of hay and join his classmates for inspection.

Snow is therapeutic to horses. It cleans their hooves and lifts mud from their coats, dappled out now in the cold clear air. It rejuvenates people as well. An odd assortment of sleds, relics of the wonder years growing up on this farm, hit the hills late this afternoon. Richard, the farm strongman, felt like a child again in the mountains of West Virginia, sitting upright on a sled two sizes too small for him, building speed, taking one hand off the sled to wave like a cowboy on a bronc, his sled popping through frozen ribs of snow compressed by tractor tires, making a sound like a tiny Gatling gun. In his slow drawl, Rich thanked Ellen for waxing the runners and giving him a push: "Why, Ellen, that was *really* fun! I haven't done that since I was a kid."

I sledded smack into the diamond-mesh fence, landed in an unharrowed dropping, cleared the snow from behind my glasses, and said the same thing.

Wednesday, December 13. Three years ago, Maryland stallions bred three thousand six hundred mares, but only two thousand two hundred foals were registered—sixty-one percent. Thus, forty percent of the one hundred-percent effort it takes to get mares to our stallions does *not* result in a registered foal. Four out of every ten mares attracted to our stallions through salesmanship and advertisements, mares we tease daily and palpate repeatedly, mares we lead to the breeding shed and ultrasound fifteen days later, mares we ship home with flannel on their halters, feet trimmed, and intestines oiled—*four out of every ten mares bred* will not produce a registered foal that just might be the next Carry Back.

Almost half our effort is in vain. This can be a very depressing statistic to folks who live with sweat on their brow, their hearts on their sleeves.

Thursday, December 14. "Why, I've never known him to be here at the bank before nine! It's just not his style. But I'll check his appointment book if you're certain he said seven-thirty," the assistant to the president told me this morning upon his arrival at eight. For the past hour, I had been the sole person in one of the largest banks in Baltimore, waiting to present my racing partnership proposal to a fellow who loves the sport. The elevator from the parking garage deposited me at the nerve center of the executive suite. "No, no, no. His daybook says ten-thirty. You must have *thought* he said seven-thirty."

Grandfather had racks of business suits, and his clients were named Belmont and Harriman and Widener. I might be caught wearing a red power-tie my mother-in-law gave me: When you squeeze the fat part, it plays Christmas carols. I felt like Rudolf the Red-Faced Reindeer by the time ten-thirty rolled around. Oh, well, just a day in the life of a salesman.

Friday, December 15. The huge top of a cedar tree blew out in the high winds of late, and Ellen chain sawed the ragged edge off the bottom. Tonight the cedar is adorned with Christmas decorations. A heavy wet snow began falling at nine o'clock. How long until the storm hits the Meadowlands, three hours to our north?

At ten o'clock tonight, twelve colts were set to break from the gate in the New Jersey Futurity. Groscar's owner said he would call if he had good news. I read until ten-thirty, wistfully computing the time necessary to run down to the winner's circle, hug the trainer, hug the jockey, accept congratulations, and then, by the way, remember to call the fellow who stands the stallion. At eleven, I gave up. Then the phone rang. My heart thumped. Visions of catapulting up the freshman sire list flashed before me. I danced to the phone.

"Well," the owner said, very matter-of-factly. "The snow hit. They canceled the card after four races. We were in the seventh."

Outside, the wind howled through the hardwoods, mocking my misplaced excitement. I crawled down under the quilt and wondered when they would put the race back up. After all, can't have a Futurity for three-year-olds, and it's already mid-December.

Saturday, December 16. It was so cold this morning, my ungloved hands froze to the galvanized steel gate. I heard a dull pounding, then saw Richard with a hammer, beating the daylights out of a water trough. Shards of ice were flying everywhere, Richard heatedly trying to reach water for the sake of the lone horse that will occupy this paddock.

"Rich, Rich, take it easy. We'll bring out a bucket of hot water and hook it to the fence. He can drink that while he's out," I told him. He looked up at me. Chunks of ice had adhered to his woolen cap and jacket. He looked as if he had been in a fight with a frozen daiquiri. I wondered how many farmers at this very moment were kicking ice in their own troughs, banging the sides of frozen buckets, blowing on their hands after lifting icebergs from water tubs.

"Horse has got to have water," said Richard the lion-hearted. "I can't stand it when a horse doesn't have water."

Sunday, December 17. I asked Mom to send out requests to the farm's Sunday crew from the warmth of her broadcast booth. She played *Carnival for the Animals* after we turned the freshman stallion out. She played the *Light Cavalry Overture* while we fed the weanlings. She played *Peter and the Wolf*, with Leonard Bernstein narrating, as I drove the truck past Rollaids. He looked at me as if his mother used to play it for *him*. What a face he has.

Sunday mornings, you can have the world to yourself if you plan a little, and your Mom is a deejay.

Monday, December 18. "The cabbies wouldn't bring him. He called them two hours ago. They just steal from him when he's like this. The man's a prince. So what if he's had a few beers?"

Uncle George had abandoned his restaurant at the dinner rush hour to deliver John King back to the farm this evening. Shorty was swaying back and forth, very sincerely repeating: "Georgie, Georgie, get some sleep...get some sleep." Uncle George took his arm to steady him across the snowy yard to his little house.

"Gotta run," said Uncle George, and he roared away in his El Dorado like the Lone Ranger on Silver.

Tuesday, December 19. Steve Earl is sitting next to me. He and I worked Sunday, just the two of us, breaking ice, throwing hay, examining livestock. Yesterday morning, though, he phoned in sick—for the third Monday in a row. It is Sunday-night carousing that laid him up. I was taught in law school that every dog gets one free bite. Clearly, Steve is dogging it, and I have already given him two free bites.

"Every time things get going good, I look for trouble to start," he said, choking up. "That's the story of my life. Maybe because my Dad never gave a hoot about my mother or me and my sisters."

His pity play worked, and I gave him a *third* chance, but as he left the office, I was torn. He's hot and cold. You can't be tired and work safely around horses. You can't be one person one day and a different person the next. The stallions won't know who you are, and will do their best to find out which personality showed up for work. I looked towards the red gates in the snow and decided to make a hard call if I get bit again.

Wednesday, December 20. Nick Carraway, the narrator in *The Great Gatsby,* describes a quiet Sunday afternoon on Fifth Avenue in New York: "Soft, almost pastoral...I wouldn't have been surprised to see a great flock of white sheep turn the corner."

This morning, I saw a great herd of white-tailed deer turn the corner near the house. They lolled in the driveway, not a care in the world. They might have been Santa's reindeer, foraging a bellyful of energy for the coming holidays. I counted nine, and looked for signs of lameness from hunting injuries. (Not long ago, Ellen saw a three-legged deer nimbly overcoming her handicap.) The biggest doe seemed uncomfortable so close to mankind. She changed directions, and climbed through the multiflora rose and grapevine to reach Winters Run. A young deer slipped and thumped loudly onto the ice, perhaps encountering the surface for the first time. Then they were gone, up through the woods to graze on young shoots of poplar pushing skyward from a timbered area. Sheep on Fifth Avenue might not have surprised Nick, but nine deer in the driveway was a new one on me.

Thursday, December 21. I revised our Stallion Contracts this week, ensuring that the owners of the six foals produced from the ten mares we breed are compelled to take full advantage of our state programs. A simple boldface sentence declares: "Owner Agrees To Nominate Resulting Foal To The Maryland Million Program, And If Eligible, To The Maryland-Bred Fund Program."

The sixty percent foal rate means that we take two steps up and almost one step back during the breeding season. That's a bite. That is why we put some teeth into the contracts.

Friday, December 22. It was zero degrees tonight at the Meadowlands when they ran the New Jersey Futurity. Groscar finished fourth and earned six thousand dollars. Carnivalay appears likely to finish the year as the leading freshman stallion in Maryland. Good thing, because this market gives up on stallions overnight, as the pool of available mares continues to shrink. It still rewards sires of good racehorses, however, and Carnivalay will hold court to a fine book of mares next spring. Following a freshman stallion is rather like watching your boy play Little League baseball. All you can do is watch from the sidelines, and wish him luck.

Saturday, December 23. "I saw an ad in the Charlottesville paper," Ellen's friend Polly Bance declared, "asking for good homes for Clydesdales that didn't have the right markings. They were out of Auggie Busch's mares, just weren't *white* where they should have been."

Behind the massive haunches of one such oddly marked Clydesdale mare— a likeness of Italy bouncing along her flank as if in a giant earthquake—I took my first-ever sleigh ride. Ellen and Polly could not resist a bubbling Christmas carol. We certainly were dashing through the snow, laughing all the way, and the spirit of Christmas rang over the beautiful Virginia countryside.

Sunday, December 24. "I think only a poet could paint Traveler," Gen. Robert E. Lee wrote to an artist who wanted to paint his famous war-horse, "a poet to imagine the horse's thoughts through the long night-marches and days of battle. He carried me through the seven-day battles around Richmond, the second Manassas, at Sharpsburg, at Fredericksburg, the last day at Chancellorsville, to Pennsylvania at Gettysburg, and back to the Rappahannock. He passed through the fire of the Wilderness, Spottsylvania, Cold Harbor, and across the James River. In daily requisition in the winter of 1864-65, he bore me to the final days at Appomattox Court House. You must know the comfort he is to me in my present retirement."

They say Traveler stepped as if conscious that he bore a king upon his back. I think he bore a poet, as well. Soon after Lee died, Traveler—milk white with age and honors—lost his master's rhythm, stepped on a nail, and died of lockjaw.

Monday, December 25. A magical Christmas, everyone home and wood on the hearth. Alice ran pine boughs along the mantles, and the racing prints in the dining room—against the fresh coat of gold paint—were as brilliant as the colorized version of Charles Dickens' *A Christmas Carol*. Beautiful black-and-white photos of August Belmont's champions, with hand-painted backgrounds and framed in chestnut, were mere identification photos for the English Jockey Club of the late eighteen-hundreds. But Mr. Belmont handed them down to Grandfather, and tonight the horses pose on pine needles, and keep vigil over the snowy landscape from the front hallway.

Grandfather died on Christmas Day, 1951, three years before I was born. His photo was on the front page of the *Daily Racing Form*, and I have read and reread the yellowed obit, a link to a man I never knew but to whom I feel so very close. I have called on him through his letters as though he were alive in his study, to be of counsel, to keep vigil over me. This business is such a humbling affair, and I have needed to know that my mistakes were not entirely my fault, that the nature of the beast called a horse invites error, that perseverance is the key, that luck will win out if you stay the course.

Well, enough of such stuff. What's on for tomorrow?

Tuesday, December 26. We stuffed our shareholders' stockings with a seven-page direct mailer today. A modest newsletter bragged on the four sires. Progeny reports were included. The birth of Country Life Racing Inc. to manage our racing partnerships was explained. We reminded breeders about opening early-foaling mares and about placing barren mares under lights. We also included the new, revised Stallion Service Contracts, and sent Maryland Million nomination blanks, with return envelopes.

Finney believed that the "absolute zero of activities on the average stud is that period between Christmas and the New Year." He lived in an agrarian society. We live in a world where information is perhaps the farm's most important fertilizer.

Wednesday, December 27. The manure man was in the middle of his last forkful when the huge claw suddenly froze up on his rig, dangling yellow straw twenty feet off the ground, as if a prehistoric bird had been captured gathering nesting material.

"Damn! Another minute and I'd be on my way to West Grove," the manure man said, scampering down from his perch. I asked him his name. "Bob. Bob. Just another damn Bob!"

From years of talking to himself above the tumult of his hydraulic fork, Damn Bob was unaccustomed to speaking in confines such as our small office. A client from D. C. phoned while Bob was shouting for help on the other line.

"If you've got problems, I can call back," the client thoughtfully offered. At that moment, Damn Bob hung up, pushed a fifty-dollar check and a box of fresh mushrooms across my desk, smiled a toothless grin, and quietly waved in thanks.

"No, go ahead. I'm listening," I said into the phone, giving Damn Bob the thumbs up as the door closed. His air of assurance, however, remained long after he had made for West Grove.

Thursday, December 28. Horse farmers are white-knuckling it through the current hay dilemma. It's not simply the fifty-three inches of rain this year (twelve above average) causing hay prices to double. As big a contributing factor is the loss of local mom-and-pop farms which annually put up thousand of bales. Such family operations made great hay. When rain threatened, moms fed the cows so granddads could climb on the tractor while brothers and cousins hustled the hay into the safety of huge bank barns. But last year *alone*, Maryland lost a record forty-four thousand acres of farmland—an area larger than Baltimore.

What lies ahead? In the first three hundred and fifty years of settlement, seven hundred thousand of this state's six million acres were developed. In the blink of the next twenty-five years, a second seven hundred thousand acres will be consumed, forming an unbroken band of urbanization from the Susquehanna to the Potomac, from the mountains of western Maryland to the Eastern Shore.

This afternoon, I stood outside the office, humming that odd song of futility from the Vietnam era: "And it's one, two, three, what are we fighting for?" But I couldn't finish the verse because I'm not that frustrated. Got a hot horse in the stud barn, a hotter one on the horizon. I didn't care that the north wind carried the sound of the loudspeaker from the Chevrolet dealership across Route 1: "Ernie, Line One...Ernie, *Line* One." I said a quick prayer for the good health of my family, and for the potential we've just begun to tap.

Friday, December 29. Heard the log splitter ripping oak as Rich and Steve worked hard all day long. Mailed stallion certificates to Kentucky for the winter sales. Grabbed the national freshman sire list: Carnivalay at seventeenth. Does that include the Jersey Futurity? The second at Philadelphia? The third in the stakes last week?

Ellen is on the phone: "I've never seen so much hay in my life. I'm looking up the hill at a hundred trucks loaded with hay. It looks like an army." But is it good enough for horses? She is at the Ephrata, Pennsylvania, hay auction. "I see five tons of second-cutting alfalfa for the studs. Should I bid?" She fights against cagey Amish dairy farmers, gets it for one hundred and seventy dollars a ton. We will wage costlier battles before winter is through.

Michael guesses that the hay will arrive just when clients are due at two o'clock. He is doubly right. Two loads of hay arrive when two sets of clients drive in, and a three thousand-gallon propane truck pulls in with an escort. Everything *always* happens at once. We push and push and get through the day, and the business of being busy relieves the mind of anxiety. It is the breeding season rehearsal—half-speed drills prepping us for game days. We throw a few freshly split logs into the office woodstove, the smell of red oak sweet in the December air, and head home after rush hour.

Saturday, December 30. They ignited the torch in the fourth race, as Desormeaux drove his mount past Shoemaker's in midstretch to win. They passed the torch in the feature, Shoemaker on the lead, trying to coax a mile and a quarter out of the old legs of the gray gelding Due North, Desormeaux again blowing past in deep stretch, getting the run of a lifetime out of a former claimer.

On his farewell tour, Shoemaker was the gracious gentleman he always has been. "Got beat by the best young rider in America," he said in a brief ceremony. Even the toughest old race trackers took their hats off to the greatest rider in history.

Sunday, December 31. "How completely a load of hay in winter revives the memory of past summers!" Thoreau declared this morning, hopping into the loaded truck. "After all, summer in us is only a little dried, like the hay."

A misting rain froze upon landing on Earth, causing a treacherous glazing, and as usual, Thoreau was too cerebrally detached from his body to be much help with physical chores. But he is good company in small doses. On the hill overlooking the watershed, he waxed: "The top of Mt. Washington is no more beautiful than this. This land should remain—for modesty and reverence's sake, or if only to suggest that Earth has higher uses than we put her to."

And he laughed heartily at our toil: "You'd been better off born in an open pasture and suckled by a wolf than to work on a farm. How many a poor immortal soul have I met well-nigh crushed and smothered under its load, creeping down the road of life, pushing before it a barn—its Augean stables never cleansed—and a hundred acres of land, tillage, mowing, pasture, and woodlot!"

I do not know how he guessed the acreage of this farm, and on any morning other than a Sunday, I might have wanted to throw Thoreau from the truck for deriding the country life. Instead, I dropped him at the office as he explained that Winter was not an Old Man, but rather a merry woodchopper, a warm-blooded youth, as blithe as summer. I had had enough.

"Fine, Mr. Woodchopper," I called to him, before shutting the door on the year, on the decade, on his foot. "Then how about feeding that woodstove! You're the one who said, 'Nothing remarkable was ever accomplished in a prosaic mood.'"

And we drove on toward the clearing dawn of a New Year.

Year Two

RULE ONE

Monday, January 1. Today, our world thawed. Newly turned yearlings sun-bathed this mild afternoon, and broodmares grazed on grass long hidden by snow. Since Thanksgiving, Winters Run has been a solid sheet of ice, two-feet thick in deep, slow-moving sections. Last night, however, heavy rains bounced off the frozen farmland, and the swollen river lifted the ice from its moorings. By this morning, giant ice floes were crashing into a fallen oak tree that formed a bridge from one bank to the other.

The oak, which had fallen in October, still had its leaves and the strength of its sap. Its branches propped the trunk against the current. It dammed up tons and tons and tons of ice. This afternoon, natural forces were in concert—the cold, the ice, the rain, the tree. They played a crashing crescendo from the watershed, an orchestra in an amphitheater designed by a glacier.

Emerson said: "Events are in the saddle, and they ride mankind."

Awed by the hollow thunder of the piling ice, feeling the temperature climb, I looked forward to the events of the coming year, as a chill of excitement swept over me.

Tuesday, January 2. Two hundred years ago, the Swiss artist Fuseli became famous for his painting entitled *Nightmare*, in which a sinister blind horse stared through bedroom curtains as his rider, a demon named "Incubus," paid a visit to a sleeping victim.

In and out of a feverish sleep all day today, as Incubus' first cousin "Influenza" visited me, I felt the black eyes of the horse in Fuseli's painting, and I felt uncertain whether excitement or exposure had occasioned yesterday's chill.

"You're not scaring me!" I said into the darkness, recalling a real horse who terrified me. "I stood a horse named Lyllos. He *was* the devil."

Wednesday, January 3. Grandfather's books are most therapeutic to the con-valescing. Under the strict rules of Mom's lending library, I signed for four items today: *The Horse in America* (1905); *Blooded Horses of Colonial Days* (1922); Vosburgh's *Racing in America 1866-1921*; and the catalogue entitled *Absolute Dispersal Sale of the Famous Nursery Stud, May 15, 1925* (with Grandfather and C. J. FitzGerald managing the sale of Man o' War's sire Fair Play and the rest of August Belmont II's stock).

It'll take me six months to read these books, and six days to forget what I learned. But it should be an interesting semester.

Thursday, January 4. The in-foal mares suddenly look as fat as houses, even lanky Auxilliary, carrying her first foal. She is due on January 11th, but maiden mares often carry late. Either that, or they foal three weeks early. The only thing for certain is nothing is certain.

Kentucky veterinarian Walter W. Zent has found that wintering pregnant mares under lights shortens gestation by an average of ten days. Some of our in-foal mares are stabled near the barren mares, whose stall lights remain on to simulate lengthening days. I wonder if the gentle rays reaching the in-foal stalls will affect gestation. We'll find out shortly.

Friday, January 5. "She's due February 1st," the mare owner told me. "I'd like to breed on her ninth day past foaling. I like *early* foals."

So do we, and we generally try to accommodate owners. But what's best for the mare? Foal-heat breeding in February seems greedy. Without April's green grass or the abundant exercise they receive in good weather, mares often haven't recovered enough in February or March to warrant foal-heat breeding. My other concern about wasting the stallion's cover is an argument certain to be lost on board-paying mare owners.

So we'll ask ourselves: Did the mare have an easy delivery? What is her history on foal-heat? Is it worth the higher rate of early embryonic loss? Has her uterus had time to rebound?

Dr. Zent's studies indicate that the mare must wait to ovulate until *at least* nine days past foaling; breeding on the seventh or eighth day is almost certainly a wasted effort. We have to play the percentages, and I would rather take one good shot with a mare on her twenty-eighth day past foaling than bruise her up on her ninth day.

That's how I feel about foal-heat breeding in the cold of winter. In May or June? Well, that is a different matter.

Saturday, January 6. Down the list I went, addressing in my miserable penmanship the fifty invitations to the inaugural Country Life Racing Seminar, at Laurel on January 18th. Our featured speaker is Sam Huff, a Hall of Famer in the National Football League, and the leading light in attracting corporate sponsorship to the West Virginia Breeders Classic, a rich event at modest little Charles Town race track. In so doing, he brought an impossible dream to fruition, and thus seemed a fitting speaker for our seminar, designed to raise the local business community's awareness of this exciting sport.

The invitations were in the mail by noon. I should be hearing RSVP's by Monday.

Sunday, January 7. Horse racing was a big deal in Maryland way before the start of the American Revolution. In those days, racehorses generally began competing when they reached six to eight years of age. Races often were contested in a series of four-mile heats. Imagine being a horse back then, needing to win the best of *five* four-mile heats just to be declared the best horse of the day. (Heck, Danzig raced only three times in his entire *career*; Carnivalay raced only four times. They could not have withstood the rigors of heat racing.)

In the pre-Revolutionary era, wrote Francis Barnum Culver in *Blooded Horses of Colonial Days*, the inhabitants of Maryland were "almost to a man, devoted to the fascinating and rational amusement of horse racing." Where there are horses, however, there is legislation. In Annapolis in 1692, "An Act for Restrayning the Unreasonable Encrease of Horses in This Province" became law. Colonial legislators were concerned about damage to cornfields from roaming bands of horses. They also worried that "the small Stature of Stallions running wild dothe Lessen and Spoyle the whole Breed and Streyne of all Horses."

Regular matched races between pedigreed horses began at Annapolis in 1745, when the Maryland Jockey Club was established. Gov. Samuel Ogle built the famous Belair Stud in Prince Georges County in 1746. The most famous record-setter before the Revolution was Maryland-bred Selim. Race courses dotted the landscape of this tiny state.

I think—on balance, over time, all things considered—no other state has contributed quite as much to the annals of the Turf as Maryland, my Maryland.

Monday, January 8. While soaking in the beauty of the Eastern Shore on a rainy afternoon, from the cozy confines of a stand-up blind on the Choptank River, as thousands upon thousands of Canada geese flew just out of range of the Kentucky shareholders on their annual visit to Maryland, I battled my conscience for skylarking while the home team worked in the wet.

To assuage my guilt at being so unproductive, at day's end I phoned an Eastern Shore horseman to inquire about his breeding plans for this season. When I called, however, all hands were at the barn, delivering a dead foal born two months premature. Unfortunately, the foal was by one of our stallions.

I should have known better. If it's your day off, *take it,* and don't ask questions.

Tuesday, January 9. "Have you received your invitation?" I asked Betty Miller, who lives a scant eight miles away, on a hill overlooking beautiful Merryland Farm. "No?"

I'm worried. I put stamps on the invitations. Businessmen will need at least a week to plan for an afternoon away from the office. Is my handwriting so bad the postman can't read it? I think about the best-laid plans of mice and men. I think about Groucho Marx, who knew something about a day at the races. An adversary challenged him with the question, "Are you a man or a mouse?" Groucho replied, "Put some cheese down and you'll find out."

Don't put any cheese down around me if the invitations aren't delivered by tomorrow.

Wednesday, January 10. One morning a few years ago, two Canada geese flew no more than twenty feet overhead as we led Allen's Prospect to his paddock. Geese, like policemen and men who lead stallions, often travel in pairs. Michael was my backup. I looked up and saw the slightly opened mouths of the geese just at the moment I walked Allen past the double-fenced paddock of Assault Landing. Sensing my inattention, or the eeriness of powerful birds so close we could hear them breathing, or the proximity of the black stallion, Allen's Prospect charged me. I did not feel it when he clamped onto my right forearm. Then my arm went limp. The stallion released his grip, his message delivered, and I fell to my knees. Michael rushed in, took the stitched loop of shank out of my left hand, and led the stallion safely to his paddock. My ego was bruised worse than my arm.

Today, while leading Allen's Prospect, I heard the songs of the beautiful white tundra swans whooping overhead. Yet I had total recall of the earlier incident—as if I was hearing the screech of a car about to crash. I did not look skywards. I kept my eye on the stud. Nothing bad happened.

Tonight, I wanted to fix this episode even more firmly in my mind. When the mayhem of May is upon me, I must remember that when I have a stallion in my hands, *nothing else is important.*

Thursday, January 11. "Rollaids has all the luck," mused Dr. Riddle as he examined the lumpy lymph nodes under the chin of our favorite patient. Fifty-three horses on the farm, and he is the only sick one. We vaccinate against strep equi, so it's unlikely he's contracted the dreaded strangles bacteria. But new horses have come onto the farm recently, and with them can travel new strains of ancient illnesses.

"The odds are ninety-nine-to-one against it," Dr. Riddle said, as he read my mind. "His lymph nodes are simply doing their job, filtering various routine bacteria. Still, the best bet is to isolate him, lance the abscess, take a culture, and apply hot compresses for several days."

As Rollaids marched away to the quarantine stalls, he whinnied in his raspy, deep-throated voice— a chuckling cartoon laugh, as if to say: "Don't worry. I'll be out in no time. Just a cold. No big deal."

I'm sorry to have to isolate him, but I can't take chances. I feel like I've become an absolute pain about viruses and bacteria. In these matters, however, the horse business does not teach you gently. I've seen the one out of ninety-nine, thank you, and so has anyone else on a commercial farm.

Friday, January 12. I thought *I* was paranoid about health records until I met a driver for H. H. Hudson Van Lines, here to pick up four two-year-olds being shipped to Florida.

"Did your vet put the *Accession Number* from the Coggins test onto the Interstate Health Certificate?" he asked me at six-thirty this morning. "They'll stop me at the Florida border, and I'll have to call a vet and wait and wait. I'm not moving one inch unless that Accession Number is on the Health Certificate."

I examined the certificate: It was timely, drawn within ten days of the horses' departure. There were four copies (two for the state, one to travel with the horses, one to file in the attending veterinarian's office). After execution, *no subsequent notation* had been permitted on a copy, not even for something as clerical as an Accession Number. Dr. Riddle and Peter Richards had seen to the proper execution of the paperwork.

I was glad the van driver didn't have to stay for breakfast. These guys are not good company when they should be somewhere else—like on the road.

Saturday, January 13. A footnote from the good news/bad news department. The culture taken from Rollaids was negative (great news), and Betty Miller received her seminar invitation, canceled by five different post offices (bad news). Businessmen might not see their invites until Tuesday, since Monday is Martin Luther King Day with no mail service. I took the matter into my own hands, phoning everyone on my list. Things happen for a reason, I said to myself, dialing away the day, running down a checklist of cliches: It will work out for the better; turn this to your advantage; every cloud has a silver lining; all is not lost. (No, just the invitations, I grumbled as I dialed.)

Sunday, January 14. By the whims of fate and family that blow us all along, Peter landed at Country Life four years ago, to be near his son Gordon and daughter-in-law Annette. (Gordon is a racing official at various East Coast tracks.) Peter can be very reserved. If my Dad had been knighted by the Queen, I'd be wearing a coat of armor and galloping Carnivalay around at state jousting tournaments. Peter has greater restraint. I asked him recently what life was like growing up the son of famous English jockey Sir Gordon Richards. Peter showed me his scrapbook. In it was a letter written to his father from 10 Downing Street on May 13, 1953:

> Dear Mr. Richards,
> I have it in mind on the occasion of the forthcoming list of Coronation Honours to submit your name to The Queen with a recommendation that Her Majesty may be graciously pleased to approve that the Honour of Knighthood be conferred upon you.
> I should be glad to know if this would be agreeable to you, and I will take no steps until I have your reply.
> Yours Sincerely,
> Winston S. Churchill

Peter carries his legacy lightly, and I feel fortunate he cast his fate with us. He regards Carnivalay paternalistically, having ridden the colt's father. He regards me, I believe, as a peer, and that is as important to me as Mr. Churchill's letter is to Peter.

Monday, January 15. Two breeding seasons ago, we bred Hal Clagett's mare Amerrico's Sphinx to Carnivalay. The stud mounted the mare several times, but I did not think he ejaculated. Carnivalay performs erratically if uncomfortable in a mare; position is everything. That evening, I phoned Mr. Clagett, and suggested we breed his mare to the taller Corridor Key, and not miss this heat period.

"We'll report both stallions," I said, "and let The Jockey Club blood-typing procedures be the final word."

He agreed. When we filed our Reports of Mares Bred, I attached an affidavit explaining why the mare was bred to two different stallions on successive days.

Mr. Clagett phoned this morning. Apparently I was mistaken about Carnivalay's carnal knowledge of Amerrico's Sphinx, but the mare's seasoned owner was magnanimous in the face of youthful error.

"The important fact is that the integrity of the Thoroughbred breed is maintained through the advances of blood-typing," lawyer Clagett said in his closing argument.

In the breeding game, the action moves fast, and the players are microscopic. Having blood-typing as a backup is similar to the benefit of instant replay, even if it did take the officials two years to make this call.

Tuesday, January 16. The Baltimore *Business Journal*, whose articles reach local corporate sanctums, phoned at 2 P.M., and asked if they could take photos at three. No problem, we replied, but every horse on the farm was covered in mud. The sixty-degree weather had the stallions feeling very springlike, and they tried to mount us as we led them past the photographer. Assault Landing's halter had slipped over his ear, and I did not notice it for all the mud. I did not want photos of a loose stallion, so I hustled him into the barn before he could act up.

"This story will end up lining some birdcage in Baltimore. Don't take chances for a stupid photo," I chided myself. Finally, I was able to pose Carnivalay against his red door, the winter sun banking off his almost piebald face. I suggested the cutline: "Just write: 'Pictured above is Carnivalay, son of legendary Northern Dancer.' Everybody in Maryland has heard of Northern Dancer."

"Don't tell me," the photographer said. "I just take the pictures. Somebody else does the story."

Why do I suddenly feel he'll select the photo of Assault Landing with his halter over his ear?

Wednesday, January 17. In the evening, the black and white glow from the closed-circuit TV monitor is the only light coming from the office. It is the signal that the foaling season is underway. Can spring be far behind? Two months from now, on nights no warmer than this, the spring peepers will venture forth from the muddy banks of the pond, to congregate on the driveway over the dike. Some nights, rushing to the foaling barn, unable to avoid their gauntlet, I've massacred dozens and dozens of tiny frogs. There is simply no alternative route. By morning, however, not a trace of a squished frog can be found. The owl, the fox, the opossum, the wild cat who lives behind the dike, all feast quickly on the carnage, gorging themselves lest they meet the same fate.

Ellen is always appalled at the frog fatalities, but I've been able to reason with her: "It's part of the food chain. It's survival of the fittest." She knows I probably couldn't even *walk* across the dike without stepping on a recent tadpole. We laugh during these arguments, and I know there is a black humor to farm life that causes me no end of amusement.

Thursday, January 18. In contented exhaustion, we drove home in the dark after today's exciting seminar. In my rearview mirror, I could see Michael's head bounce sleepily off his chest, while beside me, Ellen stared absently out at the North Star. Behind us on I-95 drove Dad with Uncle John, Alice with Andrew. I felt as if we were on a very long team bus after a high school road game.

Sam Huff kept the crowd smiling while imparting his own frustration: "In football, one team wins, one team loses. I had something to say about the outcome. Not so in racing. In a ten-horse field, there are nine losers." Local owner Frank Wright, however, was a winner during the afternoon card: "I'm fortunate. I only own this one horse. But my wife, she owns forty-two."

Leading trainer Carlos Garcia said: "No difference between stocks and horses. You buy stock and it goes down, you sell it. You buy a horse, it is slow, you get rid of it—get a new horse. No difference. Just more excitement with horses." Rich Wilcke, editor of the *Maryland Horse*, said racing opportunities in the mid-Atlantic region make it the best area in the country to own racehorses. Jay Hickey of the American Horse Council said that the tax changes were overrated: "Does anybody in this room want to go back to seventy percent tax margins just so you can write off horse losses?"

Finally, I put on the Country Life pony show: "When people in Baltimore think of tennis, they think of homegrown product Pam Shriver. We hope to be to Maryland horse racing what Pam Shriver is to tennis." Driving home, I felt as if we all had played well.

Friday, January 19. I don't know what causes a horse to chew on fences. Does anyone? We have a yearling who has taken up this nasty habit. Peter put a cribbing strap on him, which helped. Yet I can't overcome a mild feeling of disgust seeing a cribbing strap on such a young horse. His mother didn't chew fences. Where'd he learn this?

If he wasn't such a handsome colt, and the last of a line, I'd find him a new home, before another oak board is defiled.

Saturday, January 20. One of the yearlings has had a runny nose, but no fever, for several weeks. Lung damage can result if the condition persists, so I called Peyton.

"It's time to be concerned, but not alarmed," Peyton said, in a sweetly reassuring tone he perfected on distraught poodle owners who burst into the small-animal clinic where he interned.

"Various factors determine the level of concern: Duration, attitude, appetite, fever, presence or absence of coughing, and the herd health program in effect." As he spoke, he placed a pink rectal sleeve over the foal's nose and put his stethoscope to the colt's side.

"Diagnostic tests include auscultation. Depriving him of air for several seconds causes the foal to build up an oxygen debt, resulting in increased respiratory rate and effort when the sleeve is removed. However, absence of lung sounds is not conclusive, because air does not move inside tissues damaged to the point of consolidation. You may not hear any wheezing or gurgling or popping. Yet inhaled viral diseases that start in the ventral, or lower part of the lung, may be undetectable due to the progress of disease."

He released the bag from the colt's nose. The yearling coughed once, breathed deeply, but not stressfully.

"He sounds fine, but we'll take a blood count to detect infection, just to be safe."

As he drove off in his jeep, Peyton left me with peace of mind about a yearling's condition, and buzz words to keep anxiety in check. From now on, I shall be concerned, but not alarmed.

Sunday, January 21. In such perfect unison that they appear to be simultaneously yanked through the air by a powerful magnet, hundreds of blackbirds descend upon the yearlings' ground feeders. The yearlings do not seem to mind, except when the birds land at their heels. Then suddenly the horses all will startle if one youngster should bolt from the birds.

It seems so harmless, so natural, this ballet of birds and horses. It is also a beneficial relationship. The birds peck through and disperse the yearlings' droppings. Yet I recall that birds were a factor in the unfortunate death of Carnivalay's talented full brother Dance Spell. Recovering from an injured back, Dance Spell was grazing in his paddock when a covey of quail got up. He lunged forward with a start and snapped his back.

Why does experience teach you to look at everything with a jaundiced eye?

Monday, January 22. Slated for the upcoming legislative session is the satellite wagering bill, certain to cause heated discussion. History reveals the long skirmish American racing has conducted with legislators. In 1877, for instance, the American Jockey Club wrote eloquently to the New York Senate, on the eve of the Senate's decision to abolish gambling: "We stimulate interest of the general public in the development and improvement of the breed of horses...(through) a well-ordered system, under which all persons safely and fairly stake such sums of money as they think proper."

The improvement of the breed was a hard sell to legislators opposed to gambling then, and it still is today. Some elected officials imagine lice-infested OTB shops on every corner. I am ready to counter their concerns with the "Proud History of Racing in Maryland" angle, citing cosmopolitan Annapolis in 1721, offering a purse of "twelve silver spoons: The best horse according to the rules of racing wins eight spoons and the second horse four spoons."

This would tell them that Annapolis conducted prosperous racing two hundred and seventy years ago, with formal rules, and the public attended. Maryland has always been very progressive in legislation pertaining to racing, dating all the way back to the silver-spoon days.

Tuesday, January 23. "Mind if I teach you something about marketing horses?" the real estate man asked, more than a little patronizingly. He couldn't overcome my refusal to predict a rate of return on our racing partnerships. I knew I had lost him the moment he started his story.

"A horse trader says to a fellow in a bar: 'Hey, I got a real quiet horse for you. Wouldn't hurt a flea. Right out back in the barn. Whaddya say? Fifty dollars.' Fellow buys the horse, walks to the barn and finds a dead horse. A month later, he runs into the horse trader at the bar. The horse trader says, 'Hey, you got a right to be mad.' Fellow answers: 'No, matter of fact, I did right well with that horse. I held a raffle: Win a Horse. Five bucks a chance. Sold a hundred tickets. Of course, I never showed anybody the horse, except the one guy who won. And sure, he was mad, but I gave him his five dollars back.'"

The real estate guy was so pleased with himself: "See, that's how to market horses. You got to be *creative*."

I had left the dogs in the car in the parking garage. I asked Nanny and Boxy what they thought of the dead horse story as I dug around for the parking fee. "Forget dat jerk!" I felt the dogs saying to me. "We don't wan' dat guy in de club, anyway!"

Wednesday, January 24. I'd like to ghostwrite a soap script. I'd start with Joe Hamilton.

PROLOGUE: With their son Anthony in bed for the night, Joe tells his wife Betty he is going out. Nancy, the mother of Joe's maid, is waiting for him. This is the third night in a row Joe has spent with Nancy. Betty says she will not fight Joe over Nancy, so long as Nancy keeps her promise to send Anthony to college. Joe returns at dawn, makes a cup of coffee, sits down by the phone and begins dialing.

FADE IN to Joe, as the first light of day lifts the darkness: "Gooood Moooornin' Country Life! Up yet? Me and Nancy had *quite* a night. Don't worry. Everything's fine. I've got another 'maid' now. Betty doesn't know yet. I'll call you back after I get some sleep."

CUT TO a farmhouse as a sleepy young man says into the phone: "No, *please* don't call back." Then he hangs up, mumbles to his wife: "Nancy had her baby last night. Joe is very excited. I think we should get an unlisted number."

FADEOUT.

Thursday, January 25. Yesterday morning, Nancy's Scout produced a white-legged filly by Corridor Key for Joe Hamilton. Nancy's Scout is the dam of Valay Maid, Carnivalay's stakes-winning filly. Joe sold Valay Maid as a weanling, but he still has a two-year-old and now this suckling out of Nancy's

Scout. Selling Nancy's foals might help pay for Anthony's college education. Joe sleeping with Nancy is clearly a family affair.

Last night, the boys at Country Life handled the first delivery of the year. The mare was named Don't Tell Sue—appropriate since she foaled at 8 P.M. and Sue Clayton doesn't post for night watch duty until ten. The girls on the farm all went to see the movie *Steel Magnolias*, and upon their return, Ellen suggested the name Steely Dan for the new colt. I expect to see Steely Dan in concert with his mother this morning if the rain clouds break.

There is some concern that Steely Dan is straining to have a bowel movement. I remind myself he had a tight squeeze coming through the birth canal. It's possible he ruptured his bladder, the diagnosis of which is often confused with an impaction. Enemas do not help ruptured bladders, and so the veterinarian has been summoned.

Friday, January 26. Yesterday afternoon at Laurel, Carnivalay's daughter Lucky Lady Lauren won by twelve lengths. Forty-eight hours earlier, Valay Maid bounced home a five-length winner in her first start since the What a Summer Stakes, and since the flu. The *Daily Racing Form* headline proclaimed: "Still Not Beaten."

Carnivalay must read the *Form*, because every day, he acts a little more like his daddy, a Cagney cockiness, a Napoleon complex, that swagger of pride that makes little men seem big. Carny's only a shade over 15.2 hands. For conformation shots of him, Neena Ewing crouches down in the grass and shoots uphill. Northern Dancer redefined the physique of a successful stallion, and savvy horsemen began looking for a center of balance instead of size and scope. On Tuesday and Thursday of this week, Northern Dancer's granddaughters—Carnivalay's leading ladies—were the center of attention.

We interrupt this broadcast to report that Auxilliary is foaling, fifteen days overdue.

Saturday, January 27. At four forty-five this cold, windy afternoon, only minutes before ABC was to televise the Horse of the Year award, Dad phoned, said two folks were at the stud barn, and wanted to see the stallions.

Business is business, and I'm always glad of it. So Ellen taped the show and I paraded the stallions. They thought I'd come to start the breeding season. Each stud spent considerable time on two legs. Afterwards, I watched the film clip from before the Belmont Stakes, when Sunday Silence glanced his front foot across Whittingham's head. The sharp aluminum shoe drew blood.

Heading into the breeding season, all but one of our four stallions remain barefoot. I've decided to keep them barefoot, sacrificing a few points on presentation for a greater margin of safety. To the uninitiated, little difference exists between getting hit in the head by a horse with shoes on or by a horse with shoes off. Take it on faith. The former is a widow-maker. The latter might merely be a concussion.

Sunday, January 28. Mr. and Mrs. Carlyle Lancaster were beaming as we led Auxilliary's two-day-old foal to her paddock this lovely Sunday morning. This is one of the best parts of the game, when owners arrive to inspect their newborns. The Lancasters had brought their filly a brand-new, calf-brown foal halter, with a shiny brass nameplate already on it.

"I think this is a special little filly," said Mr. Lancaster, whose nickname is Jiggs. This filly jigged when Mrs. Lancaster tried to take a still photo of her. Then we all watched in delight as the mare and foal cantered away, so pretty against the blue sky and the grass tinted green now from the warm January we've had.

I was anxious to get on with my chores, but stifled such ordinary ambition and rested with my arms against the top board. Sunday is a day of observance, I rationalized, and I could observe no greater evidence of the Good Lord's bounty than to watch this brown bundle of energy twirling half circles on her hind feet, in celebration of new life.

Monday, January 29. Yesterday's rain flooded the farmlands just as it did on New Year's Eve. The oak tree in Winters Run moved last night, its buoyant crown of branches shifting to point downstream, moving like the hands of a clock set to eons instead of hours. The tree is a huge presence in the riverbed, a sleeping Gulliver.

When he first fell to Earth in October, Gulliver became a bridge for Lilliputians of all species to reach the remote territory on the other side of the river. During a moonlit walk in the snows of December, we watched a red fox elude the pointer's nose by dancing lightly across the tree, then scurrying up the steep slope on the other bank. We used Gulliver to lift us above the currents of cold water, so we too could explore the north slope of Winters Run. Gulliver's furrowed trunk provided good traction for all. Last night, Gulliver stirred in his dreams, and moved away from the bank. Our bridge is out, at least until the rains of February come to disturb the sleeping giant again.

Tuesday, January 30. Breeders must be squinting to detect the tiny superior numbers in the *Racing Form* charts, but the romps by Carnivalay's daughters last week might as well have appeared in boldface type. Every morning since, breeders have called with inquiries. Carnivalay has been a hot item since the Breeders' Cup in November, and I raised his stud fee moderately after that weekend. Now, when I tell callers he is booked full, they conceal anger at their own procrastination by insisting that their mare is worthy of the horse. I heartily and truthfully agree.

"Yes, she is a very nice mare," I say. "Could I interest you in one of our other fine young stallions? They have bright futures, as well."

A breeder with a fixation for a particular stud does not readily entertain alternatives. Yet I need to resist the temptation to overbook this young stud. My first stallion, Lyllos, taught me: Do what's right for the horse, and you will be doing what is right for the people involved. This is perhaps the cardinal element of this game—Rule One, the crisp voice that speaks above the din of self-interests: Do what's right for the horse.

I hate like hell to turn away business, and so I continue to lobby for the other three stallions. Then I laugh at myself, thinking how seriously I try to protect Carnivalay, a colt whose personality is nearer to Harpo's than Nearco's. He is a funny-faced duckling who has become a swan, somewhat akin to his father's own history:

Northern Dancer was an unmarketable yearling when Peter signed on with Windfields Farm on May 14, 1962. By the time Peter left Windfields on December 20, 1970, the unsold yearling he had broken had almost won America's Triple Crown, and the young sire's first crop included the great English Triple Crown winner Nijinsky II. In the stud, Northern Dancer was managed with uncommon restraint by Windfields Farm, a precedent I cite to breeders who won't take no for an answer when I tell them Carnivalay is booked full.

Wednesday, January 31. When I think back over January, I find myself looking forward instead of back. Three weeks ago, we said goodbye to four youngsters we raised from birth. They were bound for training centers in South Carolina and Florida. Ellen passed the van on her way to town. She recognized each two-year-old simply by their eyes, peering out of the van windows in helpless anticipation as the driver headed for I-95. We always called them by their mothers' names, never knew their given names until the new halters arrived with brass plates for Quail Hollow and Allez Prospect, for Greenwich Bay and Wannamoisett.

These young racehorses will be learning their lessons down south this spring, while we stay up here and foal their mothers and lead their fathers. When we come up for air in July, we hope to see their eyes again, peering forward from the starting gates of northern tracks.

IF THESE HANDS COULD TALK

Thursday, February 1. Races from Laurel are simulcast to Pimlico, where two floors of three thousand patrons congregate daily as if on a four-hour cruise ship. The upper deck is peopled by swells: Businessmen who pay an admittance to the Sports Palace, where chandeliers are controlled by rheostats, where during the call of the race, the lights seductively dim. When the houselights come back up, grown men crowd around automatic mutuel machines—TV screens with graphics not much different than a video game. The men stare like little boys if they punch out a wrong ticket, which, surprisingly, does not happen often.

On the lower deck, kinky reggae musicians tuck their hair into oversized woolen caps and shuffle their feet in a dance called "skanking" when their mutuel ship comes in. Down in the hold of the good ship Pimlico, the atmosphere is not softened by amenities. The lighting is harsh. When a favorite is beaten, you feel the ship lurch, and you know the waves will be felt tonight in the rough seas of the poor Pimlico neighborhood. It is a statement against my best interests to observe that, heedless of the warning slogan in the Frenchman's Kitchen, the patrons of this floor are betting with their eating money, not eating with their betting money.

Fueled by the handle from this track without horses, Maryland racing prospers. Betting translates into purses, and purses make owners happy, and happy owners breed mares to our stallions. As I walk out of the lower deck at Pimlico to a chorus of cursing as another favorite falls, I turn my head away, and I don't dare say: "What's the matter here?"

Friday, February 2. The prologue to the Nursery Stud catalogue of 1925 (to paraphrase a famous book of the same era, *The Great Gatsby*) provides a riotous excursion, with privileged glimpses into the heart of Maj. August Belmont's breeding dynasty: "There will be a high niche in the hall of hippic fame for the breeder of Tracery, Man o' War, Fair Play, Mad Hatter, Ladkin, Beldame, Ordinance ..."

146

From the turn of the century until Mr. Belmont's death, Grandfather was his secretary, in the old sense of that office: one entrusted with the confidences of a superior. Grandfather confidently negotiated the sale of classic winner Tracery to the Argentine for a quarter of a million dollars. He managed the importation of Rock Sand, who sired the filly Mahubah, who in turn produced Man o' War. At the Nursery dispersal, sixty-one Thoroughbreds sold for a total of almost eight hundred thousand dollars, an average of nearly thirteen thousand dollars—an exhorbitant sum for horses in those days.

The horse business of the nineteen-twenties reminds me of the go-go days of the nineteen-eighties. Both decades were followed by a Great Depression. So today, as we coerced winter heats from mares shipped here recently, I told my teaser: "Let's start our own Jazz Era, Jazzy. Let's go-go get the nineteen-nineties."

Saturday, February 3. A Ph.D. in biophysics from Johns Hopkins is in our office, and I am lost.

"We've configured a neural network to predict genetic breeding patterns," he tells me. He perspires as he speaks, either from the surge of long words, or because he is building up to a request for funding. "The neural network not only will discover ordinary nicking patterns, but also detect extremely subtle influences on heredity, including complex patterns imbedded among, and across, subsets of particular sire lines."

Whew! I'm really lost. Then John King shuffles past my office window. John can't read, but he knows the horse's numbers, hanging from their halters.

"Num'a tree don't look good," John informs me, shaking his head to indicate the gravity of his observation. "Better tell da 'bet." (He refers to Dr. Riddle as the "betternary.") The Ph.D. dismisses John with a wave, maybe because John is wearing his slightly soiled Goodwill windbreaker with the Greek letters of the Kappa Alpha fraternity house stitched to his chest. The Ph.D. does not understand that *little* things, not big things, make the horse business work. I am desperate to be relieved of him, to go check on Number Three.

"But I'm not sure how to proceed with my idea!" the Ph.D. calls to me, as John becomes insistent: "Num'a tree. Right out 'dere by the water tub. Call da 'bet! It's num'a tree, sick!''

Sunday, February 4. When prominent guests were invited to Leslie Combs II's house at Spendthrift Farm, so the story goes, he would put grain on the ground at the crest of the hill, so yearlings would graze in bucolic serenity in view of the porches of the great house.

For the Sunday afternoon cocktail party we held today for new investors in the horse business, Ellen took a page from Combs' book. She put the farm's friendliest foal, Auxilliary's filly, in the barn nearest the house. Ellen then escorted folks to the equine petting zoo. If Jiggs Lancaster desired a partner on his handsome filly, he could have syndicated her today faster than "Cousin Leslie" could cruise through the Keeneland boxes.

Monday, February 5. In the upcoming sale this weekend is Carnival Cruise: Bred as a two-year-old last June to Assault Landing, now a dowager at three, the first daughter of Carnivalay to help determine his promise as a broodmare sire.

I must be getting old. I foaled her. Ellen named her. She is part of Mr. Janney's legacy. Her second dam, Laughter, is a half-sister to Ruffian. Yet the question remains: Should her buyer anticipate any foaling problems when a three-year-old mare enters labor? The teen pregnancy rate in humans should assuage any misgivings, but the night watchman better not be napping around her due date.

Tuesday, February 6. Contented mares crunch hay in dry, clean stalls at dusk, and I feel such appreciation for the crew who muck these stalls every day, who carry the hay and straw (particularly those three-strand wire bales that slice your fingers).

Mares bask in the late night light of these beautiful old barns—a scene so settled and peaceful it is almost tranquilizing. But it does not just happen by itself. It takes the hard work of our good crew. While I've been busy pushing pencils, they've been pushing manure.

Wednesday, February 7. The warmest January on record has carried over into February, and we should get a good jump on the breeding season next week. Jazzman is in full glory, teasing the girls as if he is an equine Henny Youngman. The lighting program is working its magic, as well. (Yet I still grimace contemplating the electric bill during these months. Does the "shock system" of ninety minutes of light in the middle of the night *really* work? It would be an affordable alternative.)

Drs. Tom Bowman and Bill Riddle again are complemented by the relief pitching of Drs. Peyton Jones and Russ Jacobson. The four-man rotation is set for this breeding season. The stadium lights are on.

Thursday, February 8. Two mares a day have shipped in this week, reporting early for spring training. We used to prepare neckstraps with nameplates for mares arriving with no identification. But banging out brass nameplates is time-consuming, and it is *imperative* to identify the mare the moment she arrives. So, three years ago, we began attaching rubber livestock tags and numbers to *all* mares. In-foal mares receive duplicate number tags, so one can be removed and attached to her foal's halter the morning after delivery.

The orange tags can be read from fifteen feet away, even by John King's old eyes, whereas the fine print on brass nameplates is obscured easily by mud. The rubber tags look unsettling at first, like a company picnic at which participants wear labels that shout: "Hi! I'm Dawn." Yet I don't even care that they look kind of cowboyish. The orange tags are as plain as the King's English, even John King's English.

Friday, February 9. Last February, our four stallions covered a total of fourteen mares. To determine if foals were born eleven months later, Alice prepared the first batch of our Foal Postcards today, addressed to the breeder, with a postage-paid return card that reads:

"My broodmare (name) foaled a (colt) (filly) sired by (stallion) on (date). The foal was born at (farm name & full address). This mare will be bred back to (stallion). (If mare did not produce a live foal, or if foal died, please indicate so.)"

We know that four of the fourteen mares resorbed or aborted prior to January, for an attrition rate of twenty-eight percent. Of the ten remaining mares, another casualty or two will appear when the postcards arrive back in our office. By the end of the foaling season, we'll know for certain whether the attrition rate for mares bred in February is greater than for mares bred in April and May, when Mother Nature is an ally. Meanwhile, relentlessly, the attrition rate from Mares Bred to Registered Foals inches towards the constant of forty percent.

Saturday, February 10. Dr. Jones palpated a mare today and discovered a fluid-filled mass near her pelvis, on the wall of her vagina. What were we dealing with?

"First," Peyton began, "what is it? An abscess containing bacteria? A hematoma? A lymphatic cyst? If it's from an injury, three sources are likely: The stallion—trauma caused by the penis; Partruition—perhaps the foal's foot pierced the tissue during delivery (Did she have a difficult, unassisted, or unobserved foaling last year?). Lastly, iatrogenic error—a polite phrase for unintentional mistake caused by a veterinarian in diagnostic or therapeutic efforts.

"The foaling injury—a misdirected foot penetrating the vaginal wall, or delivery of a contracted foal—is most likely. I rule out the stallion because he usually traumatizes the cervix or adjacent vaginal wall. And a veterinarian too aggressive with a speculum would be more likely to damage the top of the vagina.

"The retro-peritoneal space is nature's defense against peritonitis and death after foaling injuries. An anatomic barrier, the perineum, divides the pelvic cavity from the abdominal cavity, allowing the mare's reproductive system to localize and wall off the contaminated area.

"Start this mare on antibiotics. If it is simply an abscess, we'll aspirate the mass with a needle and drain it, rather than having it rupture at breeding time, or next year when she foals."

Peyton bats cleanup for Riddle and Bowman here at Country Life, but his analytical answer to today's unsolved mystery certainly covered all the bases.

Sunday, February 11. The average price at tonight's Fasig-Tipton sale rose one hundred and seventy-eight percent. I know, sounds like a joke. It is, in part, since one mare brought three hundred and forty thousand, Laddie Dance laughing from the auctioneer's stand that the Timonium board did not go that high. But the median price was three thousand, exactly twice that of last year's sale, when half sold for more than fifteen hundred and half sold for less. We tried to buy four-thousand-dollar mares who went for eight.

American soldiers in Vietnam, upon realizing the worst part of their tour was behind them, would say: "The worm has turned, brother. The worm has

definitely turned." That's how I felt coming out of the sale tonight, and said as much to my brother Michael, who enlisted at the same time I did.

Monday, February 12. *"Aaannnddd THEY'RE OFF!"* I wanted to yell in my best imitation of famed race caller Fred Capossela, as our four stallions hollered from inside their stalls after hearing Popeye declare the breeding season underway. The moment the stallions realize breeding season has arrived, they pace frantically like caged tigers, hoping the man comes to *their* door, to lift *their* chain shank from the hook, the shank jingling in confirmation of imminent action. The stallions can build up tremendous momentum in their oversized stalls. I will never forget the morning Allen's Prospect became so frenzied he fell down. I looked in and could not see him. He was flat on the ground in the corner, stunned like a child who has spun himself dizzy with excitement. He scared me to death.

Allen's Prospect is our opening-day starter. His delivery is perfect. As he makes his way back to the dugout, the cry goes out, and the veterans like Richard and Peter, Michael and Steve Brown—certainly the other stallions—hear the ghost of Fred Caposella call from atop the stud barn cupola, his nasal voice so sharp it cuts the ribbon of the breeding season: *"Aaannnddd THEY'RE OFF!"*

Tuesday, February 13. At 8 P.M. last night, a routine foaling became a nightmare with one panicked thrust by the maiden mare Angel. Rich and I felt the foal's two front feet coming normally. Suddenly, a severe uterine contraction pulled the foal back into the mare, then violently pushed him back out again. The sudden thrust rappelled the foal's feet against the wall of the mare's vagina, exactly the area Peyton had lectured on so succinctly last Saturday. The force of the mare's contraction must have sent the foal's foot through the shelf of tissue separating the vagina from the rectum.

The next thing Rich and I knew, the foal's nose emerged from the mare's rectum, his teeth in a ghastly grimace, while his *bottom* jaw protruded from her vagina, his tongue as purple as liver. Holy smoke! His upper teeth had torn a baseball-sized hole between the vaginal and rectal walls! It was as if he was biting what was left of the mare's rectum.

Richard and I smacked Angel on the rump to get her back onto her feet. When she stood up, I cupped my palm over the foal's nose and pushed for all I was worth, stuffing him back through the hole in her rectum. We missed by inches from having a complete blowout. The young mare was scared to death, almost in shock. She gallantly finished the delivery standing up. The foal slipped from our hands and landed heavily in the deep straw of the foaling stall. He was as frightened by the bizarre turn of events as we were.

"Damn," I cursed to myself. "Damn, damn, damn. How could this have happened? We were *right here*. Everything was fine, then all of a sudden..."

Wednesday, February 14. At 2 A.M., after three hours of surgery to save Angel's reproductive future, Dr. Bowman slowly climbed into his truck for the two-hour drive back to his home on the Eastern Shore. "I've got to be at Northview Farm by six-thirty," he explained. The foaling season started in earnest today for the good doctor.

Richard walked away stiffly from the veterinary stocks. From ten-thirty last night until one-thirty this morning, he had held Angel as still as himself, never once complaining. "You okay?" I asked him as we finished. In his strong West Virginia accent, he answered, "Even a *bear* gets cold." He was dressed only in the vinyl jacket he'd grabbed when he thought he would be back in his house after routinely checking on the foaling barn.

Meanwhile, Sue washed the blood off her strong, country-girl hands, the nurse in this M.A.S.H. outfit, fetching clamps, forceps, hemostats. Dad said goodnight, as he coiled and uncoiled his fists at his sides like a gunslinger. His sixty-eight-year-old hands were cramped from holding drop cords and flashlights, from training the light ahead of Dr. Bowman's fingers.

Peyton gazed up at the post-and-beam construction of the ancient bank barn. He appeared ready to leap up and begin a set of his famous pull-ups. He can do a pull-up with one arm; ten pull-ups hardly raise his blood pressure. Peyton was pivotal to the effort, arriving moments after the foal was born. He prepped the mare for surgery, shaving her tailbone for the epidural, starting her on antibiotics and pain-killers, feverishly concentrating despite the meat locker coldness of the barn at midnight.

I walked out into the bright light of a waning moon. I heard two geese splash their wings as they lifted off the pond to feed in the cornfields, their

151

voices carrying over the water in the stillness of the night, a song so beautiful I forgot for a moment the ordeal we had just been through.

Mother Nature, human nature, animal nature, these forces circle in constant flight over this farm. Sometimes, like this morning, they all fly together at once.

Thursday, February 15. How ironic a mare named Merlin's Mistress would be Carnivalay's first date of the season. Carnivalay's dam, Obeah, is named for a form of witchcraft, you see, and his famous half-sister is Go for Wand. A wizard in his own right, Carnivalay danced like a witch doctor upon seeing Merlin's Mistress. But when he pitched forward and bucked his hind legs in the air, sending Peter scurrying away from the task of washing the stallion's penis before breeding, I said "Enough!" Carnivalay looked skeptically at me, as if deciding whether this macho display with the mistress of Merlin was in fact prudent. As is his wont, though, Carnivalay fooled around, then fell in love.

Allen's Prospect was flabbergasted at the sight of his old flame Gramatical, a Kentucky dame he dated his freshman year in the stud. He was so anxious to renew their acquaintance he forgot something. Gramatical patiently explained the fundamentals of love to her overeager suitor, and lessons learned soon returned.

Corridor Key has the blood of Polynesian coursing through his veins, and he was delighted to see Dainty Geisha presented *au naturel*. (Peter added an additional orange tag to her halter this spring after she almost decapitated us last year: "No Twitch," the tag said. Steve Earl asked: "Who's that new mare, 'No Twitch'? I've heard her name before!") Sans twitch, Dainty Geisha sparked the native dancer in Corridor Key, whose fancy two-step took no time at all.

Friday, February 16. "A lot of veterinarians would wonder why we bothered to operate on Angel," Dr. Bowman said. "The pain after such trauma often causes mares to become impacted, and if you can't get their manure to a loose consistency, the manure itself will tear out the sutures we put in. With a little luck, and a *lot* of fluids, we can get her through this stage without such a setback.

"I think it was worth the try, even if all we accomplished was to make the hole smaller, to prevent fecal material from seeping down into her vagina."

Watching Angel get up and down in the field is painful to *me*. Imagine how she feels! She kneels on her front legs, then tries to drop her hindquarters gently, but the stretching of those muscles causes her great pain, and she stands back up. Then she tries again, and again, until finally she lands hard, wincing in pain. Whew! The things we put these gallant mares through in our pursuit of racehorses.

Saturday, February 17. At noon today, Tom and Ronnie Roha, newcomers to racing, bought their first horse, a two-year-old filly by Carnivalay. The deal was done the moment the vet withdrew the endoscope and pronounced the filly sound of wind. At four-thirty this afternoon, the newcomers were in the box at Laurel when Lucky Lady Lauren won the Flirtation Stakes. Just a few minutes later, the racing secretary announced that Valay Maid had won the Cherry Blossom Stakes at Garden State: Two unrestricted stakes winners in the span of twenty minutes for Carnivalay. I thought to myself: "This is too good to be true."

The Rohas shook my hand at the end of the day. They looked as if they had just won a million dollars. I had spent so much time preparing them for the risks of this game that we were all caught unaware of such immediate rewards.

"Ah, every weekend's like this," I joked with them. "It's an easy game."

Sunday, February 18. Ever since the age of sixteen, when I first read Sherwood Anderson's story *I Want To Know Why*, I've been aware of the way people carry on when their racehorses do something exceptional. I suspect the moral of Mr. Anderson's story is that folks have a way of ruining memorable days by puffing up their chests like bantam roosters, by drinking too much in celebration, by bragging about how *they* made that horse, how *they* won the race, how *they* set the records.

Last night, I thought about how to celebrate wins. Dad went to court this past week to tell the judge that a certain jockey had been straight since last summer, that the jock couldn't do anybody any good if he was in jail, so why not let him run the backstretch meetings at Pimlico, let him tell other jocks how bad the slammer is? This jockey figured he was just a born loser, but Dad picked him up, said he just hadn't learned how to win yet. Dad figures every man's a winner, except know-it-alls.

I got goose bumps when Lucky Lady Lauren won. I almost fell down when I found out Valay Maid won. I felt like celebrating something fierce, then I remembered the moral of Mr. Anderson's story. I really did. I was damn sure glad of it late last night too, because Peyton called from the broodmare barn. He needed help oiling Angel, who still can't pass manure, and we worked on her until two o'clock this morning.

When I was sixteen, I wanted to know why this business made people drink. I was that boy in Mr. Anderson's story after a drunk trainer had spoiled a boy's image of life. That boy said: "Spring has come again and I go to the tracks mornings same as always, but things are different. The air don't taste as good or smell as good. I keep thinking and it spoils looking at horses and smelling things and laughing and everything. Sometimes I'm so mad I want to fight someone."

Now I am thirty-five, and I know why, and I am not so mad. All the same, I keep Mr. Anderson's story right on my bedside table.

Monday, February 19. Carnivalay could have run for office this President's Day. The success of his daughters lit up the switchboard in the office. All over the farm, spirits soared. Monday morning manure flew out of the stalls with vigor.

Peyton examined Angel and declared her out of the woods. We had coaxed her into eating bran laced with molasses, and only allowed her to eat the premium New Mexico alfalfa Michael bought last month for just such an emergency, alfalfa so loaded with tender leaves you cannot pick an individual flake from a bale: It falls apart in your hands. It's twenty-one percent protein, and you must be careful not to overfeed. But is such a relief to see Angel so relieved, on a Monday when the farm is so full of good vibrations.

Tuesday, February 20. "We don't need betting shops in rural Cecil County," said Dennis McCoy, the lobbyist for the Maryland Horse Breeders Association. "We need them in the next major project in downtown Baltimore, or in Washington suburbs like Rockville or Silver Spring. Think of the businessmen wagering during lunchtime. Think of the interest a nicely done betting parlor would create. Our competition is not Standardbred racing. It's pro football, college basketball, pro baseball. Purses would go through the roof with satellite wagering. But it's not on the General Assembly's agenda this session. The Thoroughbred and the Standardbred tracks can't get together on this issue."

Maryland is a tiny waterlogged state, connected by bridges and tunnels, with only one major interstate highway. Betting parlors would be like branch banks. Business would boom. But I guess we will just have to wait on such conveniences.

Wednesday, February 21. It's a long way from Tipperary, but John O'Meara made it from Ireland to Kentucky to further his knowledge of horses. He signed on today as a free agent with our club, and he appears capable of playing any position. No doubt he will be called upon for some clutch hitting soon.

Thursday, February 22. Allen's Prospect would breed a tractor if you put a halter on it, but today, aggravated by a bruise deep in the sole of his hoof, he did not perform with enthusiasm. I put him back in his stall without him covering the mare—the first such time. I called Tom Rowles to put shoes back on him (the end of my barefoot experiment), and John applied a poultice to take heat out of the stud's foot. Then I phoned Dr. Bowman.

"Could you be aroused while someone beat your fingernail with a hammer?" he asked me. Before I could answer, he said: "That's how an abscess feels to a stallion. Give him some Bute in a hot mash, do his foot up, keep him in for two days, and try him again Saturday."

Dr. Bowman is very reassuring, but I cannot shake the feeling of unease. Like it or not, our livelihood depends on the libidos of our stallions. If Allen needs Bute, I need Zantac, for there is a cause-and-effect relation between stallions who do not breed and ulcers in syndicate managers.

Friday, February 23. Trouble brewing at the foaling barn.

"Can't feel feet," Ellen says. "Can't feel head." I reach into the mare. Could be his rump, but I can't find the tail. "Get her walking, Rich. Peyton'll be here soon. Who is this mare?" I ask.

"Rainbow," Ellen says. "Shipped in with a mare named Coco." Two chestnuts. I've had a mental block trying to keep them straight. Not important now. Whoever she is, she needs help. I phone Bowman: "Doc, this foal might be backwards."

"Doubt it," he says. "Only two percent are born backwards. Likely a big foal. You're feeling his neck. His head is bent back. Peyton coming? Is your block and tackle handy?"

Yes, thanks to Rollaids. February a year ago, Steve Brown sunk an eyebolt into the oak rafter. Bought a hoist and thirty feet of rope from Bel Air Farm Supply. Never been used. First time for everything. Peyton arrives in high gear. Mare is standing. He tries to reposition the foal.

"His head is flexed. His feet are down," Peyton says. "Every time I go in her, contractions come." He runs a tube down her nose, into her trachea. It inhibits contractions. He tries again. Waits. Tries again. "What a bull. She's still straining. We'll have to do an epidural."

He shaves her tailbone, applies iodine scrub. He injects her right in her spinal column. *Must* be sterile. Surface bacteria can spell disaster. He waits, lifts her tail several times. Tail is flaccid. He tries to reach the foal again. "No good. *Still* straining. Need to knock her out. Is the hoist ready?"

Coiled a year, the ropes resist—a fishing reel backlash. Ellen phones Steve Brown. Out of bed in an instant. Imposes order on the ropes. Hooks hoist into ceiling bolt, runs ropes through pulleys. Climbs the ropes like a Marine to test the hoist. "Ready," he says.

Peyton is in the mare's jugular. "Steady her, Rich. Steady her. Don't let her get down against the wall." The mare reels, lands hard, hindquarters near the winch. She tries to raise up. Rich pins her with a wrestling move. John piles on Rich. Two men, their bodies on the mare's head. She thrashes. Please, no workman's comp claims tonight. The anesthetic takes hold. She's asleep. Peyton places a catheter in her jugular, shows John how to give anesthetic slowly. "If she quits breathing, let me know."

He ties the mare's hind legs. She's a roped calf. "Pull her up, Rich." Her hindquarters lift off the floor. Gravity now an ally, not an enemy. Kneeling, Peyton cups his hands for me to pour lubricating gel into. He carries it into the mare. Repeats. Repeats again. Then he reaches for the foal. Carries the end of a cable. "Got to get his head through first." The cable is welded to a handle. Peyton calls it his calf snare. Minutes pass. *Many* minutes. "Okay, it's over his nose. Pull gently."

A constant pull. Cable slips. I fall back. "Great, great, got his nose. Now for a foot," Peyton says. Takes the cable again and searches. Seems like hours. "Get down here. Put your hand against his nose," he tells me. "Repel him back into the uterus." Richard moves in behind me, his foot against mine, an anchor, leverage. My arm tires. How must Peyton feel, all this searching? He finds the foot, but cannot flex it to get it safely over the pelvic brim. We are reaching a point of fatigue.

"Save the mare, or save the foal?" Peyton asks, shoulder in her vagina, arm in her uterus, heart in his throat. Against the Hippocratic Oath. I consider the consequences. Are we talking Caesarean? Mare is owned by Smith. Foal is owned by Jones. A lease deal between big breeders. They both love fillies. Is this a colt or a filly? How do I know? All I have seen is a nose. Whose ox do I gore? I'm not King Solomon, dammit. Why do I have to decide this?

"Got it," Peyton gasps. He saves me from indecision. "Got a foot. Pull slow." Ellen has the handle. "That's one foot. Next." Now we have both feet and a head in the birth canal. "All right, boys, let's get this foal out." Ellen hooks nylon straps around the foal's legs, but oil makes him a slippery customer. We're pulling him uphill, the mare's legs are together. Pull harder. He should be coming, but he's not budging an inch. *Why isn't he moving?*

"Mare quit breathing!" John calls. Anesthetic and labor gang up on her. Foal only halfway out. Oxygen comes through the umbilical. If she goes, he goes; he's hip-locked. Peyton rushes to her throat, takes her pulse, grabs syringe of Dopram, a respiratory stimulant. She resumes breathing. Maybe she never quit? Maybe she just took a rest? Who knows? We begin tugging for all we're worth. If she rests again, it'll be for good.

Five people, ten hands, slipping down the foal's legs, ears, face, chest. One final pull. Foal squirts out. "Pick him up! Pick him up! Pump his chest!" Peyton orders. Foal's too slippery. He lands hard. Peyton tries mouth-to-mouth. The foal breathes. His ribs move. My eyes go down his body. At once, everyone sees what I see. Everybody's quiet, except Peyton, panting loudly in relief. He thinks the struggle is over.

"Jesus Christ!" someone finally blurts. "What happened to that foal's hind legs!"

The foal is on his side, his head in the straw. But his hind legs are not on the ground. They corkscrew out of his hips, angling up towards the ceiling, bent off the floor, the hooves curled back like the Wicked Witch's feet after Dorothy's house landed. This was no "windswept" foal, no contracted tendon victim. This foal was deformed, a photo out of a vet school textbook. These hind legs were the culprits behind the painful delivery.

"Cut the mare down," Peyton says, still the doctor. "Glad we didn't decide who to save."

The foal breathes rapidly, shallowly. I try to imagine what he has been through.

"Put him down, Peyton," I ask. "Please put that foal down."

I'm standing behind Coco, I mean Rainbow. I'm confused. She's on her feet now. I pull hard on a leather strap knotted in her tail. I am trying to steady her. Peyton pulls taut against the right side of her halter, John on her left side, Steve Earl ready to spell us. We are preventing Rainbow from crashing against the wall as the anesthetic wears off. She weaves like a drunk. "She'll be stormy for another half an hour," Peyton says. "I don't think we injured her. I never felt like we tore anything. We'll check her after the placenta passes. Won't know about the cervix for some time. Let's not let her crack her withers *now*, after all she's been through."

Three hours into it for us. Eleven months and three hours for Rainbow. She does not give up easily. In the corner, feet poking through the straw. She whinnies through the fog in her mind: Where is *my* foal? We stare with grim fascination. "A colt or a filly?" I ask. "Filly," someone says.

The crew, covered in blood and oil, is exhausted at the outcome. Where's Sue? Hospital? What happened to *her*? Cut her hand on the scalpel? Had it in her back pocket! John, will you finish the night watch, please? Thanks.

Peyton saved Rainbow's life. Of that I'm certain. But I'm still confused. A mare whose name I can't get straight delivers a foal like I've never seen. It's the second bad foaling out of five total. February has been a bitch in the foaling barn.

Saturday, February 24. Two foals were born early this morning, easy as making popcorn. Late this afternoon, Allen's Prospect ignored his sore foot and bred beautifully. In the hour before sundown, it snowed to beat the band—big wet flakes whitewashing the fencelines, a reminder that we are not yet in springtime.

The snow was only Mother Nature limbering up with a few hard serves to keep her game sharp. The purple snow clouds passed quickly overhead and immediately the sun peered out, as if from under the brim of a hat, a blazing twilight after the storm. The splendor of this Saturday afternoon, the easy rhythm of today's foaling and breeding, lifted our spirits and washed away the rough nights of late.

Sunday, February 25. Some Sunday mornings it is blackbirds. This morning, it is white birds—hundreds of seagulls dipping and diving into the fierce north wind that blew in from Canada last night. The gulls are quiet in flight, resolved to bank randomly until a downdraft lands them on the hill behind the mares. Only once did I hear that sharp screech of the seagull, that unmistakable high-pitched call that brings back memories of mornings at Maryland beaches. I can almost smell the salt water, can almost believe the sudden roar of gusting wind is the sound of waves crashing.

"Rough surf," I tell the mares as their new foals fidget to get comfortable on frozen ground, down out of the wind. "The beaches will close at noon today."

In this cold weather, the foals depend on us so totally that we are, in the literal sense, their lifeguards.

Monday, February 26. "Good afternoon, this is the State Police," the husky voice said over the phone at quitting time today. "We're informing you that a car has just run into your fence on Route 1. Took out two pine trees, and snapped some boards. We suggest you inspect the premises for damages."

Folks in Bel Air take proprietary pride in gazing up our steep hillsides as they drive along busy Route 1. However, the sight of grazing horses can be spellbinding. Today's errant driver broke only a few bottom boards, thanks to the pine trees that raised up in defense. As John patched the boards, rush hour traffic creeped in a clogged flow past us, as the harsh light of a February dusk reflected off the litter that seems to grow like an unwanted crop along this public highway. Beer bottles and plastic coffee cups, french fry boxes and disposable diapers, even pieces of the white lines the highway department glued to the pavement, all gathered on the grassy shoulders, absorbing the fading sunlight, catching the oncoming headlights, spoiling a bucolic view.

Most of these drivers appreciate this last piece of green, these solid black fences, these strong hills, this welcoming scenery at the end of their day's journey from Baltimore. But some don't give a damn. I can take the traffic. I can take the occasional loss of a pine tree. But I cannot, for the life of me, understand why anyone would throw trash onto such beautiful property.

Tuesday, February 27. The *Daily Racing Form* loves the shot of Carnivalay being led by Michael, the stallion obscuring the handler save for his sneakers. The cutline under this photo in yesterday's *Form* read: "Carnivalay: Is he the second coming of Northern Dancer?"

It is unhealthy to visit the accomplishments of the father upon the shoulders of the son, but from a public relations angle, I gave quiet thanks to the *Form*'s desk editor in Hightstown, New Jersey, for his tendency toward hyperbole. Carnivalay could cool off as fast as he heated up, but history, and the phone messages on my desk, indicate otherwise.

Wednesday, February 28. A camera filming this past month would have to be mounted on a boom, to capture the sweeping shots of a farm in a state of high gear. But a zoom lens would focus on the close-ups of foaling barn scenes: Steam rising off a mare in labor, Dad's hands in the cold early morning hours after Angel's surgery. The sound track would be music selected by the farm deejay, Mom: Uplifting music, capturing the *esprit de corps* among troops who feel they are winning the war even if losing a few battles.

Dad, as field general, often holds his hands aloft in victorious moments, whether engineered by Lucky Lady Lauren, or by lucky shots on the basketball court in the driveway. He will launch a flat shot towards the basket as if throwing a pie across the room. When it swishes, as it does with regularity, he lifts his hands high: *"If these hands could talk!"* he exclaims, in merriment so infectious his grandson Philip, an eighteen-month-old chip off the old block, clasps the rebound and raises the ball over his head in imitation: "Bee Baw! Bee Baw!"

If Dad's hands could talk, they would say to the troops: "March. Bring us another month like February, but ease up a little on the drama, willya'?"

SOMETHING SO PURE

Thursday, March 1. The voice of a single Canada goose bounced off the Winters Run watershed at dusk, echoing over the farmland, and I considered for a moment the similarities between geese and horses. For both, flight is their defense, lungs such an important organ. Stomachs are small—can't travel fast on a full belly. At seminars, equine nutritionists brandish the dried, leathered stomach of a horse, no bigger than a football, to illustrate the constant need for roughage. Both geese and horses have excellent hearing. I think tonight's lone goose was *listening* to Winters Run. Both are farsighted. To *my* eyes, the sight of either galloping horses or flying geese speaks of a marvelous freedom.

March is a month to think about such things. No time to gaze skyward in April or May; no energy to look up in June. Besides, this is the month the geese migrate, not just over my head, but in the hearts of poets. Robert Penn Warren wrote:

> Long ago, in Kentucky, I, a boy, stood
> By a dirt road, in first dark, and heard
> The great goose hoot northward.
> I could not see them, there being no moon
> And the stars sparse. I heard them.
> I did not know what was happening in my heart.

On the ground this month, we'll plow through the first third of a breeding season, enjoy the birth of foals, and awake to the quickening pace of longer days. All the while, for all to enjoy, the geese will carry on, either in solitary splendor like my friend tonight, or perhaps in a spectacle of flocks lapped one on top of another, herringboned against the wide sky.

March is cool weather, sweater weather, woodsmoke weather. March is the first day of spring. March is one of the best months to be alive on a horse farm.

Friday, March 2. The black-faced colt, born at two o'clock this morning, is contracted in both front legs, a bit like Rollaids. Today, after three hours in a small paddock, he still had not worn off the soft hoof tissue he was born with (the "Angel's Feet," I've heard it called, although no veterinarian would corroborate the term when pressed).

Rich's face couldn't get any redder, but damned if *he'd* relinquish the title as "Strongest Man on the Farm." So he kept the live weight of the one hundred-pound colt in his arms down the hill from the pasture. The foal did not improve with exercise. He knuckles forward onto his fetlocks. He will wear the hide off his ankles if we are not careful.

We might splint this colt, or wrap his legs tightly with support bandages. Peyton even suggested an inflatable air cast: Lightweight, adjustable, unlikely to cause abrasions. We need only recall Rollaids to realize the benefits of time and exercise. Still, it is unsettling to begin March on the same discordant note that rang through February: A ship-in mare delivering a package of trouble.

Oh, for an unencumbered foaling.

Saturday, March 3. This morning, on her last day of penicillin injections for a minor infection, the mare suffered a violent reaction. Peter is fastidious about aspirating the needle properly, drawing back on the plunger to confirm the absence of blood for an intramuscular injection. But either we struck a tiny vein, or the mare simply reached an allergic level of intolerance to penicillin. She went down as if she'd been shot, pinning Peter against the wall. Then she went into seizure, her eyes rolling, her legs thrashing. We called Peyton immediately. When he arrived fifteen minutes later, the mare had climbed to her feet, sweating, still in distress. He gave her Azium and Adrenalin.

"She'll be all right," said Peyton, once the storm had passed. "Whether it was an allergic reaction or an intravenous injection we don't know. But put on her health records: 'Allergic to penicillin,' just in case."

The incident brought forth stories of bad experiences with shots. John was at Arlington Park when a vet gave a shot and almost killed a daughter of Mr. Prospector. An owner bought a two-year-old in Florida; a week later the horse had a reaction to a vaccination, thrashed his way out of the stall, and broke his leg running through a fence. The consensus of advice when such a reaction occurs was simply:

"Get the hell out of the stall. Close the door tightly. Keep it dark. Leave them alone."

What a wake-up call for a Saturday morning. When you pull your boots on, hold on to your hat: You just don't know which way the wind's blowing on a horse farm.

Sunday, March 4. When Richard climbs on a tractor, he's the happiest man on Earth. He measures accomplishment in corduroy rows of neatly topped grass. This past week, Ellen bought Rich a new chain harrow, a ten-footer with its own brochure of instructions. (Yes, *instructions* for a chain harrow.) It has five-inch tines that break up thatches of grass and demolish droppings of manure. It aerates the soil, now warming to the task of producing pasture—the grass already a vibrant green from fertilizer.

The Masters golf tournament boasts no finer fairways than these fields benefiting from Rich's care. When he dismounted yesterday, a fine caking of dust on his face made his eyes shine like a raccoon's. I asked him how he liked his chain harrow.

"It's like driving a new car," he gushed in his soft West Virginia twang. Now *that's* a man who loves his work.

Monday, March 5. Allen's Prospect nudges the flank of the maiden mare—forcefully, but not violently. She leans in to him. Rich reassures her: "Good girl! Good girl!" The stallion is satisfied that he will not be kicked. He swings in behind her, maintaining contact of his chest on her rump, and gently lifts himself off the ground to mount her. He has not exposed his genitals to a sudden kick. Steve Earl releases the leg strap, and the mare stands on all fours.

Peter slips the leather breeding roll above the stallion's penis, below the mare's tail. It keeps the stud six to eight inches farther back in the mare, prevents bruising or tearing of her cervix. The stallion thrusts hard. Stops. Thrusts hard again. Stops. On his third thrust, he begins ejaculating. We keep our finger on the trigger: This is a good cover. Holding the mare steady, Rich is only a few feet from the stallion's face. He watches the stallion's eyes. "The Eyes Have It!" he says.

The stallion relaxes atop the mare. She is tiny underneath him. Her hocks bend under his weight. We give him the moment he relishes. Satisfied, he climbs down. Rich swings the mare's head towards us, which moves her rump away, in case she takes a parting kick. The stallion is washed lightly, and reinforced with "Good boy! Good boy!" as if he was a dog.

As the mare is walked away, I notice a small trickle of blood escape from her vagina. Knock on wood, I have only *heard* of stallions puncturing the vaginal wall. Our number will come up someday, but I don't think it happened this morning. Perhaps the stallion simply ruptured her hymen. However, Dr. Bowman is here. He inserts a speculum into her. "She's fine. No problem. I can see semen in her uterus."

We will repeat this breeding scene perhaps four hundred times this spring, an average of one hundred covers for each of four stallions. Still, this morning's cover was something special: A mare by Northern Dancer was bred to a son of Mr. Prospector. There are moments on a farm when you can feel the promise in the air.

Tuesday, March 6. Steve Brown built a loading chute this week. He dug down to the frost line, laid pressure-treated posts horizontally in the earth, drove reinforcing bars to tie the posts together. The top post is twenty-eight inches above ground level. The opening for van ramps is five feet wide. The wings are seven feet high, sided with oak boards spaced only an inch apart. We built an "L" chute, so that when the horse looks back, he can't see daylight, can't think about escaping. A handler can climb the wings to escape a fractious yearling, but the yearling cannot kick his foot through the boards.

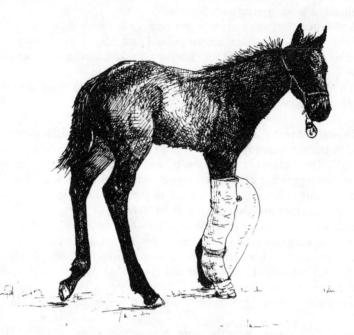

Steve worked quickly. He laid sod where the earth had been disturbed. Now the chute looks as though it's been here for years. (I just hope van drivers don't resort to their "Short Board" instead of the safer full ramp.) The old grassy bank worked fine, but the new chute is an improvement.

Wednesday, March 7. The combination of bandaging and air casting has brought vast improvement to the black colt's contracted tendons. Using an air compressor plugged into the cigarette lighter in his jeep, Peyton inflates the cast to a snug fit on the foal's worst leg, the right front. The cast stays on for four to six hours at a time. It keeps the knee in alignment as the tendons straighten. The colt's weight gradually has dropped onto the soles of his feet. He gets up and down easily in the lightweight cast. When he canters in the small field with his mother, you almost expect to hear heartwarming music from *The Other Side of the Mountain* to begin playing.

Nicknames abound for this cheery colt. Ellen calls him "Pulley," because he pulls her up the hill to his paddock, he's so anxious to play. Peyton calls him "The Boy in the Bubble." Michael calls him "Air Peyton," after popular basketball shoes pumped with air at the ankles. Me? I call the owner, tell her how encouraging the legs look. Next week, we'll experiment with glue-on shoes, an extended toe to alleviate the contracted heels the colt is developing. We don't make fifty cents a day for all the extra work this foal requires, but the satisfaction of watching Air Peyton at play is ample reward, not to mention he is sired by a home stallion.

Thursday, March 8. The window to our office sits hard by the lane where pregnant mares are led. Jennifer rapped on the window and pointed to the mare she held: "What's wrong with her face?" she asked.

The mare looked as if she had been in a fight with a rubber hose. Her features were swollen, wrinkles of flesh gathered around her jaw, her cheeks were puffy, her face full of an expression not generally available to a horse's countenance. She looked almost comical, as if she were an animation subject. I half expected the mare to say: "What's so funny? You should see the other guy!"

"Don't worry, Jennifer. I think it's just a case of the hives," I said, noticing the wheals of puffy skin along her stomach. "She must've been grazing on some spring weeds."

Nevertheless, I summoned veterinary assistance, as the mare stared at me. She is five days overdue, and I don't feel comfortable giving antihistamines to a pregnant mare. (Some common drugs in a farm's inventory, Azium, for example, might induce labor in a mare.) Mrs. Ed smiled at me through the window as I dialed Peyton's number.

Friday, March 9. When a maiden mare ships in directly from the race track, we like to pal her up with a quiet barren mare. The older mare acts as a baby-sitter, unindulgent of frantic enthusiasm. Freed from twenty-three hours a day of confinement in race track stalls, maidens prance like ballerinas upon being turned loose in a small paddock. Surprisingly, they seldom drop into a breakneck gallop, but rather prefer to flaunt freedom—tails straight up over their backs, puppets on a string, feet bouncing off soft March turf spring-loaded like a trampoline.

With excitement so contagious even the baby-sitters were acting like spring chickens, three maidens pranced about this morning, teenage girls loose on a spree at the mall. Each of these maidens was covered this past week. I crossed my fingers and hoped that hours-old embryos—microscopic Carnivalays and Allen's Prospects—were free-floating in a gelatinous bubble, safe behind the locked doors of a closed cervix.

"Play all you like," I said, watching the girls cavort. "Mother Nature will slow you down soon enough."

Saturday, March 10. It is a month of migration, and an old friend, Mike Price, visited the farm today, picking up the theme.

"Everybody's changed," he said, surprised, reflective. "I guess time doesn't stand still."

Everybody wanted to stand around and talk, but they had diapers to change and supper to fix. I shot the basketball with him, and it sure felt like old times. When I came home from Kentucky eight years ago, I put a "Help Wanted" ad in the local paper: Fifty unemployed handymen answered. It was my first great lesson in being the boss responsible for hiring capable workers. I was overmatched by the task. Then Dad walked in, said he had a fellow from the Mann House, Bel Air's halfway house, said the guy needed a second chance at life. In walked a rough young man who looked like Mick Jagger. He was sinewy—he was a sapling who'd grown up on a steep hill in life. Still, he filled the room with his presence. On Day One, he led mares. On Day Two, he led yearlings. On Day Three, he handled the teaser. On Day Four, he took the stallion shank from me. Until he left for greener pastures six years later, he treated every horse on this farm as if it was his own.

"Watch this," he said today. He picked up the basketball, rocked back, let it fly. It banged loudly off the backboard and dropped into the basket. "I haven't picked up a ball since I left home."

He still calls this farm home. We tried to keep him home three years ago: Raised his salary, helped him buy a used Monte Carlo. But he was a rolling stone. I felt like you do after a girl breaks up with you and you know it would never have worked out anyway and you are over the hurt and thankful to be stronger for it. That's how I felt, shooting the ball, letting him snatch the rebound, geese honking overhead, sun shining on the warmest afternoon of the year. Like old times.

Sunday, March 11. "You can come back to the track now," Joe Hamilton told me this evening. Even in defeat, Joe is sportsmanlike. He bred Valay Maid, and since I had never seen the filly win any of her four career starts, Joe felt he'd best keep the same socks on and, by the way, keep me away from the race track. This afternoon, Valay Maid went postward the unbeaten favorite in the Genuine Risk Stakes.

"She finished second," Joe explained. I had fallen asleep by the phone. "She was bleeding from the mouth back at the stakes barn. Had her teeth floated two weeks ago, but the jock said she fought for her mouth the whole trip. The bit scraped off a cap. That's what Carlos said. Won't meet the same bunch in the Pimlico Oaks, but I'm gonna call Billy Badgett anyway. I think the daughter can beat her aunt."

Badgett, of course, trains Go for Wand, Carnivalay's Eclipse Award-winning half-sister. The Pimlico Oaks is April 7th. The purse is two hundred grand. Every trainer from here to California will look at a purse that size. Well, Valay Maid can travel lighter now. She lifted the burden of being unbeaten from her withers. If she gets back on a winning streak in the Pimlico Oaks, I'll buy Joe a new pair of socks, and pay a little more credence to him when he suggests that Valay Maid might someday beat Go for Wand.

Monday, March 12. "Don't shed a clou't till May is out," Peter Richards cautioned in his singsong English accent, revealing his method for staying healthy in the changing weather of spring. Peter was the only member of the crew not peeled down to a T-shirt on this day when the temperature climbed to a record ninety-five degrees. By five o'clock, I had the same sleepy feeling I get when I visit the rain forest at the Baltimore Aquarium.

As I walked home at dusk, a great sound of confusion arose over the tree-tops as a hundred geese got up off the cornfield and made a low pass over the farm. A few hours later, in the bright light of a full moon, the geese could be heard circling the farm pond, groups of a dozen or more gauging the reflection off the water, discussing landing techniques. Background music to their honking came from high overhead: With the constellations on their backs, migrating tundra swans cried out—high-pitched Indians whooping from the skies. I fell asleep to this cacophonous orchestra, wondering if birds ever bump into each other in their rush to beat the heat.

Tuesday, March 13. "Put her front feet up on that grassy bank," advised Dr. Bowman as we prepared to breed Angel—twenty-nine days after her foal put his foot through her rectal wall. The emergency surgery that cold night resulted in a hole the size of a quarter instead of the size of a baseball. Fortified with semen extender in her uterus to dilute possible fecal contamination, Angel climbed up onto the grassy bank and awaited Carnivalay.

"With her front feet higher than her hindquarters, her vagina is on the same angle as the stallion's penis, reducing stress on the injured tissue," Dr. Bowman explained.

Carnivalay performed quickly, despite the forty-five-degree angle of the mare rising away from him. But the little stallion had little purchase when he ejaculated, and he withdrew suddenly. Dr. Bowman, who had guided Carnivalay's penis away from the top of the mare's vagina, was not certain the stallion had ejaculated. Carnivalay looked at me quizzically, as if to say: "Hey, that's a wrap. Ring it up, boys. What are we standing around for?"

A dismount sample would have confirmed ejaculation but, lacking such, we simply held Carny near the mare to detect unspent inspiration. He did not drop back down.

"We'll know in two weeks," Doc said. "I think we have a good chance."

Wednesday, March 14. Peyton is jamming on the subject of using Tagamet, the ulcer medication, to combat melanomas in gray horses.

"About eighty percent of gray horses over the age of fifteen have problems with melanomas around their tail, neck, or salivary glands," Peyton says, warming to the research he's read. "These tumorous masses can remain quiescent for long periods of time, then suddenly enter a stage of rapid expansion. Potentially, the tumor cells can spread to internal organs and shorten the horse's life.

"The mechanism of Tagamet, or cimetidine, is not fully understood. But it is felt that it binds to the histamine receptor sites located on the surface of supressor T cells, removing their inhibitory effect on anti-tumor cells in the bloodstream."

He's on a roll. Pay attention, class.

"The anti-tumor cells, no longer suppressed, are called up to the front line in the raging battle against the malignant melanoma cells. The stage is set. The old gray mare turned out in the lush paddock can now wage biological warfare against the threat of premature death."

I feel as if I'm at some equine Disney World, about to enter the three-dimensional "Body" exhibit.

"That's swell, Peyton, but what's it cost?"

"About eight bucks a day, ten tops. Just buy the stuff wholesale."

That's three hundred dollars more per month for a gray mare with melanomas. No wonder Tesio thought the color gray indicated a disease in a horse's coat.

Thursday, March 15. One of our teasers, Dew Burns, is by the Bold Ruler stallion Dewan. Dew is not the tail-male candidate for carrying on the sire line, but he's got heart. He jumps most of the maidens. He won't take no for an answer. Today, he took a beating from two maidens and a grouchy older mare, but he took the fire out of the girls for the safekeeping of the first-string studs.

Dew's sire won the Brooklyn Handicap the summer I was sixteen and worked at Belmont Park for Jim Maloney. With a barnful of Claiborne-raised runners, Maloney did not win a single race during the spring meeting. An exercise rider broke his neck on a stall door lintel when his mount stepped up into a stall doorway. This was a snakebit outfit. Dewan, who usually ran with his mind on the breeding shed, changed that. The morning after the Brooklyn, Maloney gently rubbed salve into the foot-long welts left on Dewan's rump by the whip of Laffit Pincay, Jr.

The heart, the studdishness, the threshold of pain exhibited by Dew today brought back memories of his sire that summer in New York. The past is always with you, I suppose. Sometimes it just needs a trigger.

Friday, March 16. Rainbow, the mare who produced the deformed foal last month, went back to her Virginia home this week. Her owner, Orme Wilson, is a grand gentleman from the old school. He believes that Rainbow would greatly benefit from a year out of production. I did not try to convince him of my optimism about getting her in foal this spring. Persuading such an experienced horseman against his will is asking for trouble.

"I'll turn her out with the old hunters for the summer," Orme said, and she boarded the van for Virginia for a year's sabbatical.

Saturday, March 17. Ellen photographed Halomer's newborn filly this morning, rushed to the one-hour developers, then returned to videotape the filly's first steps in the small paddock where foals are first turned out. At noon, I boarded a flight to Rhode Island. Late this afternoon, Tom Baylis grazed through the sports channels, catching the Remington and Florida Derbys, the Federico Tesio at Pimlico, and then, after dinner, the day's races from Aqueduct.

Time for the feature movie. Ellen had dubbed in a stirring piano sampler from a Windham Hill record. The Baylis yearlings gamboled to *Out to Play*. The broodmares grazed to *A Morning With the Roses*. This morning's foal cantered beside Halomer as a piano played *In Flight*.

Tom Baylis has loved racing all his life. But Lincoln Downs and Narragansett are gone. Suffolk Downs closed in December. Only Rockingham Park in New Hampshire carries on the tradition of New England racing. This afternoon, Tom and Marie enjoyed a smorgasbord of horses: Derbys from across the country, tapes of their race mare Homera winning the Hanna Dustin

Handicap at Suffolk, and finally, with the music so loud he turned his hearing aid off, Mr. Baylis was transported by Ellen's footage of a filly still wet behind the ears.

Sunday, March 18. On the early flight back to Maryland, we were flying low, directly over the Bethlehem Steel plant in Baltimore. I peered down into factories and blast furnaces, fires showing through vents along roofs, as if they were pilot lights on a huge gas stove. The plant was surrounded by acres and acres of ashes. Then the Chesapeake Bay ran up underneath us, bleeding red near tributaries where poor sediment control leached clay into feeder streams.

Later this morning, after the breeding session, a pair of geese flew over the stud barn. I wondered if they'd flown over the steelyards. The honking I usually love to hear sounded almost melancholy—in Thoreau's words: "A clanking chain drawn through heavy air." They flew on, uh-whongk, uh-whongking their way north, and I suddenly had a greater appreciation for the bird's-eye view they have of our world.

Monday, March 19. A quiet Monday morning is a non sequitur during breeding season. But while shuffling my feet waiting for the nine o'clock van to show, all was still. Foals slept in the grass. Birds conversed about the nice weather. It was so pleasant you could hear the grass growing, drinking in the rain from Saturday. Yellow forsythia blazed across the farm. My thoughts turned to the future, to Lucky Lady Lauren, who might be odds-on in the Forsythia Stakes at Pimlico the last day of this month.

La de da, la de da. What a life! (What's going on I don't know about?)

Tuesday, March 20. Spring roared into town today, blasting our faces with cold. Mares stood in groups in the fields, their backs to the March wind. They moved about with frayed dispositions, ready to pin their ears and snap at their foals. Today's intimidating weather took the mares' minds off cycling. We are surely approaching April 1st, the day all the unsettled mares will start to come in heat. I am anxious for this cold snap to pass. We need to finish March with a flurry of settled mares, to make way for the deluge of follicles April showers will bring.

Wednesday, March 21. A field of young mares, in foal and at ease, includes the mare by Northern Dancer, so carefully inspected after being covered by Allen's Prospect on March 5th. Last week, she tried to decrease our horse population by one. John noticed her at feeding time. She did not touch her food, and soon began sweating. John kept her from rolling by walking her. But she rapidly moved into a distressed condition. Dr. Riddle arrived quickly.

"She has some distension in the bowels," he declared, removing the stethoscope from his ears, "but nothing feels out of place."

This was good news. We weren't dealing with a twisted intestine. We canceled the van to New Bolton. Dr. Riddle administered a tranquilizer to relax her, and pumped a gallon of mineral oil into her. After a few minutes, she began lifting her tail, expelling gas in short, staccato blasts. Another good sign.

"She should be all right now," Dr. Riddle said. He returned three hours later to make certain she had recovered to his satisfaction, and I thanked him for his attentiveness in these busy days of spring.

I wondered to myself: Why is it the best horse always is the one to get sick? Can anyone explain that to me? One minute, the farm is as quiet and calm as a poetry book. The next minute, we are dealing with a million-dollar gas leak.

Thursday, March 22. Bel Air has become impatient with the triangular "slow-moving vehicle" sign displayed by law on farm equipment using public roads. We discovered this today, as we towed a thirty-foot hay elevator on a flatbed wagon, crawling slowly through town, red bandannas on the elevator blowing reproachful warnings to Mazdas and Mercedes. Yessir, we had 'em backed up as far as I could see in my rearview mirror—a ribbon of real estate agents on car phones, a parade of people whose time *must* have been important, judging by the risks they took to pass me. Yet if I had a bullhorn mounted atop the pickup, I could have told new folks to this old county what the scenery used to be.

"On your right, beneath the baby blue bunting of Bel Air's latest shopping center, is old Mr. Deaton's farm," I'd say. "Mr. Deaton died three years ago at the age of ninety, with no heirs for his nine acres that sold for three-point-five million, except for Bess, his old cow who stared at you as you were stopped at the light across from McDonald's.

"Coming up now, where the Harford Mall sign is, used to be the Bel Air Race Track. Until last year, you could still see the old water tower in the woods, but it was demolished for the new post office we needed after so many people moved out to this county. To your left, underneath the Tollgate Mall, used to be the most beautiful flat cornfield you've ever seen."

But of course I didn't have a bullhorn, and I kind of got in a hurry myself, to get the alfalfa in the barn, to get back to the business of breeding mares, so my bullhorn dream does not become a reality. I don't ever want to drive down Route 1 and think, "On my left used to be the oldest horse breeding farm in the state."

Friday, March 23. Some farms work their employees thirteen straight days, then give them a day off. But a fresh crew makes fewer mistakes, and our policy is one day off per week for everyone. We stagger the days, however, which leaves me wondering who's on and who's off. Further complicating the matter is periodic sick leave, taken by Steve Brown, who put a staple through his wrist when the pneumatic gun accidentally discharged, and Jennifer Lawler, to whom a pain in the neck is not a figure of speech. She can't look sideways to see the flying feet of foals coming at her, and so we sent her home.

Thus, slightly shorthanded, I looked around today for Bob Bowers, who has been with us only a few months. Bob is a good-natured soul who takes a ribbing but keeps on ticking. However, his conformation is not well suited for miles of required walking. Before he got fit, I teased him about his slow gait: "Do you know who Ole Bob Bowers was?" I asked him. He looked at me blankly, too breathless to answer. "He was the sire of John Henry, one of the greatest racehorses of all time."

I heard the wheels whirring in Bob's head as he picked up his step. So that's it: I was named for a famous horse. Years of youthful meanderings from job to job have been simply a prelude to my niche in life. Working on *this* farm, with *these* guys, answering to the nickname Ole Bob Bowers. This is it. This is the place for me!

Since that day, Bob has been a stalwart employee. But his stalls had to be cleaned without him today. Friday is his day off. I can't wait until tomorrow to tell him we missed him today. He'll absolutely light up.

Saturday, March 24. This first week of spring has been decidedly un-springlike. Awoke this morning to a heavy wet snow, falling straight down so hard and so fast it created slush in the water tubs in the fields. Suddenly, all the birds of spring were highlighted against the white background: Dozens of crimson cardinals, blue jays and mourning doves, titmice and chickadees, wax-wings and starlings, red-headed woodpeckers casually clinging by one foot to the feeders, Canada geese clanking through the air, pigeons pirouet-ting—all manner of birds carrying the sky upon their backs.

The wet snow blunted the yellows of forsythia and jonquils, flowering fast after the warm days of late. But who could have a long face about today's snow after the backdrop it provided for the birds? So wrapped up in daily chores are we that only a quick slap by Mother Nature can open our eyes. The snow pushed birds from the background of the woods and into the forefront of our day, and the song that came to mind was Woody Guthrie's anthem of sharing:

> *This land is your land,*
> *this land is my land.*
> *This land was made*
> *for you and me.*

Sunday, March 25. The pregnant mare from South Carolina arrived at one o'clock this morning, and appeared none the worse for having spent eighteen hours in a gooseneck trailer, the last six in a Exxon station after being towed off I-95.

"Don't worry about me," her driver said at 10 P.M. last night, before Tom Herbert of the Tamberino Van Service came to the rescue. "I've got relatives all along the East Coast. I can stay anywhere. All I need is a fan belt. Just please make sure this mare don't sleep standing up next to the Coke machine."

I'm going to sign the mare's hoofprint to the thank-you note I write to Tom the Van Man for driving through the snow at midnight to fetch this old gal. There is a spirit of cooperation during the breeding season that makes the ordeal a shared purpose.

Monday, March 26. The stallions have such distinct personalities, particularly when it comes to breeding. Carnivalay sniffs the ground near the teasing chute like a giant bloodhound, though the mare is ready and waiting less than thirty feet away. It is his game to find her. I don't like to discipline a stallion very much as he prepares to breed, as long as he is not rank or unmanageable. So I allow Carnivalay to sleuth about. He thinks breeding is a game of pinning the tail on the donkey.

Assault Landing works his jaw from side to side. His tongue turns purple and he lays his ears back and scowls at the mare. Then he bounces towards her and is up on her back with no foreplay. He bites her neck for purchase, and is excited when his bite elicits a response.

Allen's Prospect, meanwhile, is all business. When he dismounts, he utters a guttural moan that shakes a prehistoric cord, as thick as the call of a sea lion.

Corridor Key, the youngest, is the fastest draw. When this gray colt first retired, two weeks of frustration went by before I noticed him worked into a lather upon seeing a gray mare board a van near his paddock. Our old gray ghost Auntie Freeze, may she rest in peace, was available for lessons. Corridor Key has bred beautifully since meeting the Mrs. Robinson of the horsey set.

The Lord willing and the creek don't rise, these four fellows will combine for a conception rate of ninety percent by the end of July. They'll tire of the two-step by then, but this Monday morning in March, they bred true.

Tuesday, March 27. Dr. Bowman's sense of touch is generally unerring. Manually palpating a mare fourteen days after she has been bred, he will only reach for the ultrasound probe if warranted to confirm his diagnosis of pregnancy, or to detect twins. Uterine tone is his gauge. Today, examining Angel two weeks after Carnivalay covered her, uphill and all, Dr. Bowman reached for the ultrasound. I held my breath. I did not want to be optimistic. Angel, I knew, had been on Regu-Mate since we bred her. The artificial progesterone product might act like hormones of pregnancy on the mare's uterine tone, closing her cervix to further ward off fecal contamination.

When Tom spotted the black sphere indicating success, he uttered some sound between a "Yessir!" and a "Hot dog!" and everybody knew Angel was in foal. "Man, that's great," Tom kept repeating, quietly, to himself really, with deep satisfaction, as excited as he gets.

Wednesday, March 28. Tom the blacksmith was not satisfied with the glue-on shoe efforts with Air Peyton, and so he taped the extended shoes onto the foal's toes. Air Peyton was patient. He likes the way his legs are feeling nowadays.

"Tape 'em up every day. I'm gonna fashion some silicon caulks to spread his heels," Tommy said. "We'll put them on next week. We'll get this foal straightened out."

It was a quiet day, otherwise. The stallions rested. They've worked hard lately. It was great to give their libidos a day to replenish. Tomorrow's vet report will probably discover an ovulation on a mare I thought would hold, and

I will pull my hair out in frustration. But tonight I don't care. Stallion semen is a currency to be spent conservatively. We'll need it in May.

Thursday, March 29. "How old is this mare?" Dr. Bowman asked as he felt Coco's pulse during the routine post-foaling physical. Last night, Coco delivered a filly she'd been incubating for three hundred and sixty-eight days. The overdone filly was presented upside down: Soles up, head cocked, neck on the floor of the mare's vagina. Coco strained. Coco rolled. Coco was loco.

We got her up and kept her walking until Peyton arrived. He rubbed lubricating gel on his long arms and went to work. He reached for the foal's shoulder to repel it back into the uterus far enough to right the foal. The sharp pain of the twisted foal on the mare's cervix caused her to bounce her hindquarters around like the tail on a fish. Peyton, with both arms in her as she weaved, reminded me of a cartoon character who, when punched in the stomach, wraps around the fist and won't let go. Peyton lifts weights at a fitness center. He needed that conditioning tonight. He clean-jerked that heavy filly forty-five degrees to get her squarely in the birth canal, and Coco did the rest.

At eight o'clock this morning, however, Peyton examined Coco and found her pulse rate elevated to a very alarming eighty beats per minute. Coco's straining last night could have resulted in uterine hemorrhaging. Peyton pressed against the mare's gums. Good capillary refill. But he asked that Dr. Bowman follow up on Coco when Bowman arrived at ten.

I wasn't sure of Coco's age. "Eleven or twelve, I think." Dr. Bowman looked at me as if I should've known this information. He checked her teeth. "She's under ten," he said. This was valuable knowledge. Mares older than eleven have a higher incidence of uterine hemorrhage. He pressed against her gums. Again, good refill. (Vets use the term "bled out," with grim finality, when the gums stay white, when the artery has leaked all the blood into the peritoneal cavity.) He took her pulse. "Sixty-four," he said after a minute.

"Turn her out for a while," Dr. Bowman finally advised. "Take her pulse again later. It should be down in the low forties. I think she's okay, but older mares, and mares who have trouble foaling, can drop like a rock sometimes."

Coco is owned by Peter Fuller, who also owned Rainbow's deformed foal. I hope I don't have to call this fine breeder and tell him Coco dropped like a rock.

Friday, March 30. Wild horses couldn't have dragged Richard away from the farm tonight. His reasoning was simple West Virginia logic: "Last year, while y'all were at that breeders' dinner, we had two foals born. Needed the vet for one. No sir, I'm staying here tonight—make sure everything's all right."

The only bird missing from the March skies with regularity has been the stork. So while Maryland's best horse vets gobbled down crabcakes at the awards dinner at Pimlico, Mr. Stork made a few routine deliveries. Richard and Sue signed for them on the Dry-Erase Board at the foaling barn: "Affecting foaled colt 8:15. Cracking Good foaled colt 10:15. Promenador foaled filly 11:15."

Thank you, Mother Nature, for Richard's good sense, and for three easy, healthy foalings.

Saturday, March 31. Mr. Stork forgot one, it was so small, so he turned around, made a low pass over the watershed, and gently dropped a chestnut colt in Homera's stall at dawn. No time to even lead her to the foaling barn. Her blaze-faced colt has white stockings pulled high above his knees. He's quite the Jim Dandy. While everyone was at lunch today, I crawled into his stall and sat down in the deep straw with him.

"You're two weeks early," I told him as he suckled my sleeve. He has tiny white hairs on his chin and he smelled just like a puppy. I stretched his front legs out in the straw. His white stockings extend two inches up his forearm. With his face so close to mine, with his expression of complete trust, I thought there is something so pure, so perfect about a newborn foal, it makes all our hard work seem effortless.

Buoyed by four new foals in less than ten hours, the crew pushed through afternoon chores: Manure to load, stalls to bed, feed to set, hay to carry, maiden mares to breed. We were almost through when Dad called from Pimlico with news of the Forsythia Stakes: "Lucky Lady Lauren by eleven!"

We finished March in a flurry. As we steam towards April, with chestnuts in the barn and Oaks on the horizon, we are filled with the excitement of being alive on a horse farm in the spring.

IT'S ALREADY HERE

Sunday, April 1. "Cut out all this *folderol*," said the mare owner, miffed at conflicting veterinary opinions of why his mare was not pregnant. "She's a young mare. She should be pregnant. Just breed her more often."

Folderol is a refrain sung often by mare owners. Dr. Bowman believes that sixty percent of the mares conceive on the first heat cycle they are bred, but this leaves forty percent to return for subsequent breedings on later heat cycles.

Those forty percent will drive a farm manager crazy. They won't shed out. They won't ovulate. One vet says breed, another vet says wait. They won't load for the van trip. It's enough to make the Pope cuss. A sense of humor is most important when dealing with mares intent on such folderol. It helps my sanity to prepare mentally for this hectic month, when every day can seem like April Fools' Day.

Monday, April 2. Tonight, the star-crossed broodmare Angel and her foal are on their way to New Bolton, where Dr. Paul Orsini will operate on an inguinal hernia on the colt. His intestines literally drop into his scrotum. We've manually reduced the hernia the past two days, pushing intestine back up through the inguinal rings through which only the testicles are supposed to descend. But we risk strangulation of the bowel if we dare leave him unattended for more than a few hours.

Inguinal hernias are common in Standardbred colts, and castration is frequent. Surgery solves the immediate problem while also eliminating a possible defect from the gene pool. But Angel's Thoroughbred colt is only six weeks old. That's the youngest I've ever heard of gelding a colt.

We actually are sending three horses to New Bolton, because Angel is carrying a twenty-one-day-old Carnivalay foal inside her. We douse a few extra cc's of Regu-Mate in her as she boards the van. Can't let the stress of a hospital visit usurp our good fortune getting her back in foal.

Tuesday, April 3. In a wonderful assessment of self-esteem, Steve Martin in the movie *The Jerk* jumps up and down with glee seeing his name in the telephone directory, shouting: "The new phone books are here! The new phone books are here! I'm *somebody*!"

This evening, I looked up to see the Great Blue Heron flapping his giant wings purposefully towards the pond, where a dinner of tadpoles awaited him. I called to Ellen: "The heron's back! The heron's back!" No surer sign of spring than the return of the heron. We're *somebody*—caretakers of the land for wildlife to enjoy.

Wednesday, April 4. There is a period in springtime on a farm—usually late in March, this year early in April—before the weather has broken, when working in constant rain and mud might cause you to believe that no more exhausting drudgery on earth exists than to lug ninety-pound bales of wire-bound hay through fields pockmarked with hooves, or to shuttle stubborn foals to paddocks, or to hustle to muck out their stalls so they can mess them up again. Broken nights. Days on end. Simple exhaustion.

This is the period when a good crew makes all the difference.

Thursday, April 5. A day in the life of a breeding farm.

Twelve-fifteen: The foaling barn at midnight is a maternity ward with too few beds. Table Angle's six-hour-old foal is on his feet, nursing a bellyful, and I decide to move him to the main barn a hundred feet away. He gallops in my arms as we move onto the pavement. His rump drops to the ground. I wrestle him to his feet. He is as slippery as a fish, and lunges out of my grasp and into his new stall like a cork out of a bottle, crashing into the wooden wall. I find him slumped in the corner, stunned, but he raises up onto his sternum when his mother licks him. "Thank God," I say. "Eleven months in the oven were less traumatic on you than three minutes with me."

Seven-thirty: Sue's note on the Dry-Erase Board says all's well with Table Angle's colt. I watch him nurse. Looks fine.

Nine-thirty: Ship-in mare arrives for breeding. Does not stand well. She jerks as Carnivalay tries to ejaculate. Bends his penis. We try again. She jerks again. Carny gets mad, backs up quickly out of her, loses his balance, sits on his haunches like a dog, stands back up, cranes his neck. He walks around, appears fine. Relieved, we try again. "Hold her still, boys," I caution. No mood for folderol. Uncle John always said about horses: "Don't give a sucker an even break." We repeat the line for the mare's benefit. Mike keeps her leg. Third time's the charm.

Ten-thirty: Ellen says Table Angle's foal can't pee. His scrotum is filling with urine. Peyton is summoned.

Twelve-thirty: No time for lunch. Under mild anesthetic, Table Angle's strong colt is finally restrained. I name him Calamity James. Peyton inserts a needle into the colt's scrotum, swollen now to the size of a tennis ball; urine squirts out. Peyton searches in the foal's sheath for the penis. The urethra is not clogged, no adhesions prevent the penis from descending. How did urine get in his scrotum? Exercise will help the colt, and he stumbles comically, drunkenly, beside his dam in a small field.

Three-thirty: "I gave her HCG before I left home," the maiden mare's owner informs me. The clock is ticking on her ovulation, but she will not permit the teaser to jump her. Veins bulge on her neck, a bomb ready to explode. "Put her in a stall." I call Bowman. "Try a little Ace," he says, prescribing a tranquilizer. "The follicle is for real. It is a personality thing with her."

Four-fifteen: She stands. He breeds. We sigh.

Five-ten: "A geologist from the state roads department was here today," Mom says. "They want to widen Route 1 through our front fields."

Ten-thirty: I can't wait for this damn day to end. Please, Lord, no foals tonight.

Friday, April 6. The timber racing season in Maryland begins in earnest tomorrow, announces the cover on this month's *Maryland Horse* magazine. Artist Larry Wheeler's painting depicts the scene at the Elkridge-Harford point-to-point. Heavy purple clouds and bare trees are the backdrop for the white pickets of the saddling enclosure, as a jockey cinches the girth, as two huntsmen in red coats stand ready. You can almost feel the storm over the Maryland countryside, the atmosphere of the painting is so charged.

Next Saturday comes My Lady's Manor, then the Grand National near Butler, then the Hunt Cup in the Worthington Valley. Our own season here at the farm preempts our attendance, so tonight I read a favorite description of a steeplechase, from Hemingway's short story *My Old Man*:

"I fixed the glasses on the place where they would come out back of the trees and then out they came all sailing over the jump like birds. Then they went out of sight again and then they came pounding out and down the hill and all going nice and sweet and easy and taking the fence smooth in a bunch, and moving away from us all solid. Looked as though you could walk across on their backs they were all so bunched and going so smooth. They jumped the stone wall and came jammed down the stretch, and I hollered at my old man as he went by, and he was leading by about a length and riding way out, and light as a monkey, and they were racing for the water-jump."

My English teacher Mr. John Coleman taught us: "If you ever end a short story with a death or a goodbye kiss at the train station, you'll receive an *F*." Well, Hemingway never took Mr. Coleman's course, and *My Old Man* comes to a predictably crashing conclusion. But the story will sustain me through long Saturday afternoon vet lists, while nearby, the best timber horses in America rise to the occasion.

Saturday, April 7. They floated the track between races, but as soon as the tractors passed, water pooled to the surface again. Valay Maid's front end negotiated the slippery turns in the Pimlico Oaks, but her hind end fishtailed. She finished fourth, behind a couple of fillies who met her aunt, Go for Wand, in the Breeder's Cup Juvenile Fillies last fall. Well, she's keeping good company. She'll have her moment in the sun.

Meanwhile, back at the ranch, Valay Maid's mother was bred to Carnivalay. So there's bread on the water, maybe in the oven as well. Today, though, there was simply too much water for us to get the bread.

Sunday, April 8. You can't take an accurate reading of a newborn foal's absorption of colostral proteins until he is twelve to eighteen hours old. If you wait longer than twenty-four hours to administer colostrum, you achieve negligible results. The small intestine, through which proteins are absorbed, has already begun to lose its absorptive qualities.

I thought about this tonight, while stripping a foaling mare to replenish our stock of frozen colostrum. If we don't think a foal nursed well in his first few hours, or if the mare streamed milk prior to foaling, we'll thaw some colostrum and bottle feed that foal. It beats waiting around for a blood test that tells you too late that you did too little.

When I was younger, I didn't understand the finer points of the passive transfer of immunities in a mare's first milk. When a foal would come down sick, I would look towards the heavens and wonder, "Why me, Lord?" Now I realize that foals are bombarded with microorganisms the moment they hit the ground. Our skillful veterinarians, who have a better idea of the Creator's design, have taught me to look closer at the master plan.

Monday, April 9. The word for today was "Wyoming." Dang-blasted, darn-tootin' Wyoming. I never thought we'd be buying hay from way out West.

But at seven-thirty this morning, a sixty-seven-foot orange tractor and enclosed trailer sat in the van entrance, waiting to be unburdened of twenty tons of the nicest alfalfa you've ever seen.

"Convicts loaded this truck," the drivers assured us, explaining how bales were packed so tightly into the recesses of such a trailer. The dust inside the trailer certainly was murder, and Michael, John, Steve, and Rich all covered their faces with bandannas, which were soon caked with dust where the boys inhaled. They looked like Butch Cassidy's Hole-in-the-Wall Gang. The farmers out West who put up this hay never touched it. They used forklifts and pallets and then convicts. These farmers are pushing the envelope on a man's ability to transfer such weight with his hands—first from a truck, then to a hayloft, finally to the corner of a horse's stall.

Nevertheless, the dust will clear from our lungs, and we'll breathe easier knowing that mares visiting Country Life will return home fat and sassy from eating this goldarn Wyomin' hay.

Tuesday, April 10. For sixteen years, Mom drove to Baltimore to do volunteer work for WBAL's "Call for Action" program. She could solve problems of the inner city faster than they could make new ones. This spring, she volunteered for the office staff here, after Norah went back to college, after Marva took a job in town so she could drive her daughter Karianna to school. (This example of succession is, to my mind, the real strength of a family farm, even if it does exact an unnecessary measure of guilt on family members who choose paths less taken.) At Mom's birthday party tonight, I reflected on how skillfully she handles the daily crisis of a farm in spring, when phone calls are for action of an equine nature.

As I walked back home on this warm spring night, the refrain from a Rickie Lee Jones song played in my mind: "Hey, I'm a lucky guy—a real, real lucky guy."

Wednesday, April 11. "This is John, at the broodmare barn," the Irishman called at suppertime. "Gramercy's foal is pretty uncomfortable."

The first foal from this stakes-winning mare is no bigger than a German Shepherd. Born this morning at one o'clock, he strained to pass manure throughout the afternoon. Enemas had not helped. I phoned Dr. Bowman with the symptoms: Restlessness, rolling, straining to urinate. "Sounds like a high impaction, especially since he is less than twenty-four hours old," he said. I flipped through a copy of Dr. John Madigan's *Manual of Equine Neonatal Medicine*. This pocket-sized problem-solver should be in the medicine chest of every foaling barn. Under "Meconium Impaction" was the best clue: "Repeated enemas of three or more are often no longer rewarding." (Vet lingo is so succinct.)

Gramercy is owned by Orme Wilson, the same gentleman who owned Rainbow, dam of the deformed foal. His daughter, Elsie Thompson, owns Air Peyton. This family of horse breeders is keeping me very busy this spring. Orme's little colt, with limited energy reserves, could go downhill quickly. I'd rather err on the side of caution than risk losing another foal for this man. So we phoned for the van to New Bolton. I thanked John for his keen eye, and felt confident I would hear encouraging news sometime after midnight.

Thursday, April 12. "It's breeding time," Dr. Bowman declared as he withdrew the speculum from the maiden mare's vagina. "Let's see if you need that elephant juice."

For the past three heat cycles, even while tranquilized, this mare would not allow us to take her leg as the teaser approached. She would lunge forward, throwing the twitch skyward. Then she'd rear like the Mare From Snowy River, thrashing front legs at her restrainers. We have been patient. No one has been hurt. It is only April. If covered this cycle, she'll foal mid-March next year. Dr. Bowman suggested a more potent tranquilizer. Maybe the mare heard us talking, because she kept her cool with Jazzman. "Don't think we'll need it, Doc," I said.

An hour later at the stud barn, I felt I'd spoken too soon. With all four feet on the ground, she would permit the teaser to jump her. But we still could not lift her left front leg to ensure the stallion's safety. The path of least restraint would be our only route. Using the teasing chute as protection, we brought the stud in alongside the mare. When she showed heat, Richard angled her hind end towards us, away from the chute, and the stallion lifted himself up. He performed quickly, then backed down and away from the trembling mare.

I don't smoke, but if I did, I would have pulled out a cigarette after this one.

Friday, April 13. It is fitting that New Bolton Center, a place to save the lives of horses, was built in large part by bequests of dying horsemen. A plaque on the reception room wall commemorates Henry L. Collins, Jr., 1905-1961, Master of Radnor Fox Hounds, with the wonderful epitaph:

> *And there with his peers we may leave him*
> *With all the good men and the true*
> *Who have come to the last of the gateways*
> *And laughed and gone galloping through.*

New Bolton, last of the gateways for so many horses, proved to be the *first* gateway of life for Gramercy's colt, who, once relieved, simply laughed and went galloping through. The lucky fellow returned home to us on a Friday the 13th, no less.

Saturday, April 14. Rushing to the Amtrak station for the Easter train to Virginia, I carried on board the scents of home, a world away from city life. My hands smelled of antiseptics from the breeding barn. Saddle-soaped leather boots brought the warmth of a tack room to the air-conditioned car. Blue jeans were covered in the vibrant scent of newborn foals. No one chose to sit next to me, for there was no mistaking I worked on a farm.

Out the windows of the train, we passed the backside of a city—washing machines and refrigerators, tires and broken chairs, discarded right out by the tracks in west Baltimore. Well-worn paths climbed embankments towards trestles, where two boys puffed smoke and plotted conspiracies. Then came the rowhouses, too close to the tracks. So many people, so close to each other. How fortunate I am to live on a farm. From the seat in front of me, a gal with headphones pulled the speakers from her ears, and I heard the tinny voice of the singer: "Terry, take me home. 'Cause I can't remember. Terry take me home."

I am not naive about the country life. It is something of a myth. That mythmaker himself, Henry David Thoreau, often walked a scant mile and a half from Walden Pond to the city of Concord, just to better appreciate the country. And whenever I reread *The Great Gatsby*, I always stop at the first page:

"Whenever you feel like criticizing any one," Nick's father told him, "just remember that all the people in this world haven't had the advantages that you've had."

Sunday, April 15. "When oak leaves are the size of a mouse's ear," the farmers say, "it's time to plant corn."

I know nothing about general farming, but springtime in Virginia is too beautiful not to wax rural. The leaves are out in Goochland County, at Ellen's family farm. Spring is three weeks ahead here. When the corn is knee-high in July, we'll be finishing up the breeding season back home, resting under the broad leaves of our own oak trees. This refreshing weekend away from foals and follicles is a welcome recharger.

Monday, April 16. On the drive back to Maryland, we passed through the flat farmland of Hanover County, Virginia. Signs pointed towards Orange, towards Lake Caroline, towards Warrenton, and I imagined touring through the dogwoods and pear trees, stopping to read every Civil War marker I came across: "Lee's Movements," "Stuart's Ride Around McClellan," "Lee Versus Grant."

I took a long cool drink of life while the band jammed through a Monday encore performance: On studs, Irish Johnny O'Meara. On mares, the strongest man in the world, Richard Harris. On paperwork, the unsinkable Peter Richards. On foals, the talented Jennifer Lawler. On the phone, brother Michael: "Take your time," he says. "Everything's quiet."

Tuesday, April 17. Man oh man, is Calamity James a strong colt. Wary of sending this colt to New Bolton if we could patch him up here, Dr. Bowman took the matter into his own "bear" hands. His technique for dorsal recumbency exams of headstrong colts resembles a calf-roping contest. He picks the foal up by the loose skin near the stifle, kicks the foal's front legs out, and pins him. No anesthesia required, thank you.

"Don't want to be cryin' wolf to New Bolton every time something goes wrong," he is saying, examining the raw area on the colt's prepuce, withdrawing the colt's penis from its slippery sheath. John and I restrain flailing legs. "This colt's fine. Keep spraying Granulex. Keep a close eye on this sore."

We released our grip on the colt, and stood back like calf ropers do. Calamity James bucked to his feet and glared, the toughest bronc on the circuit. Hmmm. Ain't gonna be easy checking the hardware on this ornery colt twice a day.

Wednesday, April 18. Mom has five children. When the seminar announcement entitled "Alternative Mating Techniques" arrived in the farm office three weeks ago, she reserved two tickets, then wrote on the RSVP: "Too late."

The seminar began at three o'clock. I was curious to know how the speaker selected young sires whose foals haven't raced yet.

"I'd breed everything I had to Forty Niner," he glibly explained, citing a Claiborne son of Mr. Prospector inaccessible to most Marylanders. Then he mentioned Snow Chief. "There's 'good' dominant, and there's 'bad' dominant. Snow Chief is the worst-bred classic winner since Carry Back, but his yearlings look great."

He didn't have to remind me of Carry Back's humble beginnings. I admire this man's publication, and I came away with a few fresh kernels of knowledge for the fifty-dollar surcharge. School is never out for the serious student of horse breeding.

Thursday, April 19. One hour before the van was to arrive to ship Next Dance Please and her two-month-old foal home, Ellen phoned from the barn: "This foal can't use her hind end." Dr. Bowman arrived on the scene: "Could be white muscle disease, or wobbles, or even botulism. Better get her to New Bolton."

Botulism, one of the deadliest toxins known to man, exists in the soils of farms from New York to California. A handful of veterinarians fought hard to have the vaccine made available, and folks in the horse business who do not administer it in the same routine manner as tetanus and influenza should reconsider their health programs. But we vaccinate against botulism. As Next Dance Please and her foal boarded the van for New Bolton, I thought: "Damn, we were one hour away from sending this mare and foal home. What happened?"

Friday, April 20. Everybody wants to know about yesterday's foal. The mere spectre of a botulism case sets nerves on edge. Dr. Wendy Vaala phoned at lunchtime.

"We administered botulism antiserum yesterday when she arrived," Wendy informed me. (That's a cool one thousand dollars right off the bat for the foal's owner.) "But we are exploring the possibility of a cervical spinal lesion."

Late this evening, I pulled down a book of drawings by the great English painter George Stubbs. In 1766, he published *The Anatomy of the Horse*, after spending eighteen months living near a tannery, hoisting horse carcasses with a tackle, peeling away contours of muscle to reveal skeleton. A horse would hang for six or seven weeks, "or as long as they were fit for use," while Stubbs sketched and numbered every muscle and rib and vertebra he discovered.

Stubb's drawings have a foreshortened perspective. The horses' heads bear down hauntingly on the viewer. Flipping through layers of drawings is a nightmare in animation. But the cervical vertebra are so clearly linked, a chain connecting the horse's head to its shoulder, a conduit for signals to the horse's legs.

A chain is only as strong as its weakest link, I thought, shutting the light out against the hollow-eyed visages of Mr. Stubbs' cadavers.

Saturday, April 21. The owner of Next Dance Please's foal is a nurse. She is on eighteen-hour call and can only phone between emergencies at her hospital. When I'm in, she's out, and vice versa, until four-thirty this afternoon, when the phone rings at the stud barn. Jason yells to me as the breeding crew prepares for action. Carnivalay has one thing on his mind, and will not permit me to take the call.

"Tell her it's not botulism," I call to Jason, who only works weekends and has no idea what I'm talking about. "Tell her it looks more like wobbles. I'll call her tonight with an update."

What a discouraging pair of possibilities for the foal and her owner. But it's breeding season, and foals live and foals die. The prognosis for spinal column trauma is not good. Not good at all.

Sunday, April 22. It's Earth Day. Morning on the hills of the broodmare field: A red-tailed hawk swoops down and carries away a field mouse in its talons. The two fly off together, the mouse silhouetted underneath the belly of the bird like a bomb on a plane. Earth Day is not off to a good start for Mr. Mouse.

Midday on the pond: Spring peepers blare a high-pitched chorus to the rhythmic croaking of a mature bullfrog, while water in the shallow end of the pond gurgles and splashes as frogs climb on each other. The whole noisy scene is stilled as if by the wand of a maestro when, sensing an intruding presence, the frogs whistle for silence. They abide. Not a peep is heard. The silence is a warning to other wildlife: A red-winged blackbird jets by at eye level en route to the woods, where squirrels race up the barks of trees, as two Canada geese waddle off the dike and dip quietly into the deep end.

At dusk on the most beautiful day of the month, a goose lowers her head into the grass on an island in Winters Run where she has decided to nest. The long shadow of the blue heron passes over her. Fifteen feet of rushing water on either side of the island buffer the nesting goose from riverbank predators, but owls and herons in flight keep sharp eyes out for untended nests.

I can say with certainty that on the first Earth Day twenty years ago, I did not notice anything quite as significant as what I saw today. When I look back on this entry twenty years from now, I hope the improvement is equally as dramatic.

Monday, April 23. Valay Maid's dam has feet the size of serving platters. Fortunately, she's never used them on Carnivalay. Yesterday, we teased her, put a speculum in her to examine her cervix, put semen extender in her uterus, bred her, and gave her HCG. Today, she teased out. But who knows with this contrary rose? Her foal by Corridor Key is as big as a yearling, the most mature young foal we've ever seen. It's a challenge to convince this mare to make another baby.

Tomorrow, Dr. Bowman will palpate her. He'll verify ovulation with the ultrasound, infuse her with antibiotics to help her uterine tone, then suture her back up. This business of getting mares in foal is similar to making a phone call in a foreign land: It takes a lot of perseverance to get through.

Tuesday, April 24. "Uncle Merlin led by three lengths after twenty-one fences," Vivian Rall recounted the Grand National at Aintree earlier this month. "He pecked after Becher's Brook. The horse never fell—just lost his rider. I was so proud of him. And to think, I almost bred a Grand National winner."

Uncle Merlin, winner of the Maryland Hunt Cup last year, feels like a first cousin. His dam, Aunt Sheila, is booked to Corridor Key this spring, and she has an Assault Landing yearling. For a syndicate manager, a mixed feeling arises when a steeplechase breeder sends a mare to your stallion. Almost all these breeders give their horse a chance on the flat, but their training methods do not favor two-year-old racing. And by the time the horse turns three, "Well, let's just pop him over a few cavallettis to see if he picks his feet up." As four-year-olds, the horses debut in flat races at Fair Hill, and before you know it, "Well, I hunted him over the winter and I think he'll make a splendid jumper."

Keen horse people, they stay involved despite fluctuations in the commercial market. The allure of their sport is second to none. When I was a teenager and saw J. B. Secor in jockey silks, smacking his whip idly against his boots, getting a leg up, bouncing home a winner at My Lady's Manor, I thought there could be no more romantic a life than that of a jump jockey. I have since learned better, and so has J. B., but it is fun to be vicariously involved in that world through the exploits of a few home-sired foals.

Wednesday, April 25. Sitting low and still in the grass on her island, the nesting goose continues her vigil, on guard despite the hypnotizing rush of water over the rocks. Environmentalist Rachel Carson feared a silent spring. I cupped my hands to both of my ears, like an old man hard of hearing. Suddenly amplified, Winters Run roared, birds of all voices called from the trees. The woods are alive. No silent spring this year, Rachel. Just put your hands to your ears and listen, wherever you are.

Thursday, April 26. "How many people died to build this wing?" I asked Dr. Ben Martin as the buzzer sprung the door open on New Bolton's new four-million-dollar clinic. Ben has been a surgeon here for ten years. He is the nephew of Jim and Lib Maloney, and a touch of the glib Maloney humor showed through in his response. "Not enough," he said, with a gallows smile.

Next Dance Please's foal was in a padded stall. Nudged away from the feed tub by the mare, the filly careened stiffly, a tin soldier's sense of balance. If this foal were not injured, she might be worth three thousand dollars. With a damaged spinal column, she is worth nothing. Veterinarians prefer to keep animals alive, but months of treatment and therapy would be very costly. Sometimes it all comes down to a question of "never say die" versus "enough of this folderol."

I didn't have time to feel bad about this filly. Something happened to her spine. I have to look forward, not backward. I am supposed to be back home at four o'clock, to breed the same tough mare who fought us back on April 5th. As I signed the euthanasia form, Ben shook his head at the shame of a life about to be lost, but he offered no protest.

Friday, April 27. Director Robert Altman brought the film-making technique of voice overlap to a new level. His actors all seem to speak at once, yet the audience discerns essential words. This is how the farm sounds on busy mornings such as today. Key words emerge as Mom on the phone confirms afternoon bookings, Alice on the other line gives directions to van drivers, Michael and Uncle John review the mail, Peter and John set up the morning breeding schedule, Dad gives yesterday's results, today's entries, and Sunday's stakes lineup—two of five starters in the Caesar's Wish are Carnivalay fillies.

Steve Earl in the pickup, Steve Brown on the mower, Ellen in the car with the dogs, Rich on the tractor drowning us all out as the radio gently says today's high of ninety-four will seem cool to yesterday's record ninety-eight. Newborn maple leaves diffuse sunlight and bathe the scene in a soft green hue. But no movie ever smelled of the sweetness of freshly mowed grass. So I breathed deep, took my best hold, and regretted only that this very competent crew works for such a small margin of profit. I think our day will come. Yet in the happiness that comes from a healthy family, the voices tell me: It's already here.

Saturday, April 28. Scold, the Amazon Queen's dam, foaled a full brother this week, but the mare has not bounced back as strong as we'd like. When her colt rests, she joins him. When he canters in play, she trots quietly behind him. Dr. Riddle listened to her abdominal sounds. Dr. Bowman suggested continuing the bran mash given all postpartum mares. Mother Nature advised skipping the foal heat.

Often, the broodmares speak subtly of their discomfort. I am not always listening, but sister Alice, from her vantage out the office window, has my ear.

"Please keep an eye on Scold," she asked, as the mare got down in the grass again. I used to overreact to such surveillance—as if to say, I can't do everything. Now I try to listen, and observe, grateful for the help.

Sunday, April 29. Oh, what a wonderful evening on the farm. In the late afternoon sun, groups of five and six foals gathered in a huddle, called a play, then broke for their positions. The oldest foals—born in February with heavy winter coats—are muscle-bound now from growing three pounds a day. They tired quickly, and preferred grazing the seedheads of lush bluegrass. The younger foals—six-weekers in lightweight spring coats—burned off mother's milk in gleeful displays of exuberance.

As I sat in the grass and watched, Calamity James ventured forth to sniff my hair. I hoped he did not want to pay me back for my having stung him daily with Granulex, though I did chance a glance up under his belly to see how completely he had healed. Mostly, I remained still, unthreatening, and he and I made a fleeting peace with each other. When I finally spoke, it was as if he were my chauffeur: "Home, James." He darted sideways and sprinted over the hilltop in the direction of the big house, where the entire farm had gathered for a cookout, and to watch the tape Dad brought home of Valay Maid's ten-length romp in the Caesar's Wish.

Hemingway said if you talk about something too much, you'll talk all the meaning out. I sat in the back of the room, and no one said a word as Lucky Lady Lauren splashed alongside the leader, as Valay Maid splashed past them all. No voice overlap in this movie. We all heard the essentials.

Monday, April 30. With one sweeping move in midstretch, Valay Maid redeemed the entire month of April for me. The foolishness and folderol exhibited by our resident equines this past month seemed enough to cause a nervous breakdown. But a fine crew made all the difference, and my sanity levels have never been higher. I can whistle *Still Crazy After All These Years* around here and no one even notices.

So we bounce towards May on a high note, with the local favorite for the Black-Eyed Susan Stakes, with a dozen mares left to foal, with Preakness Week around the corner. Carry on, spring. We'll try to keep up.

KEEP COMING BACK

Tuesday, May 1. "The breeding season starts today."

"What do you mean? We covered our first mare February 12th."

"Doesn't matter."

"But we've already bred two-thirds of all mares booked."

"And how many are safely in foal?"

"Well, I don't know exactly. Maybe a third."

"And of the mares not yet bred, what is their excuse?"

"I'm not sure. They're all different."

"So the *real* breeding season hasn't started. Only the easy mares are settled thus far. Now come the late foalers, the resorbers, the chronic barren mares, the maidens on steroids, the return breeders. They will cycle in waves—off Prostin, off green grass and longer days, off foal heats. Now comes the challenge. This is when you really need to concentrate. Are you ready?"

"Do I have a choice?"

Wednesday, May 2. While mowing a pasture near the main road today, Rich found two beer bottles a hundred feet inside the fenceline.

"Kids must've come into the field to drink," Rich said, clearly perturbed. "What if I'd hit those bottles with the mower? There'd be glass everywhere. Damn kids."

Our night watch girl Sue hasn't reported any nocturnal trespassers. John shrugged, and recalled the gypsies in his Irish homeland: "At least you don't have the 'tinkers' turning their mares loose in your fields at night, stirrin' your h'arses all up."

Yes, at least we don't have the tinkers, I thought. But we've got folks smooching in the lane in broad daylight, beer bottles over the fences at night, and school isn't even out. I don't like the least bit of tinkering with the quiet privacy and security of this horse farm. But what can you do?

Thursday, May 3. World War II was in its first full year in 1940. The threat of German conquest caused several European breeders to export their Thoroughbreds, among them the famous gray colt Mahmoud, the Aga Khan's Epsom Derby winner of 1936. Mahmoud's brilliant display of speed in the Derby, in which he eclipsed Hyperion's course record, remained fresh in the minds of British breeders, who were incensed that such a sire prospect had left their shores. Lord Derby even banned the Aga Khan's mares from Hyperion's book.

Britain's loss was America's gain, as Mahmoud excelled in the stud. His famous daughters include Almahmoud, Northern Dancer's granddam. Mahmoud's presence is felt closer to home in the pedigree of Carnivalay, a son of Northern Dancer. And there is no mistaking his influence on the gray Corridor Key, whose third dam, Mumtaz, was a daughter of Mahmoud. Every gray foal on this farm today is a descendant of a colt imported exactly fifty years ago.

Sometimes walking this circular farm day after day, I feel constrained by the physical task of caring for horses. It is rather uplifting, like a balloon ride, to look down upon the scene, to understand where these horses have come from. I look back upstream to Mahmoud, and downstream as far as my eye can see.

Friday, May 4. The mare has gone home to Virginia now, but I think of her every time I drive out the lane. Someday soon, I will have to install formidable gates in these driveways—to keep people out of Lover's Lane, and to keep horses in. On a Sunday last month, I turned Pure Poppycock and her foal out in their small paddock behind the barn. They trotted very precisely—a hen and her chick—straight down the fenceline towards the gate at the other end of the field.

"Holy cow!" I shouted to John. "That gate's open! I forgot to shut it last night!"

The gate at the far end was ajar only two feet, but Poppycock executed a gavotte worthy of a dressage horse, and in a flash, she and her foal were free. No Hambletonian winner ever trotted as fast. John, Jennifer, even Ole Bob Bowers at full throttle, could not corral them. We would jump to the right. Poppy would jump to the left. She trotted through Shorty's backyard, trotted past Mom's birdbath, trotted down the double fence between the paddocks, and turned left out the lane towards the county road. She negotiated a maze of fencing in her dash towards disaster.

Suddenly, Dad drove up Old Joppa Road in his Volkswagen Thing. General Rommel himself in a Volkswagen of another era was never more timely to a battle. Dad swung his German staff car sideways in the road and blocked her escape route, as a dragnet of breathless pursuers closed in. I yelled to Jason, the teenager on the Sunday crew: "Jump that fence! Keep her there!" My words seemed to run up the back of his pants, snatching him aloft, up and over the fence to corner the trotting champion of Country Life, who resigned herself with an imperious snort, and nary a scratch. I phoned her owner, Mary Stokes, of Upperville.

"That's nothing,'" she said. "I've had 'em run three miles down to Route 50 before being caught."

Mary's a good sport. Still, I'm bugged. We have two heavy chains on every gate on this farm, but my human error bypassed this precaution. Gates at the ends of the driveways would have contained the mare and foal. But the gates would be a nuisance to open and close; a lot of people drive in and out of this farm. I feel sheepish admitting this occurrence, but I think everybody has had a loose horse at one time or another. And if they say they haven't, that's pure poppycock.

Saturday, May 5. We had hardly cooled out from watching the Kentucky Derby when Rich called from the foaling barn.

"Small Star is foaling but I can only feel *one* foot," he said. "I called Peyton. I'm gonna walk her."

I'm always amazed at the length of time a mare can be walked, and labor arrested, prior to the arrival of the vet. It seems as if hours pass. Yet Peyton was on the scene within twenty-five minutes. Naturally blessed with long arms (often the key to reaching a foal's entrapped legs), Peyton pushed the whole presentation back into the uterus far enough to recover the errant leg. Out popped a black colt whose first breath of life gave goose bumps to the swelling post-Derby crowd.

It was somehow a full circle, to have watched Unbridled storm down the stretch to fulfill a lifetime dream of his owner, and then to have watched a difficult delivery in the foaling barn, where dreams of such glory are born.

Sunday, May 6. Harford County once was a rural postcard of undulating cornfields and black-and-white cows. Our high school lacrosse team competed against schools from Baltimore. On our home field, visiting city slickers would smile and taunt: "You plowboys can't play this game."

Some twenty years later, to make the opening face-off at the annual alumni game, the plowboys left their fields and finished their Sunday chores early, and jogged out into a different type of field, with white lines on its edges, where memories blew like wind over wheat. The odd-year grads versus the even-years, brother against brother. A wife in the stands said: "These guys have gotten worse."

I can hardly walk tonight, but I'm so glad I came back to live in Maryland. The county has changed, but it is still home.

Monday, May 7. "What do you mean, I'm a lucky guy?" Ron Mather smiled. "If I was 'lucky,' all six of my mares would still be in foal."

Until routine palpations at Ron's farm this afternoon, all six were *believed* in foal, from covers as early as February. But the sixth mare, the maiden, had resorbed recently. Instead of a two-month-old fetus, she carried a big, soft follicle.

"Vet says she ought to be bred today," Ron told us at four o'clock. With Ron's luck, the stallion was not booked, the crew had not left, and van man Bob

Kohl dispatched a driver. This is the third mare rebred recently after incurring early fetal death. Locked up tightly in foal earlier this spring, sutured down for the eleven-month gestation, and guess what? They got lovin' on their mind again.

Tuesday, May 8. Mares should show strong heat in the grand weather we enjoyed today, but such was not the case. Dr. Bowman went so far as to suggest replacing the fellows responsible for teasing. Jazzman and I took no offense.

"I'm a management aid," Dr. Bowman lectured. "I can tell you size of the follicle, the condition of the cervix. But I *cannot* tell you to risk your stallion if the mare is not showing heat. Still, some of these gals would get in foal if you could get them bred."

We sent the A-Team to breed three mares at two o'clock. We took our time, jumped them with the teasers. (Jazzman took a kick in the forearm, but shook it off.) We were very careful. They all got bred. They all received HCG. We finished two hours later. Two weeks from today, we'll know if it was worth the risk.

Wednesday, May 9. The two-year-olds are coming north, and we hope their names will soon become familiar: Quail Hollow and Faithful Tradition, Proangle and Ameri Allen. Dozens of two-year-olds sired by the home stallions are in serious training now. We'll study the "Latest Workouts" columns in the *Daily Racing Form*, as their trainers increase their task: Three furlongs in :36, half a mile in :48. During the arduous days of the breeding season, visions of the future sustain our efforts. Some of the babies are at Belmont Park. Maybe they will debut under the trees at Saratoga. Maybe this is our year for Maryland Million success. Maybe our stallions—sons of Northern Dancer and Mr. Prospector—will carry on the great traditions of their pedigrees.

Maybe I better get back to work.

Thursday, May 10. Dean Richardson at New Bolton examined X rays of Native Summer's filly, who has the most crooked front leg we've ever seen. Viewed from the side, the filly is perfect, eyes as big and soulful as a calf's, a strong rump and shoulder. Viewed from the front, however, her right leg breaks at the knee and angles twenty-five degrees outward. When she walks, the knee springs. You wonder how she can carry her weight, yet she gallops in play without pain. Dr. Richardson says the knee is irreparable, a congenital defect involving complex bones. As a performance horse, this filly is doomed.

Her owner suggested putting her down, but a foal nursing her dam helps release the beneficial uterine hormone oxytocin. Mares whose foals have died are often difficult to get back in foal. So we bought some time for the filly. Maybe an alternative will arise. Maybe she will make a Christmas present for the local 4-H Club. Maybe some fellow on a five-acre farmette will just want a "landscape" horse to stand on the hillside. I don't know. She has no fault other than a bent leg. I can put down a sick foal. But she is not sick. 'Bout time to try the classifieds.

Friday, May 11. Late yesterday afternoon, after much wailing, Mother Goose abandoned her island nest to rising waters. Rain had turned tranquil Winters Run into an angry brown torrent. The goose had been sitting on her eggs since Easter, refusing to yield to adversity. Truant boys from across the river shot at her with a pellet gun until Ellen chased them away. A family with a pet Doberman sauntered along the banks, the black dog splashing only yards away from where she nested, her neck low, the grasses concealing her. A poplar tree lost its grip on the riverbank and landed at the end of her island, providing a footbridge for foxes. Through it all, Mother Goose remained steadfast. She was no match, though, for the raging brown water. Her eggs are gone now. She and her mate circled the watershed for hours tonight. They finally landed, a lonely childless couple, late in the evening, in the back field near the feeders, hungrily searching for grain the mares might have missed. There will be no yellow goslings taking flying lessons over our heads this spring.

Saturday, May 12. "Did you stop at the barn on your way home?" Ellen asked me tonight.

"No, what's up?" I had been at a seminar at Pimlico since early morning.

"We sent Snugged to New Bolton today," Ellen continued. Snugged had only arrived from Kentucky on Thursday, with a nine-day-old foal in tow. "She colicked around lunchtime. We worked on her all afternoon. Bowman was here. We did all we could. She was just *throwing* herself to the ground. She couldn't even walk onto the van, she was in such pain."

Nothing would have occurred any differently had I stayed here today. This crew is the best. But it sure takes the fun out of a day at the races to imagine what the home team went through today.

Sunday, May 13. Mother's Day. What an appropriate holiday for putting Snugged's foal on a nurse mare. New Bolton put down Snugged. They discovered a long scar on her stomach from a prior colic surgery. She was an accident waiting to happen, and she happened to us. But Dolly Pouska, the nurse mare lady of the East Coast, came to the rescue. By midafternoon, Snugged's foal was nursing hungrily from Trouble Sue Bar, a sixteen-hand registered Quarter Horse who stood hobbled and muzzled, while the transfer took hold.

"The color of the foal I took off her was bay, just like this filly," Dolly said about Sue's foal, evaluating subtle factors that make the switch possible. "I put Vicks Vapo-Rub up the mare's nose, and she's a third generation nurse mare. You can breed the temperament into 'em."

Dolly wears funny black shoes, like the nuns used to wear, and she spoke to me as though I was an attentive student. The education was not cheap, though. Dolly's one-page contract is weighed heavily in her favor: The nineteen-hundred-dollar lease cost is up-front, non-refundable. Snugged's foal might die two days from now, for all I know. But decisions have to be made. I was unable to contact Snugged's owner, Peter Fuller, prior to signing Dolly's contract. But Mr. Fuller regards every filly he breeds as a potential Mom's Command, who his daughter Abigail rode to victory in the New York Filly Triple Crown.

"When the mare hollers for the foal, you'll know it took," Dolly explained. "It might take a few hours, maybe a few days, but this foal is hungry enough to keep old Trouble Sue Bar busy until she decides to accept this new baby."

Dolly stayed with Trouble Sue Bar through a long rainy afternoon. When Dolly left, the mare whinnied for the foal. Mom's command was: "Heel here!" I knew the switch had took.

Monday, May 14. We've been monitoring a colt we call Hector for several weeks now. He has a chronic respiratory problem antibiotics haven't cured. Yesterday, he exploded with a fever of one hundred and five degrees. Between putting the nurse mare on, we dealt with Hector. Peyton hooked a transpirator onto the colt's head. He looked like a cart horse with the feed bag tied on. But he understood it would help him breathe. His temperature this morning was down to 101.8.

"The transpirator system delivers temperature-controlled water vapor to his airways," Peyton explained. "We use it on racehorses all the time. The warm vapors thin the mucous in his lungs and promote resolution of respiratory disease."

Why didn't we think of this two weeks ago? It's as simple as having a vaporizer in your room. Of course, this colt has the personality to accept the headgear. Most foals will resign themselves to unusual therapy once they feel their natural defenses are exhausted.

Tuesday, May 15. A mare who foaled over the weekend retained her placenta for almost seven hours. Two shots of oxytocin were ineffective. When Sue phoned us after midnight, we were ready to summon Peyton for yet another house call when the placenta finally dropped. Retention for more than three hours does not bode well for the reproductive health of the mare. Of more immediate concern, it can quickly lead to founder or toxemia. Remnants fester inside the mare's uterus, releasing toxins that rapidly spread throughout the bloodstream. We've watched this mare's temperature closely, and she is receiving penicillin injections twice daily.

These are busy days. Follicles and phones demand constant attention, but fundamentals of horse husbandry cannot be overlooked. The margin for error is unforgiving.

Wednesday, May 16. I look at the characters preparing for tomorrow's Preakness Party, and I can tell their task at a glance. Easiest is John King, alias "Shorty," who at five feet tall cannot reach the tops of the six pipe gates he painted red today. Shorty must have felt *he* needed repainting as well, judging by swatches of red on his face, pants, and sweater. Part-timer Philly Joe has shreds of hedge in his hair, from crew cutting unruly hedgerows circling the house.

Alice and Sandy have furniture wax on their hands from white tornado efforts through the major sitting rooms. Ellen's pants are covered in mulch from planting petunias at farm entrances. Mildred has an "If I ever see another egg!"

look on her face from stuffing deviled eggs all day. Mom prepares every dish; her apron is a mosaic of sauces.

Dad is on the riding mower, Richard is harrowing the fields, Steve Brown is replacing frayed screens on the porch. Neighbors arrive with flowers from their yards. Guest beds are made. Vacuums hum. Kay Tanzola writes numbers on orange tickets for car parkers. Office phones ring. Mares come and go. The horse business carries on.

The party, a mountain in the foggy distance all spring, suddenly appears closer.

Thursday, May 17. At 4 P.M., just as we finished the afternoon breeding session, the skies turned purple. If we have a downpour during the party, the front porch will flood. It has never happened, but it might.

"Don't you think these people have a leak in their own roof sometimes?" Mom gently rebuffed me. "They won't melt. You worry about the horses. Leave the house to me."

Mom has given this party since Carry Back created a surplus of hungry Turf writers in 1961. Some years, particularly when every extra dollar went into college tuitions, and before Dad joined A.A., the party was not such a production. But Henry Clark and his wife, Mary, would come every year, rain or shine—for old friends, for Mom's food. Mr. Clark trained Obeah, Carnivalay's mom. He shakes your hand and you can't really grip him because he has a bent finger, but he is to the Preakness party what Sunny Jim Fitzsimmons was to Saratoga—a wonderfully constant fixture.

This is a good year. Tomorrow, Carnivalay's filly Valay Maid is in the Black-Eyed Susan Stakes, and Allen's Prospect has his first starter, at Belmont Park no less. If Mom is not worried about rain tonight, then neither am I.

Friday, May 18. "Didn't beat a horse," the gruff reporter in the Belmont Park press box told me when I phoned about the two-year-old race. "Finished thirty lengths behind the winner."

So much for Allen's first starter. The news did not dampen the day in the least. Allen simply did not want to steal Carnivalay's thunder, because moments earlier at Pimlico, Valay Maid beat everybody but the winner. She surged between horses in deep stretch to claim second, and now Carnivalay has sired a graded stakes performer. I stayed on the farm today. I have this fear that things go wrong on busy weekends because attention is focused elsewhere. We were still breeding mares at seven-thirty when house guests returned from Pimlico.

"She ran great! Another Carny won an earlier race, paid thirty-five dollars!"

They brandished their winnings and bought steamed crabs for dinner, and a mare foaled at nine o'clock, and the Preakness weekend broke on top.

Saturday, May 19. Over the Loch Raven Reservoir, down between cornfields of the Long Green Valley, over Gunpowder Falls and back to the farm, past Shorty's red gates, past fields of mares and foals, with friends from school on their every-other-year sojourn to the Preakness, while a guest we call the "Old Pro" slipped a Louis Armstrong tape into the deck, I sang to myself: "What a wonderful world."

Sunday, May 20. While palpating mares last week, Dr. Bowman regaled us with the story line from a movie called *Weekend at Bernie's,* in which the host dies, but his friends prop him up until Monday morning. I felt as lively as Bernie today. Friends who flew in could, more or less, put careers on hold for three days. But the horses take no breaks in the height of the season. Mares cycle, foals are born, stallions are booked. With the exception of Carnivalay, who's already read the *Form* and is cockier than ever, the horses seem to mock our fatigue: Preakness! What's that? Day of rest? We're hungry.

So we plow through morning chores, feeding and breeding livestock. Richard on his day off makes a breakfast of sausage, gravy, and biscuits for fifteen, and we watch a tape of the Marx Brothers' *A Day at the Races,* and the weekend at Bernie's continues.

Monday, May 21. Last week, I watched Dr. Bowman shake his head wearily as yet another mare entered the veterinary stocks. He massaged the thick muscles of his left forearm—his "palping" arm. Bowman is a fastball pitcher: The arm is everything. With the fingers of his right hand, he felt the capillaries and veins of his left arm. Thousands of mares have resisted his intru-

sive exams, and his arm sometimes is sore from such resistance. Then he bucked up, and laughed with Ole Bob Bowers about the proper way to twitch a mare, and he rolled on, throwing fastballs, a workhorse of a man who loves working with horses.

Tuesday, May 22. "Do you recall when the statue of Lenin in Bucharest was torn from its pedestal?" asked Mrs. Madeline Kirk, wife of the former U.S. ambassador to Romania. She spoke today at a *Maryland Horse* luncheon on behalf of Georgetown University's Intercultural Training Program.

"Well, that monument was erected on the site of the most beautiful race course in Eastern Europe. The Communists tore the track down in 1960 because racing was a 'bourgeois sport' not fit for a workers' paradise. But with freedom have come plans for reestablishing the horse industry in Romania. Our exchange students could carry their country from its nascent state to a flourishing industry. After a five-month course at the Kentucky Equine Institute in Lexington, they would live on American farms, exchanging work for hands-on teaching, room and board, and a stipend of two hundred dollars a month."

When I was a boy, an Austrian couple, Johann and Julie, worked for us. They crossed the Alps in winter to flee Hitler. When Johann fixed a fence board, he straightened the used nails and hammered them back in. He was brilliantly frugal, an ingenious recycler. They worked here for thirty years. When his battle with cancer appeared hopeless, Johann knocked on the kitchen door. He had come to say goodbye. He tore at the woolen cap in his hands. Mom came back crying quietly, and Johann walked out of our lives.

I raised my hand: "Mrs. Kirk, what if these Romanians are great workers and we like them and want to keep them here?"

"They *have* to return home. We'll send you another one," she replied.

Their salary comes to a dollar and twenty-five cents an hour for a forty hour week—an employer's dream. But I don't think the Romanian students will be so excited about the goodbye part. What's there to go home to? There are only eleven Thoroughbred stallions in the whole country. And I've read that industrial pollution is very bad: "Even horses can stay here only a couple of years," Dr. Alexandru Balin told *Time* about a region. "Then they have to be taken away, or else they will die."

I see in my mind's eye a broad-shouldered Romanian lad, wearing coveralls and a Country Life baseball cap. Just as soon as everybody gets to know him, and he learns to dribble a basketball, he'll have to go back home. Johann left twenty years ago, and I still get chills thinking about his goodbye in the kitchen that day.

Wednesday, May 23. Mom said: "Pick up the phone. It's Mr. Fuller."

"Just wanted to let you know I wasn't sitting up here in Boston saying: 'What the heck is that kid doing to my horses?' " he told me. I had only spoken with his secretary Mary since Snugged died, since the expensive nurse mare was procured. "Hey, I've been in this game a long time, and luck will come. We'll stick with it until we get lucky together. Okay? Okay." And he was gone.

"What did Mr. Fuller say?" Mom asked.

"Well, it wasn't exactly: 'Good job.' " I tried to explain. "It was more like: 'Hang in there.' "

Thursday, May 24. Peyton was sitting on the tailgate of his jeep, outside a sick foal's stall, reading Madigan's manual. "You aren't going to like this chapter," he said, handing the brown book to me, opened to Chapter Thirty-Eight: Congenital Cardiac Anomalies. I knew we were in trouble.

"He has the worst heart murmur I've ever heard," Peyton said, "now further compromised by stress of a virus." He fixed his stethoscope to the colt's rib cage, up underneath the left elbow, and let me listen. Even to my untrained ears, the whoosh, whoosh, whooshing of blood roared through the colt's heart, as he strained to keep up with oxygen demand. The stethoscope had taken me to another world, and I visualized his imperfect heart, shunting blood frantically through its chambers. Peyton then put my fingers on the colt's heart, and I felt a river of blood bursting beneath my touch.

"His labored breathing has given him almost a 'heave line,' " he said, pointing to the muscle exertion of the colt's stomach, drawn now in shallow breaths. "Horses with chronic heaves will often look like this."

I picked up Dr. Madigan's outline. I read between the lines: Mild exertion could kill this foal. I knew it didn't have to be done immediately, and a second opinion from Dr. Riddle is prudent, but will be perfunctory. It breaks my heart to have to put this colt down. He is a full brother to the Amazon Queen, our best yearling. But what's a racehorse without heart?

Friday, May 25. The Bump: It is the end of May, and they come in waves. Four mares have ripe follicles—two here on the home farm, two at satellites. Two covers in one day is enough for a stallion. Don't squeeze the lemon dry. Which mares to breed? So many factors. Do what's right for the stallion. Take the mare with the best chance at conceiving. (That eliminates the old stakes producer. Well, she's been covered several times already.) Take the mare owned by the shareholder, not the outside season mare. (Three mares are owned by shareholders. Whose ox gets gored?) How are the mares teasing? Review status of follicles. Have all been palped *today*? Will any last until tomorrow?

Everyone wants the same thing—the stallion's service. Everyone wants it *today*. To miss this heat means a June cover. Make the decision. Cite your reasoning. Most mare owners will understand. If they squawk, keep cool. If they rant, politely remind them what Joe Hickey at Windfields Farm told you years ago, when the Greentree mare booked to Northern Dancer ovulated on the van ride from Country Life to there: "Sorry son, can't breed her. That's the breaks. Welcome to the broodmare business."

And don't look back. Second-guessing yourself is a waste of time. Tomorrow's coming fast.

Saturday, May 26. It takes serenity to keep perspective—when rain falls relentlessly, when the vet list is as long as your arm, when you throw off your yellow slicker because it's like wearing a sweat suit.

Everyone is soaking wet. Rugby shirts and baseball caps cling to the crew at the afternoon breeding session. Later, on the walk home from the stallion barn, I looked around at the farm. Four-panel fences seemed freshly painted (but really were only soaked black with rain), vibrant green were the mowed fields, the gates a brilliant red, the barns a perfect white. You might think on a drab rainy Saturday that the farm would not look as pretty as she does on a sunny spring day, but natural beauty needs no makeup. Through the rain comes a whisper: "This, too, will pass."

Sunday, May 27. Turned loose in the clearing day, dozens of foals bucked and cantered away beside their dams. Often it seems we concentrate so much on the few sick or lame foals that we overlook the herd of healthy babies who pose no problems. On a quiet Sunday morning on this holiday weekend, I saluted the kiddie corps, and thanked them all for staying out of harm's way.

Monday, May 28. The inherent risk to field teasing is to get blind-sided. Allegedly, a mare with an abiding dislike for Jazzman, backed up in high gear. I shouted: "Steve, look out!" Jazzman ducked. Steve Earl didn't have time. Allegedly's hind feet smacked into his upper left arm. His right arm clutched the sixteen-foot stud shank, and Jazzman pulled him to safety. No bones were broken, but the round imprint of two hooves developed inches below the tattoo of an eagle. Steve had the tattoo done in Korea, when he was in the Military Police. The eagle wouldn't be nearly as blue as the bruise coming on his arm, I grimly thought. Steve jumped up and down from the pain. He wouldn't say anything, but I expected every curse word he ever learned in the Army to come spewing forth.

"I've got Jazzy. You get to the office," I ordered him. "Get Mom to put ice on it. Get Alice to take you to Fallston Hospital for X rays."

On the way out the gate, he suddenly exploded: "Damn, this hurts. Damn!" And he cocked his good arm back and with a roundhouse swing perfected while breaking up drunken brawls among enlisted men, he slammed his fist into the oak fence. Then he jumped around some more, uncertain now which limb hurt worse, and disappeared through the trees towards the office.

Dad and Uncle John were in the Remount, but they never saw overseas duty. So Steve is our only real veteran, and what a Memorial Day he's having—cut down in the line of duty, hauled off to the emergency room.

"We've got to be more careful, Jazzy," I told the teaser, pulling his head up out of the grass. "These mares don't take holidays."

Tuesday, May 29. "I'm going to drive down to Virginia and pick up Ellen," I told Peter late yesterday afternoon. "I'll be back Wednesday morning."

Peter jumped up and down with his hands above his head, as if he just won a new car. When the folks you work with are so pleased to see you depart, maybe a break from the grind is just what the doctor ordered.

Wednesday, May 30. "Yesterday was the hardest day of the whole breeding season," Michael says guilelessly, merely a factual recap to set the scene. "Usually when it rains in late May, it's just a summer shower. But we had four inches of cold rain falling all day long. We brought everything in. A load of straw arrived, mares were coming and going off vans. Mom made a huge kettle of soup, but we didn't sit down to lunch until three o'clock."

This horse hotel is full, and the staff looks beat. Steve's arm is in a sling. John is catching a cold. Richard is apologizing because he hasn't had time to mow any grass. Jennifer's nerves are frayed from working with an all-male cast. But Peter's paperwork is perfect, and with a rested mind, I memorize yesterday's lengthy veterinary report and breeding schedule.

"And Jury's Princess foaled," Michael says, an afterthought that drew the curtain down on the long foaling season. The final score: Twenty-five colts, twenty-two fillies—forty-seven foals in all. I don't think Michael has had time to appreciate that it is all over.

Thursday, May 31. The storm blew out to sea, and under cloudless blue skies this evening, Carnivalay closed out May displaying a lusty libido that sent the ship-in mare home wondering "Who *was* that clown-faced colt?" Perhaps the little stallion overheard Michael's recap of the Spring Bonnet Stakes at River Downs in Ohio: Lucky Lady Lauren, sold in April to a Kentucky outfit, made her Midwest debut Sunday, romping home by twenty-two lengths. Our *Form* did not carry the chart. Skeptics raised eyebrows and repeatedly quizzed Michael: "Did you say twenty-two lengths?" Michael nodded wisely, patiently.

Whatever wondrous things happen on this farm—this month, next month, whenever—they are all attributable to one simple decision Dad made ten years ago this week. In the smoky basement of the old Bel Air Elementary School, seventy-five folks laughed in celebration of his tenth anniversary. "I wouldn't be sober today if it weren't for you, Joe," a young man my age said. "You made it fun to come to these meetings."

Growing up on this farm, I saw the horse business push Dad close to the edge. Maybe he should have taken off a few Tuesdays to keep things in perspective. Maybe he didn't understand this disease runs in families and creeps up on you if you aren't careful. But something brought him back to us. Now he takes life one day at a time, all in stride, whether he wins by twenty-two or loses by the same. He answered the young man at the meeting. "Just *keep coming back*," he emphasized. Then everybody joined hands—black and white, girls and boys, young and old—and said the Lord's Prayer. Richard put his strong arm on my shoulder and said in his mountain twang: "Git 'cha some cake and ice cream." So I did. "Git 'cha some more," Rich said, smiling, as happy as I've ever seen him, nodding proudly towards Dad at the head table. "Git 'cha some more."

THE STUFF OF DREAMS

Friday, June 1. "And whatever you do," cautioned the horseman years ago, "don't *ever* stick a horse with a lousy name. If he's a good colt with an unattractive name, he'll be tough to market as a stallion. If he's just an ordinary racehorse, who's going to claim him if his name is something stupid, like 'Nothing But Bills'? No thanks. A good name is a must."

We've grown to love the name Carnivalay, however you want to pronounce it. I say tomato, you say toMAto. However, Mrs. Lunger named him, and when she asks about him, she puts the accent on the third syllable—car-ni-VAL-ay. So we do, too. Breeders like to use part of the stallion's name when naming his foals, but most owners of Carny foals simply throw their hands up. Hence, Carnivalay, when mated with Grosmar, sired Groscar. Tonight at six o'clock, I phoned the Monmouth Park press box for results of the opening-day John J. Reilly Handicap.

"How'd Groscar do?" I asked, stumbling over the name.

"Gross Car? Oh, he run third," the sportswriter told me. "Hey, nice name, pal!"

Yes, I agreed. Very nice. So we start the month of June with a third in a stakes. After all, what's in a name?

Saturday, June 2. Nothing else has worked to get the stakes producer in foal, so this morning, Dr. Bowman inserted a urethral catheter to assist her in voiding urine. The mare simply has not recovered her natural conformation after delivering a whopping baby in February. Consequently, urine is pooling inside her, contaminating semen. (In the off-season, urethral extension surgery might prove necessary.)

It's not pretty, nor very assuring, to watch her stride off, the catheter acting as a constant drain for the bladder. The catheter will stay in through the heat cycle. We've coated her hind legs with Vaseline, to keep urine from scalding.

197

Yet I feel like a card player calling Mother Nature's bluff. Nine of ten mares we breed will conceive at some point. I think I see that tenth mare.

Sunday, June 3. Scold didn't holler for her foal until she was three hundred yards away from the barn. Why did she wait so long? Did she know he was sick? We turned her out with the other single mares. I am going to march up to the stud barn with the next mare and try to make another foal. I am not going to think about the colt with the heart murmur. What's done is done.

Monday, June 4. Into a tape recorder while field teasing at 8 A.M., the details (and the reasons):

"What's the story with Ocean Reef?" (Cysts surrounded embryo on ultrasound. Said Bowman: "A vet school exam might ask: 'Find the pregnancy in this photo, and explain why.' ") "Call Valley Protein to collect foal." (After 9 A.M., you miss the dispatcher. Not a smart move in hot weather.) "Get Tracie Girl in for bath." (Dried mud suffocates hair follicles. A quick hosing, then mineral oil onto coat, before skin fungus sets in.) "Order pasture seed from Farm Supply." (Sow before summer rains cause erosion.)

"Check hernia on Blue Lass' foal." (Strangulation of bowel unlikely, but Dr. Riddle's opinion should be obtained.) "Check Head Cheerleader's foal. Coughed once." (Bag appeared down, but foal can incubate virus and still nurse strong. Cough is a warning. Listen up.) "Discharge from Lisa Hackett." (Resorbed at sixty days. With ultrasound, Bowman saw bits of embryo being expelled from uterus. Wise to treat uterus with antibiotics before rebreeding.)

"Pass the word: Watch the bag on Scold." (Mastitis easy to miss, hard to cure, on mares whose foals have died.) "Jury's Princess showing early on foal heat." (Seven days past foaling and she won't leave us alone. Breed before nine, waste of time. But can we gamble in *June*?)

Back in the office later this morning, I will listen to the tape, and Peter will tell me: "Got it."

Tuesday, June 5. Race tracks are as perishable as ice sculptures. Ghost grandstands dot the Maryland landscape from Hagerstown to Havre de Grace, from Bel Air to Bowie. So I celebrate every improvement at Pimlico and Laurel. While not aesthetically beautiful tracks, such as Keeneland or Saratoga, they are comfortable sporting arenas where racehorses compete for growing purses, where breeders and owners can, with luck, float above sunken costs.

In *Racing in America 1866-1921*, Walter Vosburgh chronicled the tracks of his era. One month after the battle of Gettysburg, Saratoga held a four-day meeting, two races a day. There followed a post-war revival: Jerome Park opened in 1866, Monmouth and Pimlico in 1870, Brighton Beach in 1879, Sheepshead Bay in 1880, Gravesend in 1886, Morris Park in 1889, Aqueduct in 1894, Yonkers in 1900, Jamaica in 1903, Belmont Park in 1905.

Last week, Bob and Tom Manfuso stepped down from active management of Laurel and Pimlico. In the past five years, they revived Maryland racing, with the help of the late Frank De Francis. When the *Racing in America* series is updated to include the nineteen-eighties, these talented brothers should merit more than a footnote. They hustled hard to make the Maryland racing scene vibrant again.

Wednesday, June 6. The overflow pipe on the pond is clogged.

"Have to get a snake from Harford Rental tomorrow," I told Rich. "Can't take a chance on this pond blowing out again."

I've seen the dike blow twice. The first time, the bulldozer driver miscalculated the weight of water. The dike might as well have been dynamited. A noise like Niagara Falls roared from the valley, and we rushed down to watch foot-long fish flying through the hole. In a matter of hours, we had our own moon crater.

The bulldozer man came back and put in an overflow pipe—the same diameter as a volleyball, which was sucked down into the pipe. Water rose up and lapped over the dike in broad places. A friend convinced us to bore a hole in the dike. When the first heavy rain came, our tunnel was chewed away by rushing water. Fish went flying again. It took sixty-four truckloads of dirt to patch the dike. This time, however, we cut a gentle swath to serve as a spillway. I don't *think* it will blow a third time, but I'm going to rent a snake first thing in the morning. I bet I find that missing life jacket at the bottom of the pipe.

Thursday, June 7. Perhaps this farm is turning the corner: Offspring in stakes races, a freshman sire with potential, *some* of our clients making money. Impediments, though, still abound. The changes in this county, the increased population, the restlessness of youths gathering noisily on parking lots of the malls, these things unsettle me. We are an anachronism: Farmers in an urban society. The naturalist John Muir confronted a similar situation when mining and timber industries first plundered the West: "If every citizen could take one walk through this reserve, there would be no more trouble about its care; for only in darkness does vandalism flourish."

I'd push for an Open House Day if it would increase awareness of the last major farm within the development envelope, but I have this misgiving it would provide a blueprint for joyriders.

Friday, June 8. Over, under, sideways, down, we tried Carnivalay every way. Yet he refused to ejaculate in the obliging mare Tricky Vicky. We put him away, unsatisfied. Dr. Riddle updated Vicky's follicle: "Plus four," he said, as good as they come. He cleaned her very carefully, then inserted his arm into her reproductive channel. "She has a very short vagina. Did you use the breeding roll to keep the stallion back a bit?" Yes, but to no avail.

"What do you want to do, Mike?" I asked younger brother. "The owner doesn't want her bred later than this heat. We miss her now, it's see you next year."

"It's a cool night," Mike answered. "Carny certainly didn't breed her this morning, so if we regroup here this evening, it'll be no different than 'doubling' him. Let's try wrapping the breeding roll in a few layers of cotton, to keep him well back in her."

We did. It worked. Carny, who stands on a breeding ramp of stone dust to reach his prey, tiptoed in his little dance on the edge. The mare inched forward, *almost* out of his reach. It was long-distance love, but it was a cover for Vicky. She might now go home to Virginia in foal rather than in heat.

Saturday, June 9. It was so humid at 8 A.M., you could cut the tension in the air. Nerves are frayed in this heat. Push through the day. Watch the weather west of us. Tornadoes in Indiana, high winds in Kentucky. The stillness of the air is the calm before the storm.

Sunday, June 10. Hidden away on narrow roads named Pocock and Houcks Mill, My Lady's Manor is home to the most beautiful land in the county. Farms are concealed from the road by thick underbrush, by tall trees clinging to steep banks. When last night's wind ripped through, however, the trees lost their tenuous purchase. This morning they straddled the road. Over brown masses of entire root balls, yellow farmhouses stood exposed, as nature's curtain was pulled back. Chain saws and front-end loaders hummed through the jungle of thick trunks, through branches of leaves captured in full canopy. Farmers would enjoy no relaxation this Sunday.

New maples will quickly sprout where sunlight now pours in. In Muir's words, "Nature is ever at work, building and pulling down, creating and destroying, chasing everything in endless song out of one form into another."

Monday, June 11. A rogues gallery of impregnable mares comprise the June Hot List, a daily roster of the last twenty-five unsettled gals from the two hundred covered this season. Four mares were given Prostin on Saturday. Today they were merely condescending to Jazzman, not violently inhospitable. The breeding season is soon to be a fading light, but we must not go gently into that good night. Success from the June Hot List means too much to the ledger.

Tuesday, June 12. During the season, no one lounges over clubhouse lunches at the races. Everyone connected to Valay Maid has grown accustomed to grouping suddenly in the box minutes before her races. Today, though, moments before the Hilltop Stakes, Joe Hamilton was missing.

"His brother was killed in a tractor accident on Sunday," Betty said. "Joe flew home to Louisville the moment he heard."

The three most deadly occupations in this country each start with the letter F: Farming, Fishing, and Forestry. If you are employed in one of the three F's, you might carry a heightened awareness of life. Joe always thought Valay Maid would love the turf. All I wanted was for Joe to phone back to Maryland tonight and hear good news. In a dark moment, the filly he bred won by daylight.

Wednesday, June 13. Glorious weather continued today. Afternoon brought a visit from Achille Guest, of the Thoroughbred Genetics Laboratory. In affiliation with the Johns Hopkins Bayview Research Campus, young Mr.

Guest is assembling blood samples from hundreds of Thoroughbreds to determine genetic predisposition to disease. He asked if we would cooperate by
drawing an extra tube of blood whenever we pull blood for routine purposes,
such as for The Jockey Club testing or for Coggins tests.

"We think we can develop DNA markers that will allow equine medicine to
focus on preventive rather than therapeutic treatments," Achille explained.

Mr. Guest's father, Raymond Guest, owns a share in Assault Landing. I
think I found my first equine blood donor.

Thursday, June 14. Death is nothing at all. It does not count. I have only
slipped away into the next room.

Of course, three years later, I never notice it—the rubber shift knob on my
VW. Mrs. Clagett's two Scottish Terriers chewed the shift diagram right off the
top of it. The Clagetts had driven their van here to breed a mare, but Alden's
Ambition was in a stakes, and I insisted they use my car to reach Pimlico by
post time.

"Take the dogs, too. Wouldn't want them to miss the fun," I said, waving
as they tootled off like high schoolers on their first date. Upon returning, Mrs.
Clagett apologized over and over, pressing bits of chewed rubber into my hand,
offering to buy me a new knob.

*Nothing has happened. The old life that we lived so fondly together is
untouched, unchanged...Call me by the old familiar name.*

Weston Farm has been in the Clagett family since Capt. John Smith met
Pocahontas. This afternoon, under supple dogwoods and ancient oaks in the
family cemetery on the hill, a cool breeze stirred, pushing away the heat of the
day. Freshly planted petunias formed a border for the shallow grave where Mrs.
Clagett's ashes lay. Mr. Clagett stepped forward. I thought of advice he once
gave me: "If you are on the side of the angels, you have nothing to fear."

*Let my name be ever the household word that it always was. Let it be
spoken without an effort, without the ghost of a shadow upon it.*

"Julie asked that I read a passage she much admired," Mr. Clagett began.
He opened the book *September*, by Rosamunde Pilcher. One of the characters
battles cancer.

*I am waiting but for you, for an interval, somewhere very near, just round
the corner. All is well. All is well.*

When Mr. Clagett finished reading, I bit my lip. I felt under a spell. But I
had to rush off. The boys were waiting for me at home. We had a mare to breed
this evening at eight. My rear tires slipped on the sand driveway of Weston as I
jammed the shifter into second gear. That's when I felt the teeth marks left by
the Scotties. *I am somewhere very near...*You could have knocked me over with
a feather.

Friday, June 15. Carnivalay rolling: Once free of the shank, he rushes to
the soft dirt where topsoil washes into his field. He breaks at the knee and lands
on his side, air rushing out of him in a contented "Harrumph!" He wallows, then
up onto his back he rolls, squiggling mud deep into his coat, hooves doubled

back, legs shaking skyward—a frenzied dancer on the frat house floor. We urge him to go all the way over—like a good horse, the saying goes. Carny goes over, does his other side, rolls all the way back to the original muddy side. Then he rights himself onto his sternum like a newborn foal, and uses his front legs as leverage to drag his fanny in a circle, hind legs underneath him like a dog's haunches. Satisfied he has plastered mud to every pore of his beautiful bay coat, he lifts himself to his feet, shakes once, then trots off precisely to his territorial manure pile. He makes a morning deposit, backs up over it, double-checking that no other stallion has come in the night to announce himself.

His morning ritual takes about five minutes. But I never tire of watching. He is telling me he is fine. When he's rolling, we're rocking.

Saturday, June 16. Hot and very humid today. No air moving. Feels like summer settling in. Flies are buzzing. Tempers are short. Yet nothing could possibly be as aggravating to us as Calamity James' eye injury must be to him. Somehow he has traumatized his corneal surface, causing his eye to water profusely. Perhaps the mare's tail caught him as he nursed. Maybe a weed poked him. Whatever the cause, eye injuries to foals are not uncommon, and can quickly go from bad to worse.

Dr. Bowman stained James' cornea orange with fluorescein, to detect damage to the delicate epithelium. James was not happy about holding his head still. I'm sure he will not appreciate antibiotic ointment in his eye three times a day, nor will he welcome a needle in his fanny for penicillin injections. But we have no choice. We have to be aggressive. Seems we've been fussing over this colt from the moment he slammed the wall the night he was born, but he looks like he's worth the effort, particularly since Thursday, when Proangle, his half-sister by Allen's Prospect, won at first asking. She became Allen's *first* winner.

"Don't you be thinking about going blind, even if it isn't your rail eye," I scolded James tonight. "Just get all your bad luck used up before you get to the races."

Sunday, June 17. Ten days from now, Ellen and I fly to Colorado for a college reunion of lacrosse teammates. It's either now or never. Ellen is due to foal on October 1st, and we have free tickets from getting "bumped" off the flight to the Breeders' Cup last fall. Thanks to Go for Wand, we're going out West.

"I served in the Remount at Fort Robinson, Nebraska, just a few hours from Denver," Dad told me. "When we were discharged, we all went to Denver to whoop it up at the Brown Palace Hotel. It was just a cowboy saloon in those days."

From the various depots of Front Royal in Virginia, or Fort Reno in Oklahoma, the Remount supplied horses to the U.S. Cavalry at Fort Riley in Kansas, and furnished mules capable of pulling howitzers through the mountains of Italy, or through the difficult terrain of the China-Burma-India theater. They shipped mounts to the Coast Guard to patrol beaches on our Atlantic and Pacific coasts. They supplied attack dogs, trained on bloodstained clothing shipped back to the states from battles in the Pacific. When it was all over, Humphrey Finney was assigned to auction off some twenty-four thousand horses. After thirteen months of traveling, Finney recorded the sales average of seventy-five dollars, and a long chapter in the history of the horse had ended.

On this Father's Day, I appreciated the thought of Dad in that bygone era of boisterous men and dusty saddles. I may have to hoist one to the old boys of the Remount if I make it to the Brown Palace saloon.

Monday, June 18. The broodmare Allegedly has the disposition of a rattlesnake: If you don't bother her, she won't bother you. Today, though, she needed breeding, and so we bothered her with twitches and leg straps. Carnivalay was seconds from completing his task when I felt the mare running out of patience. Our hands were full, though. Suddenly brother Andrew came on from the sidelines. He warned the mare not to dare strike, and he steadied Carnivalay in the important moment of relaxation atop the mare. I have consciously tried to make myself a better horseman. Andrew was born with the gift.

Tuesday, June 19. This joint is jumping. We couldn't feel it, but Dr. Bowman's ultrasound machine was short-circuiting. The mare I led into the chute was a nasty-tempered liver chestnut. Her frantic behavior was not simply in anticipation of Dr. Bowman's intrusive rectal examination. She hopped up and down like a Lippizaner in fast forward. Dr. Bowman said, "Hey, let me turn this machine off. Must be some ungrounded charge we can't feel."

Dr. Bowman explained that horses hate electricity. Their four feet are encased in a horny conduit of hoof wall. I recalled the harrowing day at the Maryland Equine Research facility, after we had made the difficult decision to euthanize my first stallion Lyllos, rather than wait for him to maim yet another innocent handler. He had already put Mike Price in the hospital after carrying him down the shedrow by his shoulder. With his front feet, he had struck me above the eye, and the gash required twenty-five stitches. He was a horse without value, a bad sire and a worse actor. He bred one mare his final year in the stud, and the frustration of hearing the stall doors of the other stallions infuriated him. It was either him or us. I had polled the shareholders. I discussed the alternatives. I told them he might kill the greenhorn who buys him for a couple thousand dollars at Timonium, once the tranquilizer wears off. I said I did not want to live in fear of the phone call from that man's wife. I told them I would not wait until Lyllos had injured someone else here on our farm. Maybe some other farm could weigh the merits of a board bill versus someone's safety; it is done all the time. But not here.

At the Research facility that fateful day, I held the shank on the tranquilized stallion as a technician attached a pair of tiny jumper cables, one to the stallion's upper lip, and one to his tailbone. "Drop the shank and stand back," he said, before throwing the switch on an ordinary one hundred and ten-volt wall outlet. The charge lifted the stud off his feet. It also lifted a great burden from my shoulders.

Today, a little incident charged a whole train of thought.

Wednesday, June 20. Ole Bob Bowers does not stand the heat well, but does not lose his sense of humor. He showed for work today wearing bell-bottom pants and one of those loosely buttoned shirts dentists seem to prefer, the kind that don't tuck in at the waist. It was blue, with white sailboats serenely heading nowhere. Bob huffed and puffed behind the pace set by Jazzman as we field teased, dodging irascible mares. By 8 A.M., his sailboats were afloat on his soaking wet shirt.

"Whew! This is hard work!" Bob wheezed, like a cousin from the city who visits for a few weeks each summer.

"Imagine how Jazzman feels," I answered. "He gets kicked for a living." At that moment, my animated teaser shot us a glance that said, "Yeah! How 'bout that?"

We are near the end of a long breeding season. The heat has made us punchy. Bob started laughing his Hardy Har Har laugh, and that got me laughing, and Jazzman laid down and rolled, and you would have thought the three of us had lost our marbles, chasing mares around in hundred-degree heat. This is a silly job some days.

Thursday, June 21. My name is Ritchie Trail. Mrs. Orme Wilson named me for a section of the one hundred-mile ride near Hot Springs, Virginia, in the mountains above the Homestead Inn, where presidents and statesmen have gone for the baths. I was born in Maryland but raised in the Blue Ridge. My grandparents are Mr. Prospector and Buckpasser. I was educated in South Carolina by

Mr. Frank Whiteley. I live now at Laurel, where Mr. Charles Hadry cares for me.

I've led a charmed life—until today. Now I feel terrible. I let everybody down. I lost my first race. It seemed forever I stood in that starting gate. The other fillies all dawdled. When that bell rang, I got left behind. I got dirt thrown in my eyes. Jockeys were yelling and screaming at us. A whip hit me in the face and I had to dodge it before it hit me again. For a moment, though, deep in the stretch, I felt the wind blow past my face, and my blood coursed with the instincts of my ancestors. I was passing other fillies. I'll do better next time. I will rededicate myself. I am only two years old. Tomorrow's another day.

Friday, June 22. There are fifty foals on this farm today, and the worse-legged baby is a house horse. Theories of corrective trimming assume added significance when the foal is your own. Tonight, Peyton dropped off the May 1st issue of *Equine Veterinary Data*, a biweekly newsletter for vets. The focus was on angular limb deformities in foals.

"With few exceptions the entire career of Thoroughbred athletes lies in the hands of those making the decisions concerning foot care early in life," Dr. Ric Redden writes. He draws an imaginary dot in the center of the toe, coronary band, ankle, knee, and top of the forearm. By mentally connecting the dots, he determines where the limb deviates. My foal is so splayfooted I can't connect the dots. I finally threw the newsletter on the floor when I read the chapter entitled Rotational Deformities: "These foals walk like Charlie Chaplin. Even with today's technology, there is nothing that can be done to straighten this type of individual—no surgery, no shoes, and no trimming."

Thanks, Doc. That takes a load off my mind.

Saturday, June 23. In keeping with the tradition of superb horsemen judging the MHBA's annual yearling show, Flint S. "Scotty" Schulhofer will select champions from one hundred and seventy-five entries tomorrow. Scotty took over the Tartan Stable when John Nerud retired. The first horse he saddled for Tartan was Ta Wee—dynamite in a small package, champion sprinter two years running. A quiet, self-effacing gentleman, Scotty was dragged to the podium at the crab feast tonight. He gathered himself and thoughtfully, honestly, answered a flurry of questions:

What will you be looking for in our yearlings?

"I can live with faults in a young horse, if my first impression is good. A good horse has a presence. I wouldn't have picked Ta Wee out of a show, but one look at her on the race track and you knew she was all heart. One look at Dr. Fager, and you knew he was a champion. A horse has to have presence."

Which fault bothers you more: Toeing in or toeing out?

"A toed-in horse almost always has speed. Keeping them sound is the key. I've had some good toed-out horses, too. I won't buy one at the sales, but if a breeder sends me one, I don't mind. Nobody knows until they run what they've got in here. (He put his hand over his heart.) Of course, you have to be lucky, too."

I doubt if there's another business on earth where top people, invariably humbled by experience, so readily cite luck as a crucial element of success.

Sunday, June 24. A collective sigh of relief was heard across the state as a cool, lovely morning was greeted. In years past, the yearling show dawned as the hottest day of the year, or the most humid, or even as the absolute height of the seventeen-year locust invasion. Judge Lee Eaton had cicadas in his hair all day.

John was up at first light to fetch yearlings from the field. We do not leave them in the night before the show. They are better off moving and grazing through the night, spending energy rather than storing it up for release in the show ring. John's own yearling is the first foal by the young stallion Taylor's Special. The pages of John's program are dog-eared where the name Taylors Number One appears. This lad takes a fierce pride in his horsemanship. Buying a good horse takes only money, and luck. Breeding one takes heart.

Monday, June 25. The hot-air balloon is navy blue, with an orange stripe through its center—the farm colors. He loves to fly over our fields. Tonight, he could not have seen a lovelier sight. Richard had mowed and chain harrowed in a burst of energy brought on by cooler weather.

We heard the balloon before we saw him, as the "whooooooosh" of his tanks heated the air to rise above the tall trees of the watershed. Then he appeared, drifting low over the treetops, vibrant blue against the perfect light of late afternoon. It is always a magic moment, something out of the *Wizard of Oz*, when this balloon passes overhead. The horses pay him no mind. But we stand in amazement, and envy his perspective of our world.

Tuesday, June 26. The sun rises at five-fifty. Two white-tailed deer snort and duck into the woods upon seeing us. In the valley, mares and foals huddle together, the misty terraces of the watershed rising in the background. In the cool still air, a red-tailed hawk screeches a siren to other birds of prey. Packs of pigeons flutter by, banking in one fluid motion to land together on a hillside of grass. By six-thirty, the light of the world has already peaked.

Today, we got up at dawn to bring in Calamity James. He is turned out at dusk and brought in before sunup. He is kept in a dark stall through the day. His eye is healing nicely, because we've reduced the stress of sunlight on the delicate eye muscles. He forces us to awaken early, but once we see the beauty of the morning, we bear him no ill will.

Wednesday, June 27. We flew backwards in time today, landing at the Brown Palace saloon, in a region of this country that served as the cradle of the horse. The oldest ancestors of the horse were found in Colorado and Wyoming, dating back thirty-five to fifty million years ago. The genus Equus crossed the Bering land bridge to Asia, then mysteriously died out in North America about ten thousand years ago. Not until Cortes landed in Mexico in 1519, with ten stallions, five mares, and a foal, were horses returned to the continent of their origin. Quickly, the horse frontier jumped to the heels of the Rockies, as Plains Indians rode in retaliation against Spanish enslavement. For centuries thereafter, the horse was the ultimate weapon.

Two hours drive from Denver is Fort Robinson, the Remount depot to which Dad was assigned before the Cavalry was disbanded. It seems the horse as a weapon rose to prominence and died out in the same region.

Thursday, June 28. Before the horse, Indians transported their possessions with the help of dogs, which dragged an A-shaped frame of sticks called a travois. With the horse, lodge poles as long as thirty feet could be lashed, and possessions easily moved. Buffalo, formerly hunted on foot by Indians concealed in animal hides, were now hunted on horseback, miles away from villages. The capture of horses became the common goal of Indian war. Yet the horse also brought the Indian a mythical creature upon which to dream. A warrior could ride his horse forever across the Plains, galloping in proximity to his gods. The horse brought tribes together for the Sun Dance, and four days of fasting often led to visions of the future.

I learned these things today, and thought of how little the nature of the horse has changed. He is still the stuff of which dreams are made.

Friday, June 29. Apache was the name of the horse I rode at a dude ranch today, for fifteen dollars an hour. Apache and his remuda of other cow ponies carry passengers on mountain trails, past wildflowers such as the delicate purple lupine, through groves of Aspen trees bearing the rub marks of elk antlers, past furry groundhogs called marmots, who stand on their hind legs on rocks and whistle at us as we pass.

"Apache is good at spotting wildlife," our guide told us. "If you notice him stop and stare, watch his nostrils. If he smells deer, he'll flare. I've had five mule deer staring straight at me, and elk calves bedded down in absolute quiet not ten yards away, and if it weren't for the horses, I wouldn't have seen them."

Karen, our guide, came to Colorado via Belmont Park, where she worked for Johnny Campo in the late nineteen-sixties. Campo trained for Elmendorf Farm, for whom we stood the Prince John stallion Rash Prince. When Rash Prince sired two juvenile fillies weighted high on the Experimental Free Handicap, Uncle John and Dad syndicated him for five thousand dollars a share. I wondered if Karen rubbed the two fillies that helped put me through high school. The horse world is a very small world indeed.

Saturday, June 30. (The most famous Buddy Chilcoat story: A good-hearted but compulsive gambler who worked for us for years, Buddy once bet many pesos on a cinch at Charles Town. When it appeared the jockey had conspired to lose the race, Buddy jumped onto the track and dragged the jockey off his horse. It's on the highlight film of farm memories.)

"I thought somebody was gonna do a Buddy Chilcoat today," Michael told me when I called home tonight to see how Valay Maid fared in the Pearl Necklace Stakes at Laurel. "The jock almost fell off coming out of the gate. He got trapped twice. She made three different runs, but just couldn't get it all together. She finished third, but *Groscar* won the Lamplighter at Monmouth!"

The Lamplighter Handicap gives Carny his first graded stakes winner, a major milestone for a sire. What a wonderful phone call! Though we're miles from home, we end the month on the same note we started—with Groscar on our lips. Yet to come are the two-year-olds breaking out in July. Here I am in Colorado, no different than the Indians—riding the plains, dreaming of horses.

OH, HAPPY DAY

Sunday, July 1. Seventeenth-century French missionaries watched as Indians threw a ball into long sticks strung with hides. The Indians played their game on horseback, or they ran on foot, for miles at a time. Games between tribes lasted days. The priests named the game "La Crosse," for the shape of the sticks. By the end of our college reunion tournament today, I wished lacrosse was still played on horseback. My legs had given way, and I had collapsed on the grass next to Ray, a goalie.

"My dad was a jockey," Ray said. "Rode all up and down the East Coast. When I was a kid, everybody asked me: 'You gonna be a jockey when you grow up?' Not me. I was little, but I was smart. I saw what race riding did to Dad. All that traveling. The falls. Keeping your weight down. It killed him—that and drink."

Home life in the horse business is a roller coaster at best, but when drink is involved, hold tight. It's *really* a ride. I felt a kinship with Ray. I told him Dad found A.A. in time. He knew how lucky I was. As we walked off the field, he said: "Nope, I wouldn't be a jockey for nothing. I'm a pilot for USAir."

I picked up my gear as Ellen showed me the T-shirt she'd just bought. In small black print, the T-shirt read: "From all walks of life, they come to play lacrosse." Although I could barely walk, I had to agree.

Monday, July 2. Near the playing field was a huge bronze statue of a cowboy astride a raging bronc, the action suspended entirely on a single hind foot of the bucking horse. Among sculptors of heroes on horseback, a code hints of the rider's ultimate fate. If the horse's four feet are on the ground, the rider died of natural causes. One forefoot off the ground: Injured in battle. Both forefeet off the ground: Died in battle. I wondered how this bronze cowboy fared, on a mount with only one of four feet on the ground.

At Independence Pass today high in the Rockies, I read a historical marker:

"The Pass took its name when gold was discovered on Independence Day 1879. The perilous trip involved five changes of horses. Two lanes of horse-drawn sleighs—one going East, the other West—made an eerie winter sight. Ironically, in the present era, the road is closed by snow. The automobile cannot match the horse-powered sleigh in deep snow!"

I did not dare look out the windows of the car as we ascended the Pass to twelve thousand feet. At the summit, with both feet on the ground, I coolly decided I would rather have broken broncos than tamed the land—gold notwithstanding.

Tuesday, July 3. On today's dude ride, the Crystal River roared beneath us as Harry the Horse carried me through groves of hemlocks and stands of blue spruce, through willowy cottonwoods and fluttering aspens, up the hillside until the crimson cliffs of Redstone loomed before us.

No fear of Indian ambush on today's ride. Lapping generously over the saddle ahead of me was the chief of the Blue Knights, the fraternity of former motorcycle policemen. He rode into camp on a Harley with headphones, the bike weighing more than his horse. I wondered if he knew the latest law of the land. On July 1st, a new Colorado law became effective, limiting civil liability from "equine activities." The law recognizes that horses are dangerous.

Harry, meanwhile, was fixed to the old beaten path like a train on a track, and I did not seem a candidate to test the Caveat Equine law. Yet back at home

are four stallions who relish misbehaving while being closely inspected by visiting mare owners. It might be wise for Maryland to examine the pros and cons of the Colorado law.

Wednesday, July 4. Near Kent, England, in 1839, naturalist Richard Owen investigated the fossilized skull of a mammal no bigger than a hare, with small, uneven teeth—dissimilar to the grinding molars of the present-day horse. Naturalists were reluctant to link small leaf-eaters to modern mammals twenty years before Charles Darwin's *Origin of the Species*, published in 1858. The skull of the "Dawn Horse," which Owen named *Hyracotherium*, remained in controversy until 1876, when remains of a horse unearthed in Wyoming were found to be identical. Wyoming's "Eohippus" was established as the horse of the Eocene period fifty million years ago. When Eohippus was inserted at the beginning of the ancestral series, European naturalists conceded that the horse had evolved not in the Old World, but in North America, and the southwestern United States soon yielded a mother lode of fossils as supporting evidence.

Nowhere have I seen more fascinating documentation of this subject than at the Denver Museum of Natural History this morning. Horses are featured in almost every exhibit. Colorful wall murals display the horse in context with other prehistoric creatures; Equus roams the grasslands with herds of bison and mammoths during the Pleistocene period a mere twenty-five thousand years ago. Thick-legged, light-colored around the muzzle like a mule, Equus is being chased by wild dogs. Toes have been reduced to slender splint bones, as the exigencies of defensive flight propelled his skeletal frame upwards.

Descendants of Equus were the lifeblood of Colorado's miners. Pack trains of fifty mules, tied head-to-tail, each carrying hundreds of pounds of wooden posts to line the mines, deftly maneuvered mountain paths. Horse-drawn wagons pulled ore cars through miles of tunnels, then above ground to the mills, well into the nineteen-hundreds.

The Denver Museum is five minutes from the airport. As we flew home, pinwheels of irrigated hay fields formed green circles in the arid Plains landscape. Mentally, I stayed on those Plains, marveling as Equus outran the dogs, hushed as Indians on foot stalked the sixty million buffalo who grazed the Plains before the white man and his transportation, "The Elk Dog, The Mystery Dog, The Horse!" changed the Indian way of life. Physically, though, I was transported home, to continue in my own small way the evolution of the horse.

Thursday, July 5. Breeding farms, like baseball teams, go on hot streaks. Some days, we can't buy a pregnancy on the ultrasound screen. Other days, mares will tailgate through the vet chute one after another, each carrying fourteen-day-old embryos. On those days: Don't answer the phone, don't stop for lunch, don't change the subject.

When I came down from the office to check on Dr. Bowman's progress today, the cry arose: "Go back to Colorado. We're doing great here. We're five-for-five, on the toughest mares of the year. Go pick up some steak subs for lunch. Make yourself useful. Just stay out of the barn."

Peter understands hot streaks. His father, Sir Gordon Richards, once rode twelve consecutive winners. The year was 1933, and luck stayed with Sir Gordon on or off the horse. Flying to the race meet at Doncaster, his plane crashed. The pilot was killed, but Sir Gordon climbed out of the wreckage and rode the next day. As I took off on the sandwich run, I felt indispensable, keeping the streak alive.

Friday, July 6. "Where do you live now?" I asked a college friend after one of our lacrosse games.

"New Jersey," he said.

"What do you do in New Jersey?" I asked.

"Mostly sleep," he answered. "I commute to New York every day."

I've had days when I regretted sleeping and working at the same address. The feeling is soon forgotten and forgiven, washed away by the rushing waters of Winters Run, or swept skyward on the wings of the blue heron. My trip out West provided me the distance to observe my vocation. I rather side with Thoreau on the subject: "No amusement has worn better than farming. It tempts men just as strongly today as in the day of Cincinnatus. Healthily and properly pursued, it is not a whit more grave than huckleberrying."

As I stood on the big hill on the Fourth of July, fireworks bursting high over the watershed towards Bel Air, mares and foals brushing past, the promise that sometimes blows in the air over this farm, I could not imagine living anywhere else.

Saturday, July 7. "We bred a mare this morning on her foal heat. Can you believe it? She foaled almost on the last day of June," said a fellow stallion manager today.

We breed mares for shareholders until common sense dictates that enough is enough. After about July 10th, however, we lose the spring in our step. I asked my friend if a shareholder owned the mare he bred today.

"No, the season was purchased through a stallion auction for charity. We had to honor it."

When I first came home, I thought *any* mare was better than no mare. I routinely put stallion seasons in charity auctions, because it meant one more mare, maybe the mare who would produce the stakes winner that made the stallion, even if the season brought only a fraction of the advertised stud fee. Now, I hesitate to donate seasons to charity causes. We have no privity of contract with purchasers of donated seasons, no control over how many of his acquaintances the buyer informs of his bargain. Usually, he immediately phones the fellow who just bred a mare for the full stud fee. And these bargain shoppers usually own impregnable mares, or are contentious when presented with a board bill, or own the mare who shows up on the seventh of July for her first cover. On this minor point in stallion management, I have learned the hard way that charity generally begins at home.

Sunday, July 8. Ellen subscribes to the newsletter for the Massachusetts Society for the Prevention of Cruelty to Animals. I read the spring issue today, about how the MSPCA monitored the closing of Suffolk Downs.

"It had the potential for many problems, including abandonment, because so many animals had to be relocated. Some owners simply had their horses destroyed. And what about the barn cats left homeless?" the newsletter asked. I have tried to catch barn cats; I have an idea of the vigilance of MSPCA workers.

The animal rights movement is a gathering storm. In Silver Spring, Maryland, are the headquarters for The Fund for Animals, a two hundred thousand-member group whose national director, Wayne Pacelle, carries a passionate message. His main area of concern presently involves hunting, but any sport that utilizes an animal for profit is likely to be targeted.

Rich Wilcke of the MHBA is raising my consciousness of the animal rights issue. He is chairman of the American Thoroughbred Breeders Alliance, which recently adopted a Breeder Code of Practice to recognize prudent stewardship over horses. I looked out over the herd tonight. They appeared fat and happy. I thought about how hard we work to provide clean stalls and fresh water, about our vigilant veterinary care, about how we scold employees who are unnecessarily rough in handling our horses. I do not fear animal rights activism focusing on the breeding farms. Their target will be the race tracks.

Monday, July 9. Oppressively hot today when, all of a sudden, purple clouds rolled overhead. The sky, which minutes earlier seemed still enough for the hand of God to reach down through the clouds—as if out of a painting by Raphael—now menaced in rage. A thunderstorm delivered its own special brand of instant but violent relief. Electricity went out. We sat on the porch in the dark and discerned the headline from a Pennsylvania farming paper: "Fourteen Cows Killed in Severe Thunderstorm." Lightning struck a tree, said the veterinarian: "The cows were all lined up under the tree—almost as if they were placed side by side." As the heavens shook the earth, I asked the Lord to take it just a little easier. A teeth-rattling *Boom* silenced me, but not my prayer that our horses would weather the storm unharmed.

Tuesday, July 10. Mrs. June McKnight visited today. Her broodmare, Allegedly, produced a foal this spring with the loppiest set of ears we've ever seen. Ellen named him "Eeyore," after the donkey in *Winnie-The-Pooh*. Eeyore's ears are set low on the side of his head. At leisure, they stand parallel to the ground. At attention, they flutter up and down, like a portable TV antenna. I don't know what station Eeyore was trying to receive this morning. Mrs. McKnight said she'd never seen a foal with such ears.

Conformation faults such as bench knees or sickle hocks may be reasons for remorse, but floppy ears are hard to be chagrined about. With a smile, Mrs. McKnight concluded her inspection of the endearing Eeyore.

Wednesday, July 11. When I phoned Sue Quick this afternoon, I felt as if I had interrupted her effort to catch up on her soap operas—her feet up, iced tea in one hand, remote control in the other.

"Still breeding mares, are you?" she asked. "In this heat? I took a week off, went to Ocean City, came home. I'm sick of broodmares. But we got 'em all in foal, everyone that mattered. Lordy, when's your breeding season end?"

"I'm not sure," I answered. "A mare we worked on all spring cycled back in today. We might still be breeding her next week."

She sounded so exhausted. The week out West recharged my batteries, though, and I know I am so fortunate to have superb help, fertile stallions, and great vets. As far as the breeding season goes, I suppose it ain't over till it's over.

Thursday, July 12. "We're breeding a whole generation of bleeders," argued Billy Boniface, an accomplished trainer as well as a breeder. "Sid Watters sent me a mare from New York. 'Put her on Lasix,' he said, 'she's a bleeder.' That was eight years ago. Now I've trained three of her foals and *every one* of them is a bleeder." He was practically standing on his chair. "Horses inherit the tiniest things—the size of night eyes on the inside of their legs, the thickness of their hoof wall. You can't tell me they don't inherit bleeding, too. The mare I'm talking about won stakes because she could run better on Lasix. Being a stakes winner, she got bred to good stallions. Now her foals are stakes winners, and they'll get bred, and we'll have some more bleeders."

He paused for effect. "My favorite new word is *ergogenic*," he said, drawing out the air-go as if imitating a Latin scholar. "Know what that means? It means 'performance enhancing.' It's a fancy way to say 'hop!' We ain't supposed to use 'air-go-genic' drugs to hop horses. But ask any gambler, and he'll bet on that first-time Lasix starter."

I thought about what Billy said. I can't say I'm against Lasix, because the letter L appears with overwhelming frequency next to racehorses' names on the Pimlico program, indicating that they are racing with the aid of the Lasix medication. We derive our livelihood from horses who race, not from horses who are laid up from the effects of exercise-induced pulmonary hemorrhage—from bleeding. But I can think about Lasix. I think it is like the acid rain issue was to President Reagan. He said it needed more study. He didn't look at the trees. He just didn't want to hurt smokestack businesses. When they finish studying Lasix, I think Billy will be proven right.

Friday, July 13. Rained and rained and rained, all night long—beautiful soft rain soaking the pastures, cooling down the world. I heard the thunder of hooves this morning, and hurried to see the yearling colts in full gallop, reveling in the soft going, relieved of flies and heat and summer. Only Rollaids, he of the "Nothing Excites Me" school, stood quietly. I cocked my baseball cap rakishly between his ears. We stood as if Owner and Trainer, clocking the workouts. I was reminded of a passage in Vosburgh's *Racing in America*, and tried to recall it for Rollaids' benefit:

"Jerome Park at sunrise! There is a buoyancy in the air. The notes of the robin and the meadow lark, the thrush and the oriole blend in the anthem that rings through the woodlands. The very brooks have a joyful sound as they ripple through the mossy banks. 'The Bluff,' where the clubhouse rises among the firs, is a picture of rock and dell backed by the grove where the dome of the old Dutch church rises in quiet majesty."

Saturday, July 14. Dad loves words, and often purposely pronounces them incorrectly, just to see how they play with one another. He regards jokes as gimmickery, mere tricks of memorization. He prefers spontaneous humor. I say this for the benefit of great-grandchildren who might someday wonder. Today, at a cousin's wedding reception at Westminster Hall in Baltimore, Dad wandered among the prominent tombstones in the church cemetery, stopping to reflect at the gravesite of Edgar Allen Poe. I thought he was going to quote *The Raven*, or more appropriately, the *Cask of Amontillado*, but he said:

"How's this for the name of an Allen's Prospect colt? Edgar Allen's Poespect."

With spaces, it's twenty-two characters. Too long, eighteen is the limit. I checked the *Registered Names* book. "Edgar Allen Pony" is there. But "The Raven" is not taken. I like it. It's strong. I have just the colt in mind, the burly bay out of Promenador. I filled out the Name Claiming Form with my usual sense of futility, and mailed it away. I only send in one name at a time. I do not give The Jockey Club a chance to exercise their preference among a variety of choices. I am not optimistic about getting the name The Raven, but I will have tolled the deadline for attempting, in good faith, to name this colt without incurring a fee.

Sunday, July 15. On Winters Run today, Ellen and I drifted downstream in an inflatable boat. The stream, only a foot deep in slow-moving places, was swollen with recent rains, and we surged quickly through small rapids, Ellen steering as she called: "Pull left, pull, pull, pull." Our voices carried ahead, where two annoyed herons lumbered up from the rocks to perch in overhanging branches. We passed through a gorge of rock outcroppings on Marlene Magness' property. She is a staunch farm preservationist, and I think our children will float through this gorge someday, and that pleases me very much. The trees shaded our way, the sun popped through to bounce light off steel-gray rocks. At an abandoned bridge, Steve waited with the pickup for the ride home. On this hot July Sunday, we cooled off by going deep into the natural beauty that surrounds us.

Monday, July 16. The jogger and his dog chugged up the long hill through the woods to the wheat field, where on more seasonable nights he has startled deer, coming up on them so quickly they snort in high-pitched alarm, then rush off, brown heads bobbing above yellow wheat. Tonight, though, labored breathing from another source scattered wildlife. The hot-air balloon was sending fireballs of gas into its belly to keep it aloft in the heavy humid air. The jogger paused in the wheat.

"How do I get his attention?" he wondered. "How do I hitch a ride with this guy?"

Then the balloon drifted lazily in the other direction, and the jogger and his dog disappeared into the woods, to begin the steep descent back down to the creek.

Tuesday, July 17. At a sale at Timonium a few years back, a mare sired by a stallion named White Buffalo was offered. Her stall door placard proudly proclaimed: "The Only White Buffalo in the Sale!" Since then, whenever a modestly bred mare's credentials to throw a Derby winner are questioned, we prattle off her top line. When Filouette was led into the vet chute for the first time this spring, Dr. Bowman inquired of her pedigree.

"She's the only Wallet Lifter on the farm," I quickly cited her sire, a California stallion of regional note.

Today, a former employee of short duration was denied a license to work at Laurel. Seems he had a track record as long as your arm. We had gently relieved him of his duties here soon after he threatened to smash Ole Bob Bowers skull or, in the alternative, hire someone to do it. A few days earlier, Mom had left almost two hundred dollars in cash on her desk for a moment, as farm help came to collect paychecks. The cash was soon missing. Unspoken suspicions ran high. Today's update stirred the soup again. In the busy spring season, workers sometimes come and go before you know them very well. As the season winds down now, we are back to only family and close friends on the employee roster. I walked up to Filouette in the field tonight, and confidently proclaimed her, once again, the only wallet lifter on the farm.

Wednesday, July 18. Covered the last mare of the year today. This well-bred gal has provoked farm managers for several consecutive seasons. We had her in foal once this spring, for a fleeting thirteen-day check. Then she immediately cast out the pregnancy, and we had to retract our gleeful phone conversation with her home farm.

Last year at this time, we were burned out. Carnivalay's emergence as a sire rekindled us. Allen's Prospect's promise motivated us. We plowed through this season, with well-timed fresheners to Virginia, to the Rockies, taking full advantage of a great home team. Today's mare was barren when we led her to the stallion. Who knows? Maybe today was the ticket. Wouldn't that be great? Like hitting a home run in your last at bat of the season.

Thursday, July 19. Oh, Happy Day!

> *(Oh, Happy Day!)*
> *Oh, Happy Day!*
> *(Oh, Happy Day!)*
> *When Jesus washed,*
> *(Oh, when he washed!)*
> *My sins away.*

It is a fine day for singing spirituals. It is late afternoon, and we are sitting on a rock in Winters Run, the water rushing past in waves, and we are singing because Michael and Lisa brought another baby into the world today. They left the farm this morning at 6 A.M., and left us behind to push absently through the morning, Alice preparing Reports of Mares Bred, Peter updating vaccination records. I was busy trying to keep two flies from landing in the lap of a prospective client, who visited the office to inquire about year-round boarding. (This is the toughest nut to crack in a regional market. Not many absentee owners here. All have farms of their own, it seems, or farmer friends who will board mares for next to nothing.) Suddenly, near lunchtime, the phone rang, and Alice leapt for joy upon hearing Michael's voice imparting the good news.

"A girl!" she exclaimed. "Lisa had a girl! Everybody's fine! They've named her 'Elizabeth Ann!' "

This extended family is short on grandchildren. Dad was ecstatic. He clapped his arms around the client, who knew something about grandfathering himself.

"I've got to be off now," the client said, congratulating Dad again. "The mares will be here next Thursday. Thanks so much."

Babies anywhere are a fine thing if that's what a family wants, but bringing a baby to a farm is special. It means the possibility of a continuation of a way of life. Today is a happy day, indeed.

Friday, July 20. Almost every evening, it seems we have a brief thunderstorm that cools the farm. In the moments before the storm, the atmosphere is so charged that the telephones emit a quick chime of premonition—just a single sudden note from their ringers—then the *Ba-Bam* as the thunder cracks and the lightning leaps into the trees around us. There is such danger in these brief storms. We can do nothing but sit tight and be grateful for the rain.

Saturday, July 21. "I'll find seats in No Smoking. Marty and Joanne will pick us up at the Metro Park train station. The track is only twenty minutes down the Garden State Parkway. You'll love Monmouth Park. Flowers everywhere. The guys don't wear ties. It's Saratoga without the crowds. Valay Maid is in a famous race, the Monmouth Oaks. It's a big deal. She always gives her best on big days. Will you come with me? Will you?"

"You poor girl. He drags you all the way up here for a horse race. How are you feeling? Two months to go. You're in the homestretch."

"We'll hit the board. She's training beautifully, vanned up Thursday night, rested Friday, galloped the wrong way around the track this morning. She looked at everything. Eddie Maple will ride her. I called Santos. Can't do. I called Perret. The Atlantic City race is this afternoon, *not* tonight. I called Velasquez. Booked in Chicago. Maple will do fine. Good rider. Have lunch yet?...Watch the replay. He doesn't know she hates the rail. Watch her ears. She stops running right here. He holds her together very well. Then he comes off the rail. She runs again. I think second is very good."

"Sure you won't spend the night?"

"Thanks, but it's only two hours back home on the train. Look, Ellen, there's the sign I told you about: 'Trenton Makes. The World Takes.' Isn't Monmouth beautiful?"

"Tickets, please."

Sunday, July 22. In the cool, dusky haze, the ten single mares moved down into the deep grass, walking briskly, free of the yoke of oppressive heat which has beset them daily. This is a good opportunity for me to update their condition. Are older girls picking on maidens? Who is not tolerating the hot days well? The effort of battling flies, and the heat which brings sweat to their flanks into late evening, causes the mares to lose reserves of fat quickly. The four days of rain last week, however, has kicked up the grass, and the mares tonight all looked well.

I am especially concerned about dear old Filouette, forty days pregnant to Carnivalay's cover. She is twenty-one years old, and is the dam of Restless Con, second in this year's Ohio Derby, and entered this weekend in the rich Amory L. Haskell Handicap. We foaled a twenty-six-year-old mare three years ago: Palatch, winner of the Yorkshire Oaks in 1964. Her routine did not pamper to age. We kept her on Regu-Mate, but she stayed fit with several young mares as companions. She foaled with no difficulty. I am hoping to report the same for Filouette next spring.

Monday, July 23. The close humidity caused envelopes to stick to each other on this brutally hot day, and we prayed for quitting time to come to our oven of an office. We found solace on the banks of Winters Run. Such a cool name for a stream. As the water rushed by, Ellen fell fast asleep on a boulder in midstream, sleeping for two as the baby inside her kicked in excitement. I watched her sleep, watched the baby kick, wondered about our future—the three of us. I like to think the baby heard the river rolling by, heard the familiar sounds of water.

"Stay in there until the first day of October," I whispered. "Wait for the harvest moon."

As Ellen slept, the two of us kept the vigil, one kicking, one wondering, and we languished by the creek as the farm began cooling down from a long hot day.

Tuesday, July 24. A beautiful cool July day danced in this morning, and we took the opportunity to wean two of the January foals. At feeding time in the field, the herd of foals crowded hungrily into the creep feeder before we stealthily led the two mothers away. Surrounded by ten mares and foals munching in contentment, the two weanlings could not bring themselves to full alarm.

This evening, brief whinnying—young voices raspy and hoarse—floated across the farm, not carrying the shrill cry of panic, but an annoyed tone of bewilderment. To my mind, field weaning, executed properly, certainly must be less stressful than stall weaning.

Wednesday, July 25. "That was Proangle's owner on the phone," Peter greeted me at the office at 8 A.M. "He said she suffered a hairline fracture in her cannon bone on Saturday. They are going to drill a hole in it and rest her. She'll be away from the race track for sixty days."

Well, that's racing. Carnivalay has Valay Maid pointed for the Maryland Million Oaks on September 9th, and Proangle was Allen's Prospect's best bet in the two-year-old filly event. She ran second in the stakes at Laurel on Saturday. There is promise among some of the unraced two-year-olds, but we are only six weeks away, and a little seasoning is required to take a hundred thousand-dollar purse. I am sorry about Proangle. This is Allen's first crop, and so much depends on early success.

Thursday, July 26. After Proangle's injury, I did not think an X ray could possibly carry good news. What do I know? An Allen's Prospect two-year-old colt named Xray burst into our life today. Xray rocketed around the hard Del Mar track yesterday, winning by two lengths over a most representative sampling of California juveniles. The winner's share of the purse doubled Allen's Prospect progeny earnings, and kicked him up the ladder on the freshman sire lists.

I double-checked Xray's Maryland Million eligibility, because there are no supplementary nominations for this day. Xray's okay. I can't believe how quickly my Maryland Million fantasies have been rekindled.

Friday, July 27. A rush of excitement swept the office as Mom announced that Xray's owner was on the phone. I excitedly exchanged pleasantries, talked about the colt's future, about other babies by the sire, then I asked him: "We're curious why you named him Xray?"

"Oh, well," he began, very modestly, "I have so many horses to name, I thought maybe I'd just name them for places important to me. I'm a pilot, you know. Flown around the globe several times. I named Xray for a checkpoint over Newfoundland, if I'm not mistaken. Most checkpoints are short names, and are so unique The Jockey Club usually has the name available."

Afterwards, I sat quiet in the office, thinking about my own "checkpoints" as I navigate this farm. Suddenly I awoke, though, and felt at any moment that the jarring voice of radio personality Paul Harvey was about to declare: "And now you know the *rest* of the story."

Saturday, July 28. When Suffolk Downs closed, trainer Bill Perry thought racing at Rockingham Park in New Hampshire would show vast improvement.

"I was wrong," he told me today as he prepared to saddle Halo Flyer for the modest allowance feature. "A lot of the good outfits went down to New Jersey. Others just cut way back."

In a full card of eleven races with one hundred and two horses entered, only nine runners did not receive Lasix, only four were not on Bute. How sore are these horses? I wondered, as longshot after longshot scored. Halo Flyer was not sore, however, and his front-running victory kept alive the memory of his sire,

Lyllos, that disagreeable son of Lyphard who so painfully introduced me to the stallion management career. Halo Flyer is a homebred for Mr. Baylis, who bought the first share in the stallion. How wonderfully appropriate for Mr. Baylis to own one of Lyllos' better runners.

The "Rock" still attracts good-sized crowds, and construction of a new clubhouse entrance and paddock have added a modern charm to this old track. Vacationers pulling off I-93 enjoy a good show. As we drove out this afternoon, my last impression of the Rock was the pleasant sight of the handsome red spires of the new clubhouse, as if I was looking back at a handful of flowers.

Sunday, July 29. On Sundays, the papers arrive late to the New Hampshire camp where Ellen's family gathers every summer. The early New York *Times* edition carried no mention of the five hundred thousand dollar Haskell, much less the supporting thirty-five thousand stakes for which Groscar was favored. The Boston *Globe*, however, carried the headline: Restless Con Upsets Haskell As Rhythm Runs Third. I soon found out that Groscar finished a game second.

What a week! Proangle second in a stakes, Xray romps. Halo Flyer carries the day. Filouette, in foal to Carnivalay, is the dam of major winner Restless Con. Groscar maintains his momentum. It is all so very exciting, the sweet fruit of hard labor these past years. "Getting away" for a few days is even more enjoyable when there is so much to go back to.

Monday, July 30. Two elderly matrons, stoic keepers of this camp's traditions, were frustrated by the beeps and busy signals as they reluctantly availed themselves of the new phone shed, where three touch-tone phones stood side-by-side. I stepped to the middle phone and dialed information for Pimlico's number.

"Yes, that's right, Operator. Pimlico Race Course. Yes, it's a race track. Yes, *Pimlico Race Course*. That's right."

I felt the ladies' eyes on my back. I dialed Pimlico. The switchboard had closed. The stable gate picked up.

"Who won the ninth?" I asked quickly, quietly.

"You want results? Call the Hotline!" an irritated security guard shot back, as he hung up.

This meant I had to unfold the *Daily Racing Form* from under my arm. In my heart, I was simply a horse breeder bursting with curiosity about a first-time starter. However, the moment I spread the *Form* on the picnic table, I am certain the two gray-haired ladies transformed me into a cigar-chomping, ticket-crumpling candidate for Gamblers Anonymous. I found the Hotline advertisement and dialed the number. No luck. Can't dial it from a pay phone. I hung up in dejection.

"Would you mind, sir, helping me with my call?" one of the ladies asked me. She looked down at the *Daily Racing Form*. Then she reconsidered. "On second thought, I'm afraid I can't tell you my calling card number. Wouldn't be a secret any longer. I'm sure you understand."

At breakfast in the dining hall this morning, she smiled knowingly at me.

Then she sat down with her family, and I imagined her whispering: "See that young man over there by the cereal? He's a professional *gambler*! Right here in the camp!"

Tuesday, July 31. Without uttering a word, my new niece has clamored for recognition in a month full of memorable characters to me, from Eohippus to Harry the Horse. Yet she should know she was not born in a vacuum, that she came into being at a wonderful juncture in the farm's travels through time. She arrived between the last mare of a fine breeding season and the first flight of a promising two-year-old crop. She was sandwiched by Xrays and Valays and Rollaids, between hot streaks and cool storms, between the aspens of the Rockies and the birches of New England.

To Elizabeth Ann: May you read this someday, and know this month's for you, baby.

SUMMER MUSIC

Wednesday, August 1. It's August, and there is water, water, everywhere. To reach a New England lake, and respite from routines of a horse farm, we had crossed the Hudson River, and were stalled at rush hour on the Tappan Zee Bridge. Underneath us, the great gray river flowed silently, yet echoes of its past could be heard through the haze shrouding the canyons of Manhattan downstream and the blue mountains of the Adirondack upstream. Henry Hudson sailed here in search of the Northwest Passage, shortcut to the Orient's riches.

Legend says he chained an elephant to the foredeck of his ship, a present for the Lord of the Indies, but the quest was halted by the narrowing river near Saratoga. I stared at the flat water beneath the escarpment of rock at Tappan Zee and imagined an elephant rounding the bend, splitting the still air with his scream—that bellowing warped horn-note leaping across the centuries.

The English and French battled here for control of the "Empire Valley," and generals such as Washington were drawn by war in the seventeen-eighties, to return again upon discovery of recuperative springs near Saratoga. Robert Fulton's side-wheel steamer churned the river from New York City to Albany in 1807, and in the Clermont's wake, empire builders traveled the river to gather on porches of Saratoga's great hotels, near High Rock Spring, near Congress Spring, near Empire Spring, clutches of millionaires around whom the sport of horse racing evolved in this country.

We idled on the bridge as if captured on canvas by a Hudson River School painter. In due time, of course, we made it to the New England lake, where this evening a steamboat whistled across the water at dusk—not a whistle, really, but a baritone call, from the belly of the boat, the sound Hudson's elephant might have made.

Thursday, August 2. "You're a nerd, man, you know that?" Carl said. "Nobody but a nerd reads an encyclopedia on vacation!"

222

My brother-in-law loves the ambiance of a race track. You can almost picture him on the wooden porch of the racing secretary's office at Rockingham, a cigar in his mouth, mugging for an invisible camera. "You don't think I got a beef about this encyclopedia on the dock? You bet I got a beef!"

But Carl, I admonished, this is *The History of Thoroughbred Racing in America*, a highly readable history of the sport from which I derive my livelihood. Did you know, for instance, that Eddie Arcaro, ol' Banana Nose himself, was suspended from racing—with no assurance of ever being reinstated—when he told the stewards he was trying to *kill* another jockey who bumped him during the 1942 Cowdin Stakes? In that race he was riding Occupation, whose full brother Occupy was champion two-year-old colt in 1943. Occupy stood at Country Life. We stood a champion, Carl. We once stood a champion.

"Hmm, well, no, I didn't know all *that*," Carl said, suspended in reflection. "But hey, you're still a nerd, and that's still an encyclopedia. And this dock ain't no library!"

Friday, August 3. In the southeast corner of China this summer, American and Mongolian archeologists are celebrating a historic fossil expedition, toasting each other with fermented mare's milk. This ritual began with the hunters of the Ice Age, who offered mare's milk to the gods for good luck. The horse was a prominent animal to early man, as evidenced in superb cave drawings discovered a mere fifty years ago in Lascaux, France. The drawings did not decorate living spaces, for no remains of fires or sleeping quarters are nearby. The drawings were religious in nature; the caves were places of worship, where mare's milk was like altar wine.

When I get home, I think I'll head for the caves of rock near Winters Run, pack some crayons and a Thermos of mare's milk. I'll draw stick-legged horses and drink milk till the cows come home, filing an appeal to the gods of the Maryland Million for good hunting on September 9th.

Saturday, August 4. Today's sports pages bemoaned the loss of racing's two biggest stars, Easy Goer and Sunday Silence—history now—injured in preparation for the rich challenge race at Arlington. History, though, is replete with the broken promise of such races. Three stars of yesteryear—Triple Crown winner Assault, handicap champion Armed, and rags-to-riches Stymie—battled through the 1947 racing season exchanging the distinction as the world's leading money-winning Thoroughbred, while track promoters tried to lure them together for a rich race.

Atlantic City offered fifty thousand dollars in May, but Stymie's poor form early that season squeezed him out, and a revised offer extended only to Assault and Armed. Washington Park and Belmont Park entered the action. When Ben Lindheimer at Arlington upped the ante to one hundred thousand, the Chicago track set the date for August 30th. Assault suffered a minor setback, however, and the race was rescheduled for Belmont on September 27th.

By this time, Stymie had forged to a one hundred thousand-dollar lead in the money race, and his absence from the Belmont match drew protests, to no avail. Five days before the match, Armed was defeated by Polynesian, and Assault aggravated a tender splint. By race day, the bloom was off the rose, the event was declared a betless exhibition, and the purse was donated to charity. Armed won easily by eight lengths; Assault was retired for the season. The three gallant campaigners never again reached the lofty heights of their 1947 season.

These racehorses were owned by top stables: Armed by Calumet, Assault by King Ranch, Stymie by Hirsch Jacobs. This year's stars are similarly well connected: Easy Goer is owned by the Phipps family, Criminal Type races in the Calumet silks, Sunday Silence carries on the Hancock family tradition. The historical parallel ends here, for Armed and Stymie were geldings, and Assault proved sterile at stud. Easy Goer and Criminal Type might reproduce in the lusty manner of their sire Alydar, Sunday Silence with the fire of Halo. One fine summer in the future, battles might be rejoined by offspring of these champions. Such a balming thought takes the sting out of today's racing news.

Sunday, August 5. On Church Island today, the sermon was from the Book of Luke, about a fellow whose crop is so large that he needs to tear down the old barn and build a bigger one. Instead of sharing his crop with his neighbors, the greedy fellow builds a new barn, but foolishly erects it atop his most fertile cropland.

It doesn't take an extension agent to recognize the moral of this story. Lord knows I should understand a little Scripture. Dad went to Notre Dame. Andrew was named for a monsignor. I had Jesuit teachers through the twelfth grade. My first few years in the horse business, I threw good seeds on rocky ground. No crop. Yet my family kept farming with me. A bumper crop is coming, sprouting as I speak. I don't want a new barn. I want to catch back a little from the poor harvest of the past. I owe it to my family.

Monday, August 6. It is pouring, but Ellen's mom, Doris, finds the silver lining: "Rain before seven, clear by eleven." The actuarial uncertainty of that statement has evidenced itself, and it is now three-thirty in the afternoon, still raining on the lake. Waves slap on the rocks, and I am dreaming the afternoon away—a young boy back home on the farm.

The whitewash man is coming today, Mom is saying. *Stay out of his way.*

At the first sound of the truck in the fields, I rush down the long porch and out through the yard. The truck is completely white, catching so much of the powdered lime spray that blows in the air from the hose aimed at the fences. Even the tires are white. The man with the hose is a ghost. The hose is white. His hands are white. He is an apparition. Gusting wind carries the spray in a fog to nearby telephone poles, water spigots, privet hedges. Chains on gates are coated. Green grass underneath the fence is laced with white. While the grass is still wet, I walk through it in dusty sneakers, and my shoes suddenly look new again.

Don't get the creosote on you, Mom is saying, just a few years later. There is no magic. Creosote is cheaper. Times are bad. *It'll burn your skin.* I wake up with my face on fire from the fumes of the black paint.

"I just knew it couldn't rain all day," Ellen's mom is saying as I wander out onto the porch. I stand on the dock and look out over the thick mist on the lake and think to myself: The whitewash man has come.

Tuesday, August 7. In the Boston *Globe*, more column inches are devoted to afternoon and evening cards at Raynham and Wonderland Greyhound tracks than are allotted to Thoroughbreds at Rockingham or Saratoga. The July 30th issue of *Sports Illustrated* quotes a dog owner: "Put a dog track next to a horse track, we'll beat them every time. Right now in Florida, there are seventeen dog tracks to four Thoroughbred tracks." Most gamblers prefer dog racing. It is conveniently conducted at night, unlike daytime Thoroughbred racing. It is less expensive to attend. The tracks are smaller, therefore, the action is easier to follow. The races are only ten minutes apart, instead of twenty-five. (On August 4th at Raynham, *twenty-four* races were carded.) Greyhound racing ranks sixth nationally in sports attendance.

I am beginning to realize that while I am looking down my nose at the dog tracks, that long-nosed Greyhound is hot on my trail.

Wednesday, August 8. "Why don't we rent a cabin up near the lakes? You can rest and get well," the jockey's wife had said. It rained the first day, the second day, the third day. He could feel his ribs healing, but he couldn't feel his nerve growing back. It was always like this after a fall. The mind had to heal with the body. The thunderstorm above the lake rattled him the way fireworks shake war veterans. He flashbacked, horses pounding towards him, jockeys frozen in the saddle, with a dead aim on him. A hoof clipped his side—just a taste of death, like getting slapped by the side mirror on a hit-and-run car.

"Well, I'm not the first to fall. Poor ol' Gamby went down too. He's three hundred wins away from the magic six thousand, and he's bouncing along on the turf course like a doll. Carl Gambardella, fifty years old, best rider at the Rock. Then Baze goes down. Broken collarbone. Then the McCarrons, Boston boys. Chris gets trampled. Gregg crushes his foot in the gate. Those guys ride good horses. Me? I'm a two-bit Rock jock. Who cares about us saps who have to ride in New Hampshire in *January*?"

He thinks about the vodka he put in his Fresca at lunch yesterday. His wife didn't know. "Can't do that again today. Jocks can't drink, never could," he says to himself. His eyes are closed but his mind is awake at 4 A.M. His ribs are like knives as he breathes. He listens as the hemlocks gather the rain and drop it like gunfire on the roof.

"Don't you hear it? Sounds like horses, running right at me, and I can't move. Don't they see me? *Can't* they see me?"

"Honey, wake up," his wife says. "You were having a nightmare."

Thursday, August 9. Dad left a note on the dresser in the room at Saratoga: "Went racing. Am leaving after the races for Sis'. Xray 3rd in De Anza S at Del Mar to Iroquois Park, Lukas' Chief's Crown, in 1:09 and change. See you at home."

I met an acquaintance at the track: "Your Dad just left. He was dressed in a floppy porkpie lid he called his Henry Fonda hat. He said he was 'On Golden Pons.' I told him he was running that joke back pretty fast, since he told me the same thing yesterday. He's a bird. I love him."

Friday, August 10. When you drive through the white gates to Greentree's private area of the Saratoga backstretch, you might as well be in the nineteenth-century. Two barns as long as trains face out onto a private training track. A hill of grass rises in the center of the track so all you see are the straightaways like paths into the clouds. It is a magical kingdom of stately pines, where a filly named Go for Wand reigns as queen. Billy Badgett is the Queen's guard.

"I feel sorry for the other two if it rains," Billy says with genuine sympathy. "This filly's a monster in the mud."

After the ninth race this afternoon, Marshall Cassidy makes the announcement: "Attention horsemen. The main track will be sealed tonight due to a forecast of rain. Please be prepared to take your horses to the training track in the morning."

Tonight I looked out over the track, rolled smooth as hard-top. They want everything just right for the big race, for the Queen. Tonight, Go for Wand has Saratoga all to herself. Tomorrow, flanked by only two opponents in the Alabama Stakes, she should have it all to herself again.

Saturday, August 11. The movie *Race America* at the National Museum of Racing has no narration. The sounds are the drumbeats of hooves, clangs of starting gates, cheers of crowds, a mare's labor, a foal's first cry—this sport told in split screen, with wide-angle lenses, in circular sound.

Today was like the movie, no narrator, just images, just sounds: From the high walls of the museum, Stubbs' anatomical models, gray horses with no skin, overlays of muscles and bones, coming straight at you, lost, no eyes to guide them back to the tannery from where Stubbs commandeered them. A jazz band under the trees near the ice cream stand. Music mixes with calls of taped races from TVs in branches. Replay of the Monmouth Oaks: "Valay Maid is making a strong run to finish second." Ellen and I hear her name through the din.

Fifty feet away, Badgett saddles Go for Wand very near the white fence separating fans from horses. He wants to share her, I believe. She resists the tongue-tie. He takes her head in the vise of his arms, his hands as deft as a blind man's, close together in her mouth. He stares out at the crowd as he makes a knot in the cloth of the tongue-tie. He might just as well be playing a harp, his stance is so similar. Randy Romero's boyish face is tense with concentration. Electricity surges through the crowd. The filly passes, Romero's back straight; he sits high in the saddle.

Now the race is over.

"She's a freak," a horseman behind me says. "Ten days ago, she wins the Test Stakes, goes seven-eighths in 1:21. Today, she goes a mile and a quarter in 2:00⅘, over just a 'good' track. Who's the best filly you ever saw?"

"Carny's sister," I tell him.

"Who?"

"Never mind. Buy you a beer?"

At sunset, flocks of migrating tundra swans lift off Adirondack lakes and fly south over the green slate roof of Saratoga's grandstand, swans glittering like sequins, as they capture the last light of a memorable day.

Sunday, August 12. In the gold braided silks of the *real* Queen, the black velvet of his cap bunched with excess material, Angel Cordero, Jr., somehow looked like a kid playing dress-up. He is the King of Saratoga, and he coaxed a furious stretch run out of the Queen's horse to win by a scant nose, as the Queen's purple silks shown rich against the green turf course.

Mack Miller stepped right off the silver screen from his appearance in *Race America*, porkpie hat and down home Kentucky voice; only now he is saddling the very fractious Who's to Pay for the Bernard Baruch Handicap against champion Steinlen. Who's to Pay rears skyward. His groom refuses to let go, gets lifted two feet off the ground, dangles from the halter—an isolated act of courage necessary to get the job done. They both glide back to earth as Miller approaches with the overgirth. Who's to Pay is ready to blow again. (I wonder how he was to handle *before* being gelded.) Miller's horsemanship pays off when Who's to Pay wins going away.

Earlier this week, I saw famous faces of a pride of Hall of Fame jockeys, all together, Atkinson and Arcaro, Ycaza and Rotz. I felt as if I was at a wax museum, and the statues were moving. Saratoga is a fairyland. I can never get enough of her magic.

Monday, August 13. Sally Lundy nails down details concerning other horses in her stable, then turns her mind to Pavia, an Allen's Prospect two-year-old. The best youngster in the barn, says the farm, says the trainer. Brought east from Del Mar. Better than Xray? Yep, think so.

Pavia looks like his daddy, I think—all sinewy muscle, flat head intent on getting somewhere in a hurry, a dim instinct of purpose even before he is asked to really run. He coils next to Sally on the stable pony. They make their way to the starting gate. Pavia does not like the gate, refuses to enter, sits back just as the gate springs, gets left badly, even though he broke alone, but soon his brown hindquarters are propelling him down the wide stretch and over to the grandstand, where the terrace breakfast crowd does not notice how easily he is working.

"Bring him back up a few more times," the starter tells Sally.

"He's usually better than this," she answers. "I think the pony scared him. We'll bring company next time."

Driving home this afternoon, down the Northway, onto the Garden State Parkway, over to the Jersey Turnpike, I feel as if I have been underwater these past two weeks, seeing with the clear vision of a diver's mask, permitting focus on only a very few things—the characters in short stories from summer reading, the white hairs on the inside of Go for Wand's front leg, the rough-cut frame of the juvenile Pavia. Buoyant with the promise of the future, we splash back to the surface tonight, where details of farm life float in a circle on the quiet waters of August.

Tuesday, August 14. On holiday, news of the non-racing world can be gleaned through the Reuters dispatches in the *Daily Racing Form*. At breakfast one morning, half a paragraph was devoted to the signing of a peace treaty in the Far East. The tiny headline cried out: "China, Indonesia Pals."

Tonight we saw our first footage of the military buildup in the Middle East. Steaming towards the Persian Gulf was the U.S. warship *Saratoga*. This juxtaposition between the cool waters of the Spa and the roiling sea beneath the battleship of the same name disturbed me.

World War II was no picnic for horsemen. Grandfather was left without the help of his sons, called to the Remount. Uncle John had been a young farm manager before the war. He and Grandfather foaled Elkridge in 1938, and the veterinarian rubbed whiskey and powdered mustard onto the colt to revive weak lungs. While Elkridge was jumping over fences and into the Hall of Fame in the nineteen-forties, Grandfather—his heart weakened by a stroke—was trailering his mares about the countryside to be mated to stallions such as Case Ace at Joe Roebling's Harbourton Stud in New Jersey. (From one such trip came Raise You, dam of Raise a Native, Allen's Prospect's *grandfather*.)

During the war, infields of race tracks were converted into barracks, and grandstands became defense plants. War bonds were sold with mutuel tickets, and casualty reports included horsemen from many families. I know it's selfish, but I think it might be just my luck to have two hot stallions just as war breaks out. I fear it will be quite some time before the *Racing Form* reports: "U.S., Iraq Pals."

Wednesday, August 15. "What are you going to do about Rollaids?" Peter asked me, bluntly, directly, parentally. "He is developing symptoms of a wobbler. He is much worse than when you left for vacation. He is losing action in his hind end."

Rollaids' elevator has never gone to the top floor. He has no utility as a Thoroughbred, a fact pointed out to me by the business-minded. Though still as a stone next to me as together we watch the other yearlings gallop in play, Rollaids is not a pet rock. He eats. He costs money. I have procrastinated in determining his fate. I have put a fedora on his head and laughed until I fell on the ground. But if Rollaids is a wobbler, he's as good as dead.

"Dr. Riddle just left," Peter said late this afternoon. "He thinks Rollaids may have an abscess in his front foot, causing him to move awkwardly behind. Still, he thinks you will have to put him down someday."

As I watched Ellen prepare the Ichthammol poultice to draw the heat out of Rollaids' foot, I was just glad someday wasn't today.

Thursday, August 16. Put the last Kentucky mare and foal on the van this afternoon, white flannel on halters, coats dappled even through the sun-bleached hair of midsummer. Our job completed, we headed to the creek. Sitting on a rock with my feet in the swirling water, I am transported by the steady white sound, punctuated occasionally by a distinct *galumph* as an extraordinary surge of water drops its weight off supporting rocks. A twig from a willow tree is snagged in the rapids. With no roots, green leaves bud tenderly down the length of the branch, a flute with fingers.

Four days of rain from last week have swollen the creek unusually high for this time of year. Everything is running higher than normal—emotions, morale, energy. I have seen Augusts when the stress of the recently completed breeding season poisoned the water, when hard work did not bring reward, only frustration. Years of breeding basement mares to basement stallions almost defeated us. Miserable farm wages did not cover living expenses. I have seen Augusts when the feeling has been one of tenacious clinging to a blind belief in a better day to come. Developers tempted us: Cut off your thumb. Save your hand.

We greet this August with open arms, strong hands to play.

Friday, August 17. On weekdays at Pimlico in good weather, trainers, owners, jocks' agents, and sundry professionals who make a living from the Turf all congregate in a section of outdoor grandstand boxes near the finish line. It is all very chummy. It is all very sporting.

No one knows how the two-year-olds will run. They have no form, only whispers of workouts. At the finish of today's race, one box cheered. Nine others sat quiet. You could feel the immediate polarization between the winner and the losers, but it soon passed, a wave receding back into the surf. Nine losing two-year-olds still have their maiden conditions. There is only one non-winner of two. There is no rancor among trainers when two-year-olds run. There is only hope, and disappointment, and hope again. Back to the barn, cool out the horse, begin again tomorrow.

Saturday, August 18. "I went to a sale of stallion seasons last fall," said Bob Manfuso, "and John Mayer and I began playing a game between ourselves of 'over and under.' We bet on every season, him winning, me winning, and the sale went by so fast. Afterwards, I explained to John: 'You see, that's the same feeling that makes a day at the races exciting—the rooting interest.'"

Today at Pimlico, the management kept the action coming fast and furious. At two o'clock, we drifted down to the paddock to see Valay Maid. She never turned a hair in the hot confines of the enclosed paddock. She turned heads, however, when she galloped in the Twixt Stakes by six lengths. Hank Allen and Maryland-bred Northern Wolf upset Maryland-bred champion Safely Kept in the inaugural Frank J. De Francis Memorial Sprint. Minutes later, the simulcast of the Travers Stakes had us glued to the tube, but Filouette's son Restless Con could not heist the million-dollar purse. The action was not over for the day until we dialed the Monmouth Hotline: "As they turn for home in the Ulysses S. Grant Stakes, it's *Groscar* taking a commanding lead."

This evening, we offered steamed crabs to the gods of the Maryland Million.

Sunday, August 19. At a restaurant named Michael's, near the sale pavilion, the Sunday night crowd ordered steamed shrimp with Old Bay seasoning. It wasn't because anyone was feeling lavish; but after a long day, you feel entitled to a decent meal.

Les Salzman took off his tie. His company Equivest is trying to provide an alternative forum for local horses to be sold, but he is fighting the market, the timing, the reluctance of breeders to put their best stock in anything other than Fasig-Tipton's select sale in September.

The dearth of good yearlings compels the two sales companies to compete hungrily. A few breeders simply cross-entered their yearlings: "If I don't get a big price here, I'll run her right back at Fasig-Tipton." I watched as a lady who had paid sixty-five thousand dollars for a no-guarantee season sold the resulting foal for sixty-five hundred dollars. She smiled bravely, jumped in her Mercedes, and roared off. At twelve-thirty in the morning, shrimp salad sandwiches arrived, as a feeling of camaraderie, of shared hardship, swept the bar, no one anxious to leave.

Monday, August 20. As if he'd been served with a literal gag order, The Raven appeared at the gate of the yearling field this morning, the noseband of his halter stuffed snugly into his mouth. He had rubbed the skin raw on the corners of his mouth. He looked as if he'd been tied to the railroad tracks. Andrew managed to slip the halter off, but when we attempted to put a new one on the colt, he careened away, head shy now after a night of choking on leather.

What to do? Hunger is the way to a horse's brain. We placed his halter inside the feed bucket, opened the halter up wide. The colt put his head into the bucket. We pretended we were not interested in catching him. He was absorbed in the pellets when we eased the halter over his nose, slid the headstall over his ears, gently buckled him up. The Raven picked his head up, surprised. This colt will knock you off your feet. We harnessed him today. He looks caged in his new halter, and I dream of the day his energy will be unleashed.

Tuesday, August 21. Peyton studied under Dr. Albert Gabel at Ohio State. Down the rows of hospital stalls, Dr. Gabel would pull open each door to reveal the morning's maladies—strained hocks, tapped joints, moon blindness—questioning the young vet students in Socratic method. Peyton respectfully mimicked Dr. Gabel this morning.

"So, you've seen estrus in pregnant mares," he began. "What could explain that?"

I answered that sometimes mares who are in foal display heat, but progesterone overrides the estrogen, and they stay in foal.

"And what would cause excessive estrogen production?" Dr. Jones asked.

I have not seen a pattern to it, I responded, but several years ago, we actually covered a ship-in mare who was already in foal from an earlier heat cycle. She was fussy, but not violent.

Professor Jones reflected. "What's inside that mare that might cause additional estrogen?"

Just as in school, I fumbled for an answer as a faint knot began in my stomach. The foal? I ventured.

"And why would a foal produce estrogen?"

Again, the knot. Interminable moments passed.

"Because it's a filly foal!" Dr. Jones finally concluded for me. "Studies indicate that a female conceptus produces more estrogens than its male counterpart—the reason for signs of estrus in pregnant mares. If a mare you think is pregnant is horsing another mare over the fence this summer, don't cry. Just make a note. See if she doesn't produce a filly next spring. Class dismissed."

Wednesday, August 22. Back in school again tonight. At childbirthing class, the nurse instructed us to rest on sleeping bags on the floor. She turned out the lights and played a tape of soothing music—maybe waves on the shore, maybe wind.

"Now, coaches," she addressed the men. "Help mom imagine a relaxing scene."

I whispered in Ellen's ear: "Think of sandy beaches at Del Mar, where Xray gallops. Think of pine trees at Saratoga, where Go for Wand sleeps. Think of the mornings in the mist, as Pavia canters through the fog." I dreamily recalled the custom of Southern women not to be pregnant in summer—to be "light in August." This month has been heavy for Ellen, but I can see the light in our lives, and on this farm.

Dreaming of these images, I felt a presence near my feet, the way you sense an alarm clock before it rings. "Coach, coach!" the nurse said. "You aren't supposed to go to sleep. You are supposed to help mom relax."

Thursday, August 23. Today belonged to Xray, second yesterday in the Balboa Stakes, an important prep for the rich Del Mar Futurity on September 12. Allen's Prospect now has his first graded stakes performer, and we mailed memos to shareholders to spread the good news.

When you stand a stallion, you are married to him—for better or for worse. I am still on the wrong side of for richer or for poorer, but it's definitely for better when two-year-olds run early.

Friday, August 24. Spent time yesterday tracking down trainers to make certain they knew of the pre-entry deadline for the Maryland Million. Carnivalay's Lucky Lady Lauren lives in Kentucky now, but her trainer was not at Churchill Downs. He was with the Arlington division.

"I'm leaning towards a stakes at Turfway that weekend," he said, "but I'll keep her eligible." Xray's trainer was not at Del Mar, but with the Saratoga division, where apparently there are no phones in the beautiful old barns. I left a message at the stable gate.

Dad came home with names hastily typed in the racing secretary's office. Allen's Prospect might also be represented by Calledon's Prospect, a West Virginia speedball who won his first race by nine at Charles Town, his second by ten at Delaware Park. Under Vince Moscarelli's West Virginia farm run some of the finest limestone deposits on the East Coast. Vince trains Calledon for California song-writer Burt Bacharach. Raindrops were falling on Calledon's head at Delaware. I thought it ironic that Mr. Bacharach bred a mare by Stop the Music, but that staying influence in Calledon's pedigree should mean he'll carry his speed. And of course Valay Maid is listed for the Oaks. Groscar, unfortunately, again wins the award for most-conspicuous ineligible offspring, and the sting of having a good runner drop through the cracks of eligibility still rankles me.

Saturday, August 25. The first task for new employees is to master historically esoteric names for a dozen different fields. Steve Earl quickly discovered the method in the madness, after I explained to him that there once was an apple orchard—russet apples, best after the first frost—in the Orchard Field, even though not a single tree remains; that the Ball Diamond Field no longer contains well-worn base paths or a backstop; and that the Bleacher Field is named for its proximity to the Ball Diamond. Steve now understands the heritage of field names. I heard him tell a rookie employee this spring:

"Turn that mare out in Roger's field."

"Who's Roger?" the rookie asked. Steve looked at me. I nodded.

"Well," Steve began patiently, "Roger was Ellen's horse, a great guy, a real character. He died last year. Strangest thing. I noticed him just getting up and getting down real fast. I got Ellen right away. Had to put him down that night. But he was turned out with the single mares behind the pond before he died. *That's* Roger's field."

Steve's delivery was perfect, just the right amount of whimsy to flavor the facts, a natural historian. I would name a field for him, but there are too many Steves around. It could lead to confusion.

Sunday, August 26. Driving a Subaru Brat we took in exchange for a board bill, we use the leisurely Sunday pace to examine the nineteen foals in Mike's field (eight are weaned); the six yearling colts (six and a half, counting Rollaids, who is baby-sitting a sick weanling); the six yearling fillies (the Amazon Queen is so regal); the eleven single mares in Roger's field (add two more for Coco with the hoof abscess and newcomer Rosie Rooz—by Run the Gantlet out of Allegedly, recalling the gal who claimed victory in the Boston Marathon after jumping in at the end of the race); the two weaned mothers (no grain, just hay, dry them up); and lastly, the three teasers.

Richard and Steve Earl do this chore twice a day. By the look of the horses, they are doing a splendid job.

Monday, August 27. It's close to a hundred degrees today, and not a leaf is stirring. The pond brews algae on days such as this. Horses sweat in their stalls, sweat in their fields. We double-check availability of salt and mineral blocks in pastures, and mentally prepare to endure a heat wave: The five-day forecast promises no relief.

On the walk home, a brown snake with a black stripe straddles the lane, and I do not see him until I am about to step on him. "*Holy Cow,*" I scream, and jump into the air. I hate snakes. Ellen, however, does not, and routinely untangles black snakes caught in mesh netting that keeps birds off the blueberry bushes.

"What kind of snake was it?" she asks.

"I think it was a copperhead." (Any snake that isn't black or green is a copperhead to me.)

"What did you do with him?"

"I didn't do anything *with* him. I jumped higher than Elkridge and ran home."

"Who's Elkridge?"

"Never mind. Hot today, wasn't it?"

Tuesday, August 28. Very overcast at 8 A.M., but soon as the clouds burn off, it will be another day of wilting temperatures.

"Nothing like the heat in the Middle East," Peter declares. "I know what our soldiers are going through. You may not believe this, but I saw it snow in Egypt once."

"If you say so, Peter," I answer, restraining myself from mentioning elephants on the Hudson. In the early nineteen-fifties, Peter served two years in the British Army when England and France controlled the Suez Canal.

"The Egyptians didn't know what snow was," Peter continued, "but we all saw it."

Peter worked in General Headquarters, where on workdays when it did not snow, they started at dawn and clocked out at noon.

"No one can work through the heat of midday over there."

The GHQ staff then would cruise down to the Bitter Lakes, careful not to step on sharp fossils in the lake bed as they pushed off in sailboats. The Korean War put an end to afternoons at the lake. Warships steamed through the Suez en route to the Far East. Peter's tour was up, fortunately, and he sailed back to Great Britain, to begin his apprenticeship as a trainer with Paddy Prendergast.

Enlisted here five years ago, Peter performs with fastidious brilliance the task of managing health records for a revolving herd of ship-ins and year-round boarders. I have watched him in the heat of the breeding season coolly maintain his policy of administering complete vaccinations to any and all incoming horses before they leave isolation areas. He accepts no excuses for absent paperwork. When he tells me something, I listen. If he says it snowed in Egypt, it snowed in Egypt.

Wednesday, August 29. The burly bay colt will not do business as "The Raven," after all. I had used the *Registered Names* book current through last year; The Raven in the meantime became one of six hundred and fifty thousand active or protected names in the data base. Roger L. Shook, director of Registration Services, believes a happy customer exists for every unhappy one, that the owner of a filly named The Raven would be pretty upset if Roger had also granted The Raven to us.

"The quickest way to check availability of names," Mr. Shook suggested, "is to use an IBM-compatible personal computer with a modem, use our Remote Name Claiming Software to submit names by telephone. You'll have an answer within twenty-four hours."

Since The Jockey Club moved to Lexington under new guidance, the entire institution has taken on the personality of a top Bluegrass horse farm: Efficient, knows clients' names, same gal on Tuesday you spoke with on Monday. It's no longer any fun to bash The Jockey Club.

This time, I sent in the name "Edgar Allen Pro" for the colt. I don't have a modem, and I haven't sprung twenty dollars (add three dollars more for postage and handling) for a current *Registered Names* book. Maybe like Edgar Allen himself, I like the mystery of waiting for an answer to come back from the dark.

Thursday, August 30. Nice cool day. A rush of wind blew the humidity out to sea and tipped its hat with a twenty-minute downpour to make the world great to live in again. Mares trotted in play rather than in escape from horseflies. The gray Corridor Key, often seen doing handstands across his paddock to dislodge vicious flies, luxuriated in the relief from the heat. It is Mr. Summer, not the horses, on the run today.

Friday, August 31. Leaves wash down from walnut trees in the cool breeze. Shadows of two herons form dark chevrons across the ground as I think: They can't be this low! But they are—not fifteen feet overhead, flying dinosaurs in the rich, blue plumage of fall, not the steel-gray feathers of spring.

In the Office Field, fifty Canada geese, joined in stark whiteness by two snow geese, presage autumn, the flock feeding hungrily on seedheads of grass as two sentries post the lookout—necks extended, heads turning like periscopes scanning the sea.

The tan Labrador, Ozzie, escorting Richard and Steve on their rounds, cannot believe he sees dozens of birds. He dashes into their midst, and the flock is airborne in chaos—five to the left, fifteen banking to the right, perhaps twenty-five straight overhead. The sound of their honking echoes against the hillside like music from an amphitheater, a nervous orchestra tuning up, woodwinds running wild.

Mom places an album on my desk, Arnold Bax: Tone Poems.

"Read the liner notes about the pieces. One is named *Summer Music.*"

The composer struck a chord with me. He wrote: "A lovely day is full of the enchantment of sun and sea and that strange feeling that there is only a thin veil between this world and some faeryland where no-one grows old and no beauty can fade."

Not since I was a young boy, in the era of whitewashed fences, have I enjoyed a summer on the farm so much. The horses are running on all fronts, and the coming of sweater weather means the debuts of whispered two-year-olds. All the while, I am aware that the horse business will treat me harshly if I step through the thin veil between this world and the fairyland that Mother Nature, and my own imagination, create on these one hundred acres every day. So we'll cross our t's and dot our i's and carry on the business. Farewell to one fine season, hello to the promise of another.

SO MOVED

Saturday, September 1. "How the hayburners doin'?" asked George Cogan, steeped in racing lore by his father the New York lawyer—who pauses every morning at OTB shops, and who retires to his study every evening for the 7 P.M. replays, armed with mutuel tickets and the *Daily Racing Form*, sequestered in solitary splendor after announcing to his family: "Riders Up!"

George's annual visits always include a trip to the race track. He appears in a winner's circle photo like Woody Allen's character *Zelig*: Big glasses, shock of black hair, grinning above the last row of excited racegoers crammed into the photo. "And look, there's George!"

Ellen rushed from the house as George and I pulled in the lane this afternoon: "Groscar just won! Marty called before it was even official. He was in the secretary's office at Monmouth." I mentioned that Groscar apparently just won the Choice Handicap at wonderful Monmouth. George punched the air in excitement and launched the Labor Day weekend with his father's famous battle cry: "Riders Up!"

Sunday, September 2. This morning at feeding time, a pilot from the neighborhood airport in Fallston circled overhead, buzzing loudly in the still, summer air. I suppose he imagined himself as Denys Finch-Hatton in *Out of Africa*, hoping to frighten the wildlife into a spectacular run. Yet as I stood in the field of mares and foals in growing annoyance at the pilot, the wildebeests around me paid no heed.

The horses do not stir for even the lumbering troop transport planes that bank like flying Sequoia trees on evening reconnaissance flights from nearby Aberdeen Proving Grounds. The Shock Trauma Helicopter, slicing the air into ribbons of sound miles ahead of its path, does not disturb them, either. They graze placidly while sirens scream past on Route 1 en route to the Fallston Hospital. I asked Tesio for an explanation.

"An animal's senses serve the requirements of preserving the life of the individual, and thus the species. I witnessed during the bombings of Milan that racehorses at San Siro were never particularly upset. They apparently concluded from individual experience that the uproar involved no bodily harm, no cause for worry."

A modest man, humbled by initial failures, Tesio is a superb companion for student horsemen. "I thought I knew," he prefaced his book, "but the truth was—I had not yet learned to reflect—to reflect, that is, on the whys and the wherefores."

Monday, September 3. In one last hurrah of adventure before the baby arrives, Ellen and I climbed into the orange raft and ran the rapids of Winters Run. It is not the Zambesi; we often grind to a halt on the smooth rocks of the river bottom. Still, the serenity of being alone on the silver water is matched perhaps only by the sight of mares silhouetted against an orange sky, a scene truly out of Africa.

As we climbed in the pickup for the ride back to the farm, I popped a Lyle Lovett tape into the deck: "Now, if I had a pony, I'd go out on the ocean, I'd ride up on my pony on my boat." Floating on top of the world, we closed out the Labor Day weekend.

Tuesday, September 4. "Here's some good news," Cricket Goodall called from the Maryland Million offices today. "Lucky Lady Lauren is entered, after all. The preentry form was delivered to the wrong address. With the holiday, it just arrived at Pimlico this morning."

Seems like old times. Lucky Lady Lauren was foaled here, the product of an immaculate conception inside the tough old war-horse War Exchange. I remember the evening I arrived home from town after Dr. Bowman confirmed pregnancy by ultrasound, Mike Price cheering like a Rebel who had just routed the Yanks: "Hot damn, they said she wouldn't get in foal, but we got her, didn't we! Yessir, Carnivalay got himself a stakes-winning mare in foal!"

Her owner so enjoyed the videotape Ellen made of the foaling, was so certain of success, that he named the filly for his daughter—never an advisable decision. Carnivalay stood for twenty-five hundred dollars that season; last spring, Lucky Lady Lauren was sold for exactly one hundred times that amount. She will ensure a fair pace for Valay Maid. I picture a Country Life-sired exacta, a perfect one-two punch.

Wednesday, September 5. Met with attorney Dick Abrams in Wilmington today, to discuss racing partnerships. On his wall are win photos from several ventures. On his coffee table are horse books. "Take them home with you. I've read them," he offered.

Not since law school have I carried as many books under my arm. *Racing in Art*, by John Fairley, laments what he perceives to be the back burner status of sporting art in the museums in Louisville and Saratoga, and the sad confinement of the Woodward collection (the Baltimore Museum of Art cordons off this gallery so frequently I hesitate to supply directions to inquiring visitors). The book makes amends, from Stubbs to Skeaping.

Racing Days—black and white photos by Henry Horenstein, essays by Brendan Boyd—is a duotone world where the subculture of this sport gets a call. My hair will smell like cigar smoke when I've finished this book, but I can't put it down.

The Body Language of Horses, well, Tesio told it better fifty years ago. I read until I came across a "frisky foal," then went on to *Thoroughbred, A Celebration of the Breed*, where John Denny Ashley said in pictures what takes me a thousand words. I wanted to ask Dick if he'd ever read *This Was Racing*, by Joe Palmer, edited by Red Smith. ("Yes, I was implicated in this," Smith wrote, "but only as the driver of the getaway car.") No Turf library is complete without it. Dick's books enforced my belief that there is not much middle ground where horses are concerned: You either couldn't care less about the hayburners, or you love them so much you buy four books at a time.

Thursday, September 6. "Lucky Lady Lauren? Why, she on her way to Pimlico!"

The groom who answered the phone at Churchill Downs this morning might as well have been Will Harbut, who declared Man o' War "de mostest hoss!" or Clem Brooks, the famed Spendthrift Farm stud groom who recited by rote the bronze plaques flanking Nashua's stall. Tonight, I reached to the bookshelves for John Taintor Foote's 1915 story *The Look of Eagles*, originally a magazine piece, about "Ole Man Sanford" in Kentucky, about a little black colt who outran the bay Postman. I read this passage:

"Suddenly he (Blister, the groom) disappeared through the doorway and there came to me the words of the song, which were these:

> *Bay colt wuck in fo'ty-eight,*
> *Goin' to de races goin' to de races;*
> *Bay colt wuck in fo'ty eight,*
> *Goin' to de races now.''*

I report simply that the sound of the groom's voice this morning reminded me of another era, and that Lucky Lady Lauren is en route to Pimlico.

Friday, September 7. "You have sand in your clock," Carlos Garcia teased his assistant, timing Valay Maid's three-furlong blowout for Sunday's race. The *Racing Form* will report :34 3/5, but will not impart the ease with which this leggy filly swallows up ground. She will be the odds-on favorite in the Maryland Oaks. Alice and I visited the barns early this morning. Lucky Lady Lauren has arrived from Churchill. Xray is due in later this morning from Belmont. In four previous Maryland Millions, the best showing by the offspring of a Country Life stallion is a lonely second-place finish. It's about time for us to light up the board.

Saturday, September 8. On the backroads near here is a long, green valley the first settlers cleverly named the Long Green Valley. Such fertile land is precious, and farmers concede only a few unplanted feet for the road shoulder before corn begins its tasseled ascent. Drivers dropping down from the plateau on Hydes Road find themselves in a green-walled bobsled run of corn, awash in the rustle of air between stalks, in the smell of corn ripening in topsoil so soft you can push your fingers into it.

The road affords city-bound travelers the benchmarks of changing seasons: Foals in spring, hay in summer, corn in fall. Today the corn is as tall as I've ever seen it, but I know the cornfields soon will be harvested, the land open again to the horizon. For these farmers, and for myself, it's time to see how good is this year's crop.

Sunday, September 9. Yellow "Post-it" stickers are affixed to passes and parking stickers on the big desk in the office: Getting twenty-four folks to the races is quite a production. Steve Earl is off early for the airport in the Cadillac on loan from Jeff Vaughn for the day, to pick up the Baylises from Rhode Island, who have Halo Flyer in the starter handicap. The "Will Call" window is alerted. Mom is coming over later, after her radio show ends at noon, with Ellen, who is pacing herself, and Dad, who has the parking sticker. We tease Uncle John as he climbs into Kay's van: "We like to preload the elders," we tell him as he giggles into the back seat. Downey Gray from Louisville and Tommy and Angie Brannock from Charlottesville, and who am I missing? I know I counted twenty-four tickets. It's eleven-thirty. Let's roll.

The rest is a blur. Hal Clagett's Ameri Allen is rubber-legged, but on the lead, as Ritchie Trail storms down the stretch. The two daughters of Allen's Prospect finish a scant head apart to take first and second in the Maryland Lassie, and we fly out of the box seats and dash to the winner's circle, Uncle John gently reminding Ellen to take her time down the steps, please, please, *be careful.*

Valay Maid is the three-to-five favorite in the Oaks, but is chasing some speed horse, while Lucky Lady Lauren bides her time in third place. Turning for home, LLL squeezes Valay Maid, who props a bit, just like in the Monmouth Oaks. She hates the rail, but the announcer screams: "The two daughters of Carnivalay are head-and-head." But they have set it up for McKilts, who overtakes them both at the wire to win by a head.

No time for remorse. Back to the paddock, to see Xray, to see Calledons Prospect. Xray rears skyward as Sally Lundy puts the saddle on him. If he flips over in this paddock, he's dead: Concrete floor, lose a couple of horses in here every year. Calledon, meanwhile, is acting like a kid wearing 3-D glasses: "Let me outta here!" he seems to be thinking.

Out of the gate, Calledon pitches forward, jockey clinging, then regains his feet, rushes up, is blocked, finds room next to Xray, as the announcer—still reeling from the one-two finish in the filly race—tries to be prophetic: "And the two sons of Allen's Prospect are on the move." He quickly retracts his enthusiasm: "Xray is being asked to run early, and he is not responding."

I switch my binoculars to Calledon. The cavalry charge is too much for him. He drops from third. Where is Xray?

"Look!" Ellen yells. "On the rail, the red cap!" Allen Paulson's red, white, and blue silks are clear now, Craig Perret gunning Xray through a hole no wider than a man. The colt muscles through to win, ears pricked, ready for more. In the span of an hour, Allen's Prospect's career is assured. A freshman stallion has been launched.

Now the whole entourage—deputizing for the absent owner—flies down familiar steps, crosses over the track surface to the infield, to the replica of the old Pimlico clubhouse porch, where Jim McKay is patient with the presentation. Xray buys us some time when he unceremoniously fertilizes the turf in the winner's circle. McKay under his breath asks Dad: "Joe, you working the broom today?" We all hear him, laugh, the shutter clicks, Dad and Uncle John climb up onto the porch. They both glow.

"Ten years," I think to myself. Ten years ago they kissed the booze good-bye. Today, serenity prayers are answered. Maybe it took ten years to climb that porch, but they did it. There is hope. There is always hope. No finer nor truer example exists than the Old Boys up on the porch. Don't anybody ever doubt it: People *can* stop drinking. There must always be hope.

Then Cappy Jackson gathers us all together to take a family portrait of sorts for the *Maryland Horse*, and I am so excited I forget to smile at our great good fortune. I am stunned.

Monday, September 10. Went to bed at eleven, slept until three. Woke up. Smiled. Replayed races in my head. Recalled the horses in the paddock, Ritchie Trail in her blinkers for the first time, Ameri Allen tall, gaunt, drawn, ready for the race of her life, tight as whipcord—only a week since her start at Philadelphia from which seven hundred in earnings lifted her into the last slot in the Lassie field. Their sentimental Lassie exacta paid two hundred thirty-nine dollars, and more than a few farm hands had it. Xray, my little pro, was patient, but ready to blow. Calledon was a wild-eyed bundle of nerves. The track handicapper had cryptically noted of Calledon's Charles Town romps: "Laughed at cheaper." He wasn't laughing in the noisy Pimlico paddock.

I couldn't sleep. Yesterday wouldn't wear off. Andy Beyer of the Washington *Post* had stopped Dad, asked what it meant economically to a farm to have such a day. Dad fudged, reminded Andy that it should come as no surprise: The studs are well-bred, by Northern Dancer, by Mr. Prospector. I stood to the side, rapidly adding up our own yearlings, weanlings, and mares in foal. In Roger's Field, in the Office Field, in the Orchard Field, in the Baseball Diamond and Bleacher Fields, an across-the-board appreciation in livestock can be recorded. That's what it means, Andy, I wanted to say, but it was Dad's moment. The promise displayed by the freshman sire will make it easier to feed his get through the winter, to get them to the races, to take the chance that *someday*, we might own the good race-horse, not merely stand the sire.

At dawn, I walked the long gravel driveway to the barns, where I hopped in the pickup and drove to the Bowling Alley, to eat a hearty breakfast, savoring the splendid performance of our babies. In the office at 8 A.M., to answer the phone. Henry Rathbun, a veteran shareholder, called: "I bet you didn't sleep a wink last night, did you?"

The Old Boys know what a young man is still learning.

Tuesday, September 11. In a fling of irresponsibility before fatherhood claims my full attention, I went with Downey to a Grateful Dead concert tonight. I think these musicians would have made good horse farmers. They persevere. Instead of stallions, booked full are their concerts. For twenty-five years, they've played homebred music, and have accumulated a faithful clientele to weather ups and downs of the trade.

Distributed free in the village of fans surrounding the concert hall was an eight-page newspaper called *NWR*. Inside was the story of Crazy Horse, from the book *Bury My Heart at Wounded Knee*, by Dee Brown. Anyone in the horse business could appreciate Crazy Horse. We are the Indians, the economy is the white man:

"In the Black Hills seeking vision, he asked for secret powers so he could lead the Oglalas to victory if the white man ever came again to make war. Crazy Horse knew the world the white man lived in was only a shadow. To get into the real world, he had to dream, and when he was in the real world, everything seemed to float or dance. In this real world his horse danced as if it were wild or crazy, and this was why he called himself Crazy Horse. He had learned that if he dreamed himself into the real world before going into a fight, he could endure anything."

Tapping my feet and clapping to the music, I thought about the farm. Twelve years ago, when the kids were in college and the Old Boys had misplaced hope, we were dead. The question over the land was not whether it could raise horses, but whether it would percolate for development. Well, gratefully, we are back from the dead, all of us a little like Crazy Horse, dreaming before going into the fight, enduring. We clapped until midnight and wished the music would never end.

Wednesday, September 12. How do you set a stud fee on a freshman stallion who just jumped from twelfth to third on the national lists? (Uncle John once told me that for a fleeting moment in 1961, right after Carry Back won the Preakness, our stallion Saggy was the leading sire in the country.) Michael, the business graduate brother, knows the value of the bird in the hand. I see two in the bush, for there are one hundred days left before the crop turns three. Three key months. Time for Calledon to fly at the West Virginia Breeders Classic on September 29th; for New Jersey-bred Promising Lad to fulfill promise in the rich New Jersey Futurity; for shin-bucked Pavia to debut; for convalescing Proangle to return; for Xray to carry another day. Thirty two-year-olds haven't even started.

Poker-face it, Michael. Play until the hand is done. In for a penny, in for a pound.

Thursday, September 13. Last week, Steve Earl kicked the dust near a gate and unearthed a buried coin. A numismatics expert declared it from the Colonial days, and now Steve has a five-dollar quarter. Nothing that happens in the paddocks surprises me. This morning, Turn to Money missed the coin, but found a prominent knot in a locust post and hooked her fanny on it. Peyton arrived, looped a few big sutures around a plastic drain in the wound, and double-checked her tetanus status. I wanted to mention the old saying, "It's a long way from her heart," but I suppose a broodmare's heart is closer to her reproductive channel than is a racehorse's, so I kept my mouth shut.

Friday, September 14. I completely forgot a local radio talk show was due to call at six-fifteen this evening.

"Quite a Maryland Million Day for you folks!" the talk show host began before I realized I was on the radio.

"That's right, Dave. We're all very excited."

"It's Doug, but that's okay. Everyone expects the stud fee will go up next year. Whaddya' think? Any ideas on that for our listening audience?"

I paused. In my mind's eye, I thought I knew his audience: Grizzled grand-stand veterans, sitting on the famous marble steps of Baltimore neighborhoods, a beer in one hand, tomorrow's *Form* in the other, the radio in the window, waiting for the Orioles game to start.

"Well, Da, umm, Doug, I think I'll just sit tight for now. Hate to paint myself in a corner."

"I understand. Hey, listen, thanks for being by the phone. In other racing news, Horse of the Year Sunday Silence has been sold to Japan."

I hung the phone up, and reached for tomorrow's *Form*. Mom called. "You sounded just fine on the radio." So much for knowing the audience.

Saturday, September 15. Richard chain harrowed and mowed the dormant Route 1 field, and today we led over five mares with foals still at side, obediently followed by thirteen weaned babies. One of the weanlings is showing early signs of epiphysitis, and I am annoyed at myself for weaning him when the ground was hard. I did him no favor, and should have waited for rains to soften the turf before subjecting him to a stressful day of wandering in search of mom. I'm reminded of an interview in which bloodstock expert Russell Jones outlined various strategies for success. He concluded glibly that: "Unfortunately, nothing works all the time. Half of all your decisions will probably be wrong."

This thought did not soothe the aching ankles of the weanling, but it made me feel better.

Sunday, September 16. In the basement of the big house is a small room cluttered with Grandfather's old files. The feeling is the same as sitting in remote stacks of a huge library, pulling dusty books from shelves. Today, I tripped back in time to the last days of the Jazz Age, to the last days of the great breeder John E. Madden. In that era, before deals were made over the phone, files contained the complete understandings.

Western Union Telegram: 1929 Mar 28 AM 10:55 Gave You Very Low Price On Those Two Mares Stop You Were To Get Ten Percent If Sale Made Within Reasonable Length Of Time Stop I Guaranteed Them In Foal And If Not Will Return The Money Stop Please Make Decision FitzGerald Here And Likes Those

Mares And Hancock Is Coming Tomorrow=John E. Madden.

Grandfather bought the mares for clients, then added a postscript to his letter of acceptance.

"I want to thank you for your kind offer of shipping a pony for my children. I do not like to criticize or be too particular about a gift horse, but as my children are growing fast, I would prefer a Welsh pony or a small horse, about 14.2, and remember he must be very quiet and safe. With kindest regards."

The file ends abruptly, on April 8, 1930, with a letter from Grandfather to Miss Daysie Proctor, Estate of John E. Madden.

> Dear Miss Proctor,
>
> I have received your letter of March 28th, enclosing the check for $500 being refund on the mare Crystal River, which proved barren, and which was the understanding with the late John E. Madden when this mare was purchased. Thanking you,
> Sincerely yours,
> Adolphe Pons

Grandfather died before I was born, but he comes alive when I study his methods and learn from his letters, in those quiet hours alone in the stacks of his library.

Monday, September 17. Peter Fuller won a Kentucky Derby, won the New York Filly Triple Crown with a homebred his daughter rode, knows something about the horse world.

"All the hoopla happens, then it stops as quick as it started, and you wonder where everybody went," he cautioned me. The horse business giveth, and it taketh away. Nobody on this farm is getting a big head, I told Mr. Fuller. Too much history. Just happy for the moment. Roof will fall in if I turn my head.

"No, don't be afraid of it," he said, and suddenly, I wasn't.

Tuesday, September 18. How could the fellows in the Remount *not* have been good horsemen? They lived the game; to wit, from the yellowed pages of Uncle John's Army handbook *Horsemanship and Horsemastership* tonight fell a folded sheet of paper marked VIRGINIA LAUNDRY, Chester Street, Front Royal, Va., Phone 126. Date 4/4/1941.

Scrawled in pencil on the front of the laundry ticket was a homemade prescription for Red Iodide Mercury Blister, for Reducine Blister, and directions for mixing the blister paint. On the back, ingredients for colic medicine were itemized. Without veterinarians or tack shops, most of today's young horsemen, including myself, would be at a pronounced disadvantage. The Old Boys, on the other hand, would manage just fine.

Wednesday, September 19. Alice, Marva, and Ellen are the motivating forces behind the beautiful gardens on the farm this fall. Zinnias, gourds, sunflowers, tomatoes, herbs, cucumbers—all variety of flower and vegetable have burst forth in full bloom after the well-spaced rains of summer. Splashes of color highlight stone walls, illuminate farm signs, border the stallion show ring. No visitor to the farm has failed to admire the bountiful flora. Kipling understood that there is more than meets the eye on this score:

> Our country is a garden,
> and such gardens are not made
> By singing: "Oh, how beautiful!"
> and sitting in the shade.

Thursday, September 20. Steam rises from the pond on sudden cool mornings, and the fish must feel safe in the mist, as they rise to the surface, daring the blue heron to pluck them from the waters. At the slightest motion, however, the fish turn tail and dive, breaking the water in audible waves across the glass surface. Clapped hands set off a chain reaction across the entire breadth of the pond. The heron is not amused at my behavior.

This quick cold snap has stirred up all the animals. Dogs run in circles past us to the stud barn, where stallions stand on hind legs and challenge handlers. In the paddock, Corridor Key pulls his head away before the chain is released. Steve's arm stretches, but he does not let go: "You stand! You stand!" The stud obeys, but only until the chain slips over his nose, then he's off to the races, hindquarters like pistons, frog-jumping in unison. Like a boat, though, he soon planes off, gallops along the surface of the dew-laden paddock. We watch him until he is played out.

No time to dawdle. Yearlings to break, fields to mow. Feeders to feed, seeds to sow.

Friday, September 21. "I did some research on that stallion," said Tom Roha today, curious about examining yearlings at Sunday's upcoming sale. "He injured his stifle in training. Rather than operate, they retired him. Would a stifle problem be passed on to his foals?"

Dr. Riddle did not think so: "The stifle joint is a hinge, really—similar to the human knee. Located at the forward part of the hind leg, it extends when the hind leg propels the horse forward. If aggravated, injection with a corticosteroid could reduce inflammation. If the stifle locks up, perhaps we'd cut the ligament which holds the patella—or kneecap—in place. This is a fairly common surgery and does not impair the horse's athletic ability. I would not expect stifle problems to be passed on, except it might indicate poor conformation behind, which of course is heritable."

Armed with such knowledge, we will look closely at the hind legs before we buy a yearling by this sire for Tom.

Saturday, September 22. Riding in a car bothers her back, so Ellen and I did not drive to Philadelphia Park to see Valay Maid run in the Cotillion Stakes, which holds her Breeders' Cup fate. The simulcast to Pimlico would suffice. En route, we stopped in the Long Green Valley, where two hundred Canada geese rose out of the cornfields in a spectacle straight out of *National Geographic*. The sound of their wings cut the air like sheets blowing on a clothesline. (I keep notions of omens to myself, but felt the beauty of the geese to be a good omen.)

We wagered and waited. The screen showed the horses coming out onto the track, then the monitors went dark and a message appeared: "Signal from some areas *Blacked Out*." Fans grumbled, windows closed. Over the discordant din, a tiny voice came over a ceiling speaker. In the darkness we huddled near the voice, like in *Radio Days*, the family on the edge of their seat waiting for news of the war. "Dance Colony is not firing. Jefforee is beaten. It's Valay Maid making her move. She's drawing clear. She's lengthening her lead. It's Valay Maid—by five!"

We erupted in joy. Next stop: Belmont Park.

Sunday, September 23. By the time the third of three Allen's Prospect yearlings reached the sales ring tonight, the auctioneering crew knew by heart the announcement slips I had submitted, updating the sire statistics through the fabulous Maryland Million. Terence Collier and Reiley McDonald read the slips with excitement that spilled over to the prospective buyers, concluding with the zinger that Xray is being pointed for the Champagne Stakes on October 6th. Our clients have so much more invested in these yearlings than the sales average of eleven thousand dollars, and they deserve consideration during the two minutes required to sell a horse which took two years—and twenty-five thousand dollars—to produce.

Monday, September 24. Horses named for people seldom run well, unless of course you are John Nerud and your surgeon is Dr. Fager, and your favorite New York *Times* racing columnist is Joe Nichols, nee Fappiano. For a spell during the nineteen-sixties, Dad and Uncle John named fillies for daughters: Nebulous Norah, Fair Carol. They stopped short of naming colts for their sons, although the unsound sprinter Prince of Space could arguably have been any one of four boys.

Our guest at lunch today had been Michael's eight-week-old daughter Elizabeth, big-eyed and smiling, a little person now. Tonight, we bought a yearling, just to protect the stud. Maybe the jinx of naming horses for people doesn't apply if someone else named her. Only time will tell if Liz's Prospect, a May foal with some growing to do, overcomes historical precedent.

Tuesday, September 25. It's not baby Elizabeth crying this morning; it's the *horse* who has tears rolling down her cheeks. What's the problem here?

"It's conjunctivitis, of unknown origin, in both eyes, aggravated by grooms at the sale wiping her face dry to show her," Dr. Riddle opined. "It is *not* the flu, as one might imagine. Any number of bacterial or viral agents can cause inflammation to the membrane that lines the eye, causing discharge. Perhaps some abrasive object—a seed hull, a piece of hay—or irritation by flies is the culprit. However, it is generally not contagious. Treat her with antibiotic ointment a few times a day and it should clear right up."

Poor Liz is getting a little head shy from having her eyes fooled with, but at least she can see us coming. We will be aggressive until these eyes clear up. Madden said: No Foot, No Horse. He might have added: No Eyes, No Horse, either.

Wednesday, September 26. "Once more unto the breach, dear friends, once more..."

Symptomatic of my relief today is this refrain from Shakespeare's *Henry V*, playing over and over in my mind all day. Thank goodness, we are not "breeched," with an *"e,"* as Ellen's doctor feared on Monday. Yesterday afternoon, we packed our bags, tidied our affairs, phoned the in-laws, and headed for the hospital for an ultrasound exam to confirm the doctor's suspicion. If this kid is coming feet first, the doctors might be compelled to perform the operation named for another of Shakespeare's subjects, Julius Caesar.

"Oh, yes, I drive by the farm every day on my way to work," the ultrasound technician pleasantly chatted. "I never seem to have a camera when the sun is just right on the horses' backs, or when the foals are up near the fence."

She asked us whether we used ultrasound on our mares as she slapped keys to photograph the image on her screen.

"There is nothing breeched about this baby," she finally pronounced, as the color came back to Ellen's face. "He's in the vertex position, coming headfirst. I say *he* generically. I can't tell if it's a boy or girl. You're due October 1st?"

"Yes," Ellen said.

They say babies can discern words even in the womb. I will not be quoting Shakespeare while the moon is on the rise this week.

Thursday, September 27. A few years after the track at Bel Air closed—mortally wounded by conclusions of a comprehensive racing report submitted in 1960—I rode my bike to the abandoned site, climbed up onto the inside rail, and walked the entire half-mile circumference of the track on the six-inch plank of wood. At the end of the grandstand, I passed a sign that read: "Colored Women's Toilet."

I found a copy of the damning legislative report while rummaging through old files in the basement library. It noted:

"The track in Harford County has facilities in such condition, inadequate and unsanitary, that hundreds of thousands of dollars would be required to put the premises in suitable condition."

I have bemoaned the loss of this nearby race track, often wondering how many fans racing might have developed had such fairs not passed away. Yet tonight, reflecting on what the track might *really* have been like for some people, I am not sure my nostalgic sentiments are founded on hard rock.

Friday, September 28. "But the catalogue says in boldface: Fully Nominated to Maryland Million/Registered Maryland-bred!"

"The catalogue is not controlling. The announcements as she entered the ring are controlling. We announced she was only *provisionally* nominated to the Maryland Million, and that she was bred in Maryland. We did not say she was *registered* as a Maryland-bred."

"I understand the announcements are the final word. But why did you put that information in the catalogue if the fees hadn't been paid? I was busy thinking about bidding, which is what a buyer is supposed to do. It appealed to me that these two items had already been paid. I relied on the catalogue. Her consignor missed registering her as a Maryland-bred by May 31st of her yearling year. Now the fee is not fifty dollars, but two hundred dollars. An additional one hundred and fifty dollars completes the Million nomination. I'm out three hundred and fifty dollars.

"Lordy, what if I hadn't double-checked? I wouldn't know this filly was ineligible—for life—for the Million until she broke her maiden by ten lengths and I went to enter her. How many other folks who bought yearlings at this sale think the catalogue page was correct? I think you should write a letter to every buyer telling him to double-check the accuracy of the Maryland Million and Maryland-bred information on that catalogue page. Cover your flank. The Little Bold Johns and Safely Kepts of the future might have just walked through your ring."

"I understand your position. I'll talk it over here in the office."

Saturday, September 29. At nine o'clock this evening, moments before Slim Pickens was set to kick open the bomb bay doors in the movie *Dr. Strangelove*, I paused the VCR tape. I wanted to be ready to fetch my own cowboy hat and run around the room yelling "Yee Hah! Yee Hah!" in anticipation of Dad's phone call from West By God Virginia. By now, Calledons Prospect should have rocketed around the bull ring tonight in the two-year-old stakes, part of the West Virginia Breeders Classic extravaganza. The phone did not ring.

"No news is *not* good news," I told Ellen, as I hit the Play button, and the bomb bay doors opened with Slim Pickens mounted atop the hydrogen bomb, and he went bronc riding off into space, waving his hat and hollering "Yee Hah!"

Darn, I was so hoping to do my Slim Pickens imitation tonight for Ellen.

Sunday, September 30. Precipitated perhaps by my having mowed the lawn at a *single* request, Ellen entered labor this evening. As coach, I timed her contractions, following the childbirth class outline for this stage. The instructions told me: "Coach, get the car ready. EAT." I did both, twice.

There being no business so paramount as the matter at hand, I made a motion to adjourn September. Ellen seconded. So moved.

GIVEN AND TAKEN

Monday, October 1. Thanks to Damon Runyon, we have an account of the birth:

"I consider this a most embarrassing situation. I'm chewing a cigar in the waiting room while Hot Horse Herbie takes a few stitches in the kisser, and I see this guy I know from Pimlico. He and his doll pull up in a hurry, and I can tell from the look on her face—lest I remind you I am something of an enthusiast when it comes to dolls—that she is definitely ready to deliver a screaming infant.

"Now I am not such a guy as never thinks of anything else except betting on horses, but I figure even-money is not a bad price that this guy sees maybe a thousand horses bein' born. He is not such a guy I would figure as will ever get high blood pressure over a doll in labor. Irregardless, I follow them to the delivery room and am most helpful, 'cause I do not want him sored up at me and give me no more horses to bet.

"These two are breathing like a Belmont longshot, and I see it coming. The guy's gonna go white-eyed and roll out any second now. In comes this doctor, who come to find out later is a vet in Texas before he comes up here to the land of pleasant living. He sticks a needle the size of a stiletto in the doll's back, while he's calling the guy 'Coach.' The vet says to her, 'You should be feeling some tingling.' I look at Coach, and he's leg-locked from holding her still so as she don't move, but his eyes go glassy, and before you can blink, the vet is catching Coach and dragging him over to the broom closet out of the way. The nurse takes the guy's cheaters, so now he's blind. She drops smelling salts in his lap, and generally wishes him a most speedy recovery.

"Meanwhile, the doll is feeling better about this predicament she's in. Before you know it, out comes a baby boy. The nurses are making book on his weight. It's like fight night. He weighs in at eight pounds, nine ounces. Now, I can put two and two together as well as anybody in this town, and judging from the lungs on the kid, Coach and his doll are in for a few changes.

"I find Herbie in the hallway, and I start in, but then figure, nah, he will not believe me if I tell him. Truth is, I consider it a most embarrassing situation, the way Coach went white-eyed and all, right in the middle of all 'dose dames."

Tuesday, October 2. In *Lancaster Farming* last week, Ellen found a classified ad requesting a quiet horse for a learning-disabled daughter. In response, Jim and Charlotte arrived this morning to inspect Rollaids. The old saw about not looking a gift horse in the mouth does not apply to my yearling Rollaids, whose teeth are his only normal attribute. His head is too big for his body, he has suffered mild laminitis since August, his growth rate has slowed to the point where he is indistinguishable from the weanlings. He is, however, the calmest horse I have ever met.

"If his laminitis worsens, he may have to be put down," I cautioned, in the first of my many disclaimers.

"We had a foundered pony," Charlotte countered, Rollaids' head cradled in her arms. "We can work around it."

Fueled by the unforgiving ledger, sentiment is building that the Rollaids saga come to a humane end, here where it started, Dr. Riddle doing the dirty work. But old loves die hard. Charlotte and Jim persevere, fully apprised of Rollaids' faults. They have looked at numerous horses. They know their daughter. They have dealt with founder before. Jim showed me a tattered wallet-sized photo of their thirty-acre farm near Red Lion, Pennsylvania, and I could almost see the learning-disabled daughter and the physically-disabled horse walking side-by-side past the red bank barn toward the white house on the hill. Rollaids has been a horse only a father could love, but I see a girl in his life now.

Wednesday, October 3. The world was back in focus today as we loaded the new Master of the Universe backwards in the rear seat for the trundle home on Route 7, past the produce stand and the rainbow of autumn color—orange pumpkins, yellow squash, Indian corn, brown stalks—past the Dew Drop Inn, past the cedar trees of Old Joppa Road, to the lane of Country Life. We settled in, Ellen's Mom making dinner, Ellen and our new son asleep, as the late afternoon sun warmed the house.

I walked to the top of the hill overlooking the watershed, opened the beer I wanted to have at the Dew Drop Inn, and laid back to stare up at the blue sky. I gave quiet thanks for all that has happened. Either I was dead tired, or on the verge of discovering something big about life, but I have never felt as happy as I felt today on the big hill.

Thursday, October 4. Ogden Nash had trouble with names, and so do we. We tried J. P., but initials are not really a name, I suppose, unless you are O. J. Simpson. We tried Joe, but there is only one Joe P., and that's Pop. Ellen asked: "Why not Hoss? He's big enough." But the Ponserosa is not ready for a Hoss rather than a Little Joe. So we affectionately called him Josh, and it stuck. Now there are two of us. This would certainly confuse Mr. Nash, who didn't know what to call a lot of people:

*I can remember the races between Man o' War
and Sir Barton,
and Epinard and Zev,*

*But do you address a minister
as Mr. or Dr.,
or simply Rev.?*

Friday, October 5. A leisurely influence floats in the air of an Eastern Shore autumn, settling over Dr. Bowman's Dance Forth Farm. Burnout certainly exists in the American Association of Equine Practitioners, and for nine months of the year, Dr. Bowman is the prototype from an AAEP press release:

"One must travel thousands of miles in hazardous traffic, subject to frequent disabling injury from the most dangerous domestic animal. Night calls, unbelievable overhead, poor clients, uncollectable fees. Why do equine practitioners stay with it?"

At Dr. Bowman's today to pay a bill, this poor client witnessed the good vet's response. He hopped off the Quad Runner from showing yearlings, and hurried to coach soccer practice at nearby Washington College in order to be home for dinner with five bright children, a wife who outshines them all, and a caged crow he found limping in a Darlington driveway. An Eastern Shore autumn is his answer.

Saturday, October 6. I wanted to tease my mother-in-law Doris about the movie *Throw Momma From the Train,* but restrained myself as I drove her to the Baltimore station to catch the Amtrak for Richmond. I dawdled on the ride home, stopping at Pimlico to catch the simulcast action. Allen's Prospect had three two-year-olds racing today, and one tonight at the Meadowlands. Allez Prospect ran fifth, first-timer Altastar ran second. I dialed the Belmont hotline for Xray's fate in the Champagne Stakes—the first time a Country Life-sired runner has ever competed in a Grade One event: Darn, off the board. It's up to Find a Penny now, making her debut under lights to our north. Not too much progress on the freshman sire list so far today.

Sunday, October 7. Maybe the last true warm Sunday of the year, I think. The Earth is tilting almost imperceptibly towards winter. The trees have more shape and stand away from each other. Tiny bugs hatch out; their wings catch the afternoon light. I hear leaves falling in the woods behind me, see yellow jackets working the flowers, but there are no flies—thank goodness, no flies. The foxtail grasses are buffed brown on the hillsides, sassafras trees are orange. Poison ivy and sumac turn red faster than the hardwoods to which they cling. The poplars hint of the flame that will burn bright before the leaf falls. The maples are holding green. Honeysuckle, blooming from tropical influences of late, is thick on the diamond-mesh fence.

This is what I saw from the yard on the first Sunday of your life, young man.

Monday, October 8. Mention "pin-firing," and you are likely to create controversy. Faithful Tradition, a two-year-old filly by Carnivalay, is here for some R and R, and Dr. Riddle fired her ankles today. Nerves are blocked, of course, but the smell of burning hair and flesh as the firing iron pierces the skin over and over is part of the element that fuels controversy. There is no arguing with success, however, and many runners have resumed fruitful careers after having been fired. The inflammation artificially produced by the firing iron is thought to benefit the healing process.

This is a talented filly, but her mother, Steady Wind, had ankles calcified in old age to the size of softballs. Her sister, Turn to Money, won stakes and earned more than a hundred thousand dollars; she had sore ankles. It is a faithful tradition of this family to overcome bad ankles.

Tuesday, October 9. The yearlings are feeling their first human weight as our riders start "bellying" them around the stalls, and I am reminded of the Moscow Circus. The Russian horseback riders are called djigits. Their philosophy is not simply to perform daredevil riding feats, but to impart a narrative message: The horse represents spirit, the sword clears the path for creativity.

"The djigits are peaceful, with noble character," said Tamerlan Nugzarov, the most famous equestrian in the Soviet Union. "When we were forced to fight, we hid behind our horses as we rode."

The obstreperous Edgar Allen Pro is testing the peaceful nature of the Country Life djigits. Later this month, when the yearlings parade to the large paddock in full view of the office, a bit of circus riding might be required. I am considering Dr. Riddle's emasculating sword to clear the path of Edgar's creativity.

Wednesday, October 10. "Valay Maid worked yesterday," Joe Hamilton informed us bright and early this morning. "She went seven furlongs, and Carlos said she was blowing hard when she came back—just what he wanted. One more long work maybe, then some sharpening. Carlos is dedicated to this filly, and I know we will not be embarrassed in New York."

Thursday, October 11. Tropical Storm Hughie blew hot air in our faces today, as a steamy drizzle soaked the crew leading yearlings. This evening, the kid, Ellen, the dogs, and I caravaned to the creek, where *in utero* young Josh had kicked in response to the beat of water over rocks.

Through the woods an owl called to us, perhaps a youngster himself, trying out his own lungs. How vividly I recall the sounds of owls mating in the trees this past spring—louder than cats in an alley—a chorus of high-pitched hoots bursting through the still night air. Maybe tonight's owl was conceived in that cauldron of sound that had me perched at the second-floor window, with a shoe aimed at interrupting the carnal splendor of my big-eyed neighbors in the woods.

A light rain is falling now as I write this at 8 P.M., the kid asleep, dinner almost ready, shirt buttons clicking against the dryer. What a warm feeling.

Friday, October 12. Here today, gone tomorrow, I suppose. So Ellen and I had visited the farm lawyer several days before the baby was born to have wills prepared. The horse business often is such an informal paperwork game that perhaps it carries over to other aspects of life. Now, however, should I take a fatal kick in the pants, the kid and his mother will be protected:

"I, blah blah blah, being of sound and disposing mind, memory and understanding, do make this my Last Will and Testament. I give and bequeath all automobiles (my VW wagon has logged a hundred and fifty thousand miles), jewelry (a National Turf Writers pin admits its wearer free to any track in the United States), silverware (a pewter julep cup for Maryland-bred Salohcin—Nicholas spelled backwards—winner of the 1989 Broad Brush Handicap at Turfway), household goods (would a hundred-dollar reconditioned Parianni saddle with stitches by Dr. Frankenstein qualify?), works of art (my print of Richard Stone Reeves' portrait of Travelling Music will be priceless as soon as I get it framed), and all other similar articles...to my wife."

For the benefit of other similarly high-toned young horsemen who have not yet considered their own mortality, where there's a will, there's a way to keep valuable heirlooms looming about, and out of Uncle Sam's hands. Do it, now.

Saturday, October 13. Red Smith's obit on Damon Runyon:

One remembers, for instance, (Runyon's) report of the deathbed advice an old gambler gave to his son. It went, more or less, like this:

"Son," the old man said, "as you go around and about this world, some day you will come upon a man who will lay down in front of you a new deck of cards with the seal unbroken and offer to bet he can make the jack of spades jump out of the deck and squirt cider in your ear.

"Son," said the old man, "do not bet him because just as sure as you do you are going to get an earful of cider."

Mr. Runyon hit on the key to the horse business. There are no sure things. Sooner or later, you get an earful of cider. I hope to never lose sight of that, and in my old age, I hope my son reminds me.

Sunday, October 14. The new models were out for a spin this morning—cruising over soft turf in the Office Field. The eight yearling fillies crowded their quality-control inspector. Exposed to a cow kick from close quarters, I waved the girls away. They peeled out, throwing short, choppy kicks skyward at a run—mud flying from tires. I goaded them into a gallop, and was reminded of John Madden's grandson, Preston Madden—famous for escorting farm visitors in a Cadillac through the fields, bouncing over the vast hills of Hamburg Place in gentle pursuit of yearlings.

Inside the dappled hides of our sporty gals, a delicate and complex motor system is evolving. Sports medicine has revealed that the bone of young horses increases in density with the stimulus of only a few high-concussion loads. On this morning's romp, bone was being made, connective tissues—

tendons, ligaments, cartilage—developing strength through repeated stretching. The fillies played bumper cars in the mud: Cadillacs in the rough, not quite off the assembly line, horsepower untested, Mother Nature fine-tuning. Come next summer, jockeys will stand in line to hit the throttle on this year's models.

Monday, October 15. Today was the foal nomination deadline for the Breeders' Cup. Michael and I anguished over which of four homebreds to grace with a five-hundred-dollar vote of confidence.

"Peter, who's the best weanling out there?"

Peter looked at me as if I'd just asked who to bet in next year's Derby. I wanted to yank some Breeders' Cup representative out from the closet and ask: "How the heck do you expect folks in the regional markets to tack on five hundred dollars to homebreds we can't sell for a thousand dollars?"

I can say with certainty that if the Breeders' Cup only cost two hundred and fifty dollars, like the Maryland Million, we would have entered all four homebreds, because for that thousand dollars, this guesswork would be eliminated. How many other small breeders would concur? Would it be enough to offset the loss of two hundred and fifty dollars revenue for each nomination? As it is, we just threw our hands up, picked a filly instead of a colt, and prayed we had not just thrown good money away on the wrong horse.

The irony, of course, is that entries for this year's Breeders' Cup closed today. Carlos said: "Valay Maid worked seven furlongs Friday in 1:26. She get tired, sure, but she get fit, too. We go to New York. We have a chance."

Carlos thinks it will be a small field. Maybe Go for Wand and Valay Maid will be the only three-year-old fillies? Imagine that: Twenty-five thousand fillies in their foal crop, and perhaps only two—Carnivalay's sister and Carnivalay's daughter—make it to the top female race of the year. I am a fool not to have dug deeper in our pockets for the three homebred weanlings excluded tonight. You have to play to win.

Tuesday, October 16. An industry newsletter received today calculated that only 16.4 percent of yearlings in a broad study group made a profit for their consignors at major sales last year, mostly because stud fees are still too high. They didn't include Maryland sales—not major league. The odds must certainly be greater in these regional markets. If 83.6 percent of consignors to major Kentucky sales are losing money, why sell?

In the Maryland market, breeders race. The best local yearlings were not offered at the Fasig-Tipton Midlantic select sale, which averaged eleven thousand dollars. Instead, they are under tack, being broken, spending nights in run-in sheds, destined to carry their breeders' silks until they are claimed for twenty-five grand, or sold after a maiden win for sixty, or raced in the lucrative Maryland-bred stakes program, earning an average of fourteen percent breeder bonuses and an average of seventeen percent owner bonuses on top of the purse. Naturally, not all fare so well, but the opportunity is there.

Without the Maryland racing scene, Country Life would be a housing development. We would have no market for our horses.

Wednesday, October 17. Familiar mental gymnastics were performed today, as Michael and I tried to arrive at a presentation to the owner of a spanking-new stallion prospect, a brilliantly fast three-year-old colt named Citidancer. He outran his legs, and fractured a sesamoid in only his fourth start. He won his first three races by a total of thirty-six lengths, then he foolishly dueled champion Housebuster in the Jerome Stakes. He lost the duel and his racing career.

Through a painful process of trial-and-error, we have blundered our way through six stallion syndications since 1981, retooling the ailing factory with some Edsels and some Cadillacs. Now the plant is up and running, but to maintain momentum means taking risks.

Tonight might be the first night since the baby came home that he sleeps more than I do.

Thursday, October 18. With Michael quarterbacking the negotiations, I loaded Wailing Jennings into the back seat, turned on the country music station, and set off with Ellen for Goochland County, Virginia, where her brothers and sisters awaited their first glimpse of the new kid on the block. He was a good shipper, and we landed before the rains hit. We emerged from the farmhouse at dusk to witness a magnificent skyscape. To the west, the sun peered out from behind clouds racing ahead of the cold front. To the east, heat lightning exploded over Richmond, as if the old city were burning again in the fury of the Civil War. The hayfields on the farm shone luminescent green.

Ellen held the boy in her arms as a rainbow climbed out of Beaver Dam Creek, framing the scene at sunset with a portent of good fortune. A painter would have rubbed his eyes at nature's revelation of colors. I froze the scene on film, but will not need the photograph to recall the first evening the kid set foot on the red soil of his mother's homeland.

Friday, October 19. "I suppose it's either fish or cut bait," I responded to Michael's phone call at noon.

"The timing's right. They've answered all our questions. They've been very fair, but they need an answer," Michael said.

"Okay. I'll call them now. I'll tell them if the colt passes a physical exam of his testicles, and is insurable for mortality, we'll do it."

"That's fine. He fits our program."

You should not buy hay or horses sight unseen, but today I committed to buy a horse without first looking at him. I asked enough questions, however, and relied on trusted agents' advice. When a calm settles over you after a decision is made, your instincts are probably telling you something. Still, I won't be certain about cider in my ear until I have seen him in the flesh.

Saturday, October 20. "He passed," Michael said over the phone at eleven this morning. "Sallee has a van leaving Belmont at 7 A.M. tomorrow. I'll get Richard to bed down the empty stall in the stud barn."

Here we go again.

Sunday, October 21. "Okay! Okay! Okay!" The kid is lip-quivering crying right now, and Ellen is at the barn. Nothing I say has any effect. *"Shhh! Shhh! Shhh!"*

Barrie Reightler of the *Maryland Horse* told me: "If the kid won't hush, put him in the 'football' hold." About magazines and mothering, Barrie knows best. I cradle the screaming football in my left arm, his head in the crook of my elbow. He has gone right to sleep.

Monday, October 22. Uncle John's house is a hundred feet from the stud barn, and I motioned for him to come across the yard and inspect the new stallion, Citidancer.

"My, he's a well-turned colt," said the old cavalryman, as John held the new recruit in the center of the stall. "A deep chestnut, like Equipoise. A rich color."

In thick white support bandages over his injured leg, the colt stood at attention for Uncle John, forgetting for the moment the playful proximity of the Irishman's coatsleeve. In the gathering dusk, the colt's coat bounced stall light against my eyes. After dinner tonight, Ellen returned from her first inspection.

"Yep, he'll fit right in at the stud barn," she said. "When I turned on the light, he was standing in the corner, dreaming of girls. I turned the light back off."

This news warmed my heart, because a young stallion is asked a great deal in his first season at stud. If he doesn't get the numbers his first year, it's uphill the rest of his career. The casual interest he displayed tonight bodes well for his task next spring.

Tuesday, October 23. "So," Marge Dance began, mischief in her eyes, flanked by members of the almost exclusively female Press Club that comprises the *Maryland Horse* staff, "tell us about passing out in the delivery room."

Mothers and would-be mothers gathered in the hallway as I squirmed. Not since October 1st have I felt so light-headed. Oh, for days gone by, when Dads read *Field and Stream* in the waiting room, and Coaches were on football fields.

Wednesday, October 24. Most breeders need another stallion share like a hole in the head, but an intriguing "What if?" element blankets Citidancer, and the response to my phone calls today was favorable. Most good stallions have three elements in common: A good sire line, a better bottom line, and speed. We've done stallions with only two out of three: They didn't work. The new guy seems to have the right elements. If he is fertile and we can attract good mares, with luck, the big three will take over.

Thursday, October 25. The *Form* listed Saturday's post positions this morning, but we hardly had time to notice. We had momentum, and we concentrated on the new horse without distraction—plowing through cover letters, photocopying race charts and sports page articles, addressing, stamping, logging in the mail. Finally, at quitting time, with Alice gone to the post office, we stopped to contemplate our girls Valay Maid and Go for Wand in the Breeders' Cup.

I looked ahead to Saturday, imagining Gregg McCarron's TV narration as the Distaff field enters the starting gate. "Go for Wand loads easily into the number two post. What a special filly! She could be Horse of the Year after this race. Her brother, Carnivalay, is the sire of Valay Maid, who is now loading perfectly into the number seven post. We're almost set to go. Back upstairs to you."

On the entire card, *only two* three-year-old fillies are entered: Go for Wand and Valay Maid. It is clearly a family affair for us.

Friday, October 26. "Don't look for any help from Xray," Brookside Farm manager Ted Carr said when I asked about his Allen's Prospect two-year-olds. "He broke his cannon bone in the Champagne. Won't race again until next year."

The freshman sire's leading flagbearer is on the shelf. We had scanned the Belmont Park daily workouts in the *Form* without seeing Xray's name since the Champagne, and Ted's update confirmed a dark suspicion. There is safety in numbers, however, and more two-year-olds are preparing to carry the torch, battling the attrition exacted by racing wars.

Saturday, October 27. When Go for Wand won the Alabama Stakes at Saratoga, I was the happiest fellow at the track. Carny's half-sister, the best filly in the land. Wow! Then suddenly a roar arose in the grandstand next to me. A fellow had just hit a hundred-thousand-dollar Pic Six. He eclipsed me as the happiest man at the track. I introduced myself, sharing mutual good fortune. Turns out he was from Pennsylvania, same as Go for Wand, who he loved. He knew all about Carnivalay, too, and said he would come visit us this fall.

After Go for Wand died today, I thought about this fellow, about all the people who had embraced this filly. Her brilliance reflected light on the comic piebald face of our favorite equine clown Carnivalay, a colt who went from being just another son of Northern Dancer to being half-brother to Miss Universe. Only a filly of Go for Wand's talents could have relegated the great Northern Dancer to a supporting role in a pedigree. The day last fall when she won the Breeders' Cup Juvenile Fillies was a watershed day in the history of this farm; Carnivalay would not be upstaged, his daughters Valay Maid and Lucky Lady Lauren running one-two in the stakes at Laurel. After almost a decade of mud in our eyes, we turned the corner in a single afternoon.

I want young Josh to read this diary someday and understand that whatever Go for Wand meant to his family, she meant just as much or more to the family of racing, from Belmont to Boise. We all lost a sister today.

Sunday, October 28. It is absolute nonsense to think for one minute that Carnivalay understands what happened yesterday. Nevertheless, I found myself talking to him tonight in the stallion barn.

"It's all right, old boy," I said, rubbing his nose through the bars of his stall. "Everything's all right."

A few hours later, I woke up in a nightmare: All the stallions had broken their legs at the ankle.

"It's not all right," I told myself. "It is not going to be all right for a very long time."

Monday, October 29. "Can't sell this sport for a while," a fellow who sells racing partnerships for a living told me today. "I pulled the guys in this morning and told them to lay off trying to sell anything. Pretty big chill went through everybody Saturday."

Tuesday, October 30. The kid awoke every hour on the hour last night, and I felt like a zombie all day. At quitting time, I gathered in a huddle with the farm crew and caught up on various horse care items. John is removing the new stallion's bandages for an hour each day, to let the leg breathe a bit. The colt is feeling good about himself, and does not know his sesamoid is cracked. Peter is poulticing a yearling filly who graveled, but who nevertheless felt good enough today to launch Andrew skyward: "I landed on my feet, then on my rear. Softest landing I've ever had." (I stub my toe and look for a cane; Andrew cracks his ribs and is cracking jokes about it.)

A broodmare in the December 4th sale has developed a skin fungus that Steve Earl will tend to promptly. A weanling was sent to Dr. Riddle's for an abscess on her neck. Richard left Corridor Key in today because of a loose shoe; Tom Rowles our blacksmith is due tomorrow, and so is Dr. Riddle, to perform pregnancy checks on all the mares. (I dread the thought of who might not still be in foal.)

The horses are not the only ones getting checkups. For his one-month examination, the kid weighed in today at eleven pounds eleven ounces, a gain of thirty-one ounces in fourteen days. No jockeying in his future. They pricked his heel to take a blood test, and he screamed unforgivingly at me.

On the phone lines, Go for Wand lingers. I found myself sitting at my desk, chin in hand, shaking my head. It felt good to walk out on the farm, where details of life occupy the mind. Tonight, the moon is on the rise, the world spinning us towards the future and away from the tragedy of the past weekend.

Wednesday, October 31. October was a month of give-and-take: I felt the greatest joy I've ever known for the new member of the family Ellen and I received, and a very deep sadness for the family member we *all* lost. The cycles of the horse business do not pause long for joy or grief, however, and our eyes looked forward today upon hearing Dr. Riddle's report that all eighteen mares were confirmed in foal. Yet I can't let go just yet.

Sometimes when the credits roll at the end of a movie, they show you clips of the memorable scenes, reviews instead of previews—the smiling and the clapping, the warm "feel good" parts of the movie. That's all I can see now, the Greentree barns at Saratoga last August, Billy Badgett dropping his last carrot on the ground, rinsing it off carefully, walking over to Go for Wand under the trees. He cuts the carrot with a hoof knife, flattens his hand out, feeds the carrot to his champion. The music builds, then stops. A single credit line rolls up the screen, the epitaph:

Go for Wand was buried at Saratoga, scene of her greatest triumphs.

Filmed entirely on location in the real world. That is the rub, as Damon Runyon might say. The screen fades to black, and we gather ourselves to face each other as the houselights come up.

FROM DARLEY TO CHARLIE

Thursday, November 1. Maj. Goss L. Stryker enlisted in the Spanish-American War, but he was too late to charge up San Juan Hill with the Rough Riders.

"Just as well," he intimated to me this morning. "Our rifles hadn't been fired since the Civil War."

A keen student of horses, Maj. Stryker became a steward in the big Jockey Club in the sky at the age of ninety-three in 1971. Many of the horse books from his extensive library were presented to me by the executrix of the Stryker estate, and my early mornings of late have been spent sequestered in the basement study with a screaming month-old boy—out of earshot of his exhausted mother. At wit's end myself, I summoned the good Major's memory to take over the reins of the Dawn Lecture Series.

"I'd be most happy to educate your son on the origins of the Thoroughbred," he began his first lesson this morning, feeling right at home among his books in my study. "We'll begin at the beginning."

Friday, November 2. "The first horse left North America, crossed the Bering Strait, and roamed into Asia. The climate of the world was changing, continents were still being formed. At this stage of evolution, the horse was but a small agile pony, in search of grazing land.

"Some went west through Russia and Poland, down into Turkey, over into Europe as far as Spain. Others migrated southwest, down through the Middle East, evolving from ponies into the taller, lighter horses of the Arabian Peninsula and the Barbary Coast of Africa.

"Enter Julius Caesar, who was very fond of his ponies. When he attacked England in 55 B.C., however, he ran into the javelin-throwing mounted cavalry of the Celts. The stout Celt war-horses were from the Barbary Coast, brought to the Celt homeland of Gaul by the Spanish. The Romans quickly began interbreeding the Barb with horses from Turkey—gelding inferior offspring to transport vast armies, while pitting the superior colts in chariot races.

"Centuries passed until the Crusades of the twelfth and thirteenth centuries, when knights returning from the Middle East recognized the need for fast, strong horses. The constant wars of the fourteenth and fifteenth centuries depended on cavalry. By the sixteen-hundreds, the native stock of Britain did not lack for strength, but for speed and endurance. The stage was set for the infusion of blood from the 'Eastern Sires'—the Barbs, the Turks, the Arabians."

Saturday, November 3. Maj. Stryker does not teach on weekends, so Ellen and I packed the freshman into the car and headed for Ocean City to celebrate our anniversary today. En route, we drove into Standardbred country, the flatland boundary between Maryland and Delaware, where weathervanes on barns reveal the business of the farm: A trotting horse headed into the wind. The Standardbred business in this region is headed into the wind as well. The owner of the Maryland trotting tracks is barred from his place of business after being arrested on cocaine charges; his building developments are being rocked by the recession; and problems facing his racing monopoly threaten mom-and-pop operations forming the lifeblood of this industry.

It is precarious enough to weather changing winds of the Thoroughbred game, but our problems pale when compared with our kinfolk in the local Standardbred game.

Sunday, November 4. On approach from the causeway, the brown grasses of Assateague Island blanket a brackish savannah rolling towards the ocean, and the eye scans the russet landscape for signs of wildlife movement: Deer in daylight, wild ponies the color of mangrove.

On a bayside point, two mute swans fly close by, white wings whistling. Behind us, a chestnut pony is indiscernible in the brush. It is hard to imagine how horses survive on this shifting barrier island. The ponies are said to have been stranded after escaping from sinking Spanish galleons. This is a controversial point, for the barrier islands have not always been entirely isolated from the mainland by bays. The ponies might have walked here, then been cut off by a powerful hurricane, waves obliterating the route of retreat.

These charming little tourist attractions represent such a rich heritage— Eohippus of Wyoming come full circle (well, almost). They cannot make it back to the American West because they will not cross the concrete causeway. That barrier is all that prevents them from migrating to the lands from where they mysteriously died out twenty-five thousand years ago.

Monday, November 5. His hocks break over when he walks, his breathing is labored, his back is swayed. He is twenty-nine years old—and not long for this world. I take three quick photos of Northern Dancer, as Ellen holds the baby boy close to the fence—a young man on the way into the world, an old man on the way out. The stallion does not raise his head higher than his knees. I have one picture left, and focus carefully on his nameplate. Suddenly his head

rises, his ears prick, and he gazes into the distance. The shutter clicks. He is regal once more, frozen forever in time.

When I arrive home, I tell Michael if he wants his two-year-old son Philip photographed with the legendary Northern Dancer, he'd better be quick. I do not think the horse will live past the first of the year. Cold weather will be his undoing.

Tuesday, November 6. In the aftermath of Go for Wand's fall in the Breeders' Cup Distaff, no one felt like rejoicing when Valay Maid inherited third place. I only fully considered the importance of a Grade One performer while proofreading Carnivalay's stallion register page today.

The tragedy of Breeders' Cup Day resists my efforts to forget. Worse still is that, however unconscionable it sounds, Valay Maid's third-place finish added a hundred thousand dollars to Carnivalay's progeny earnings, lifting him well above the million-dollar milestone for this year. What a game. What a crazy, ironic game.

Wednesday, November 7. Exactly one hundred years ago, a rich chestnut colt with four white feet performed so brilliantly that no owners would challenge him. Accordingly (wrote Walter Vosburgh), "The race Salvator vs. Time (1:39¼) followed over the straight course at Monmouth Park on Aug. 28, 1890." Salvator warmed up with a mile in 1:50, then raced a mile in 1:35½.

Imagine a colt under absolutely no pressure from opponents slicing 3¾ seconds from the track record—after warming up a full mile. So brilliant was Salvator.

In August at Saratoga this year, another deep chestnut with four white stockings went postward. He might as well have raced only against time, since he won by fourteen lengths in 1:09, Jose Santos never raising his whip. I have no delusion that the new stallion was of Salvator's ilk, but Henry Stull's painting of the famous nineteenth-century racehorse in Vosburgh's *Racing in America* bears an uncanny resemblance to the twentieth-century colt in our stud barn.

I hope the resemblance ends there, for Mr. J. B. Haggin sent Salvator into the stud with great prestige. Wrote Vosburgh: "He was a triumph of breeding. The dams of the great racehorses of the country were brought to mate with him. Great things were expected; but he was an utter failure."

It is humbling to realize that no matter how carefully we lay the groundwork for a successful sire, history says the odds are against us.

Thursday, November 8. From 8 A.M. to 8 P.M., we aided Bernie and Bill as they shot footage for a video on the stallions. No longer will I watch the simplest comedy on television without appreciating the actor's ability to do long takes. Fortunately, Michael's enthusiasm carried the voice-over, and the stallions carried themselves. We stressed the Maryland Million, from Ameri Allen to Xray.

I am amazed that a hundred and fifty man-hours are required for a fifteen-minute video. Yet if it enables prospective breeders to patronize these stallions without having to navigate the beltways of Washington and Baltimore, the construction on I-95, and local congestion on Route 1, the time will have been well spent.

Friday, November 9. Prepared the syndicate agreement for the new stallion today, and found two publications of my alma mater, the University of Kentucky College of Law, to be very valuable.

Legal Aspects of Horse Farm Operations, by James H. Newberry, Jr., is a concise handbook for lawyer and layman alike, while the binder entitled *Annual Equine Legal Seminar* provides sample syndicate agreements, and a boarding agreement that every farm should execute with their clients. Even with shipping, Newberry's book and the Seminar binder cost less than one billable hour in the office of a skilled equine practitioner.

Saturday, November 10. At five-thirty this rainy Saturday morning, the kid was bawling, the cat was crying, Boxy the pointer was barking to beat the band. I swaddled the kid warmly and carried him under the porch overhang, where we discovered the culprit behind a sleepless night. A black and brown hound with feet the size of pancakes was nestled in Boxy's bed, exhausted after a night of treeing raccoons under our window. When I approached, he slinked away with the cowardly crawl of a dog who had been beaten too much. He wore a two-inch-wide black collar with a portable electric shock box the size of a radio. I put a bowl of kibble on the ground, and he slinked close enough for me to read the phone number of his owner.

"That must be Boomer," answered John Cornett, who lives ten miles away. "I paid twenty-five hundred for him three weeks ago. I'm glad you found him. I just figured he'd got run over in Route 1.

"He's a Walker Hound, sired by Yadkin River—the best coonhound blood there is," Mr. Cornett explained upon arrival, as he worked a fresh chew of tobacco. "Still haven't found the bluetick hound I lost last night."

It wasn't so bad after all, getting up before dawn, geese on the wing, rain slapping on leaves, kid asleep in my arms. I almost made the mistake of asking Mr. Cornett in for coffee, but then we would've been fast friends, and he'd ask me if I'd let him hunt here some more, but truth is, sleep's been a little hard to come by lately, what with Boomer and the baby and all, and the bluetick still on the loose.

Sunday, November 11. Fussy boy awoke before the sun again this morning. Looking out the window from the basement study, we watch the creek flooding from the two inches of rain yesterday. The sun's first rays pick their way through the forest understory.

I can't tell whether these mornings in the study are establishing a schedule for the kid that I can not possibly maintain into the coming foaling season. Yet these mornings are so special in their splendor, in their quiet solitude, that I am more at peace at dawn than at dusk. The kid, unfortunately, is exactly the opposite.

Monday, November 12. Vanished generations built themselves into the barns on this farm. The oldest bank barn is a three-story structure whose depth is legendary. Two silos standing as sentinels do not reach the top rafter of the immense structure. Abandoned when the last Black Angus cattle left in the nineteen-fifties, the silos became vertical tunnels for exploring children. From inside their heights, we peered out and down into the courtyard of the broodmare barn, where Johann and Dad and Uncle John and Bill Magness would be supervising the breeding of mares. We did not know why the black horse jumped up on the backs of the mares, only that it was a very quick ritual—men scurrying about with a sense of purpose, water from steel buckets thrown on the gravel as horses were led away. I never dreamed the success of the whole farm revolved around this unfathomable little ritual.

Tuesday, November 13. "During the late sixteen-hundreds and early seventeen-hundreds, a number of Arabian, Barbary Coast, and Turkish sires were imported to the British Isles. The distinction of the breeds was often confused: Barbs were called Arabians; Arabians were called Barbs.

"The three most significant of these imported stallions were the Darley Arabian, born in the desert of Palmyra, a product of thirteen hundred years of selective breeding by the Bedouins, imported to England about 1705, most famous as the ancestor of Eclipse;

"The Byerley Turk, imported by Capt. Byerley to Ireland in 1685 to serve as his charger in King William III's campaigns, a horse with scant opportunity at stud, yet to whom Herod traces;

"And the Godolphin Arabian, sometimes called the Godolphin Barb, a teaser called into service when another stallion refused to perform, who, when put to the mare Roxana, sired Cade, sire of Matchem.

"From the Arabian, the Turk, and the Barb descend Eclipse, Herod, and Matchem—the three foundation sires of the Thoroughbred breed."

Maj. Stryker paused to allow his students time to grasp a firm footing on the infancy of the breed.

"Am I going too fast?" he asked. The infant's head bobbed in sleep. How marvelous! I've stumbled on a way to keep the kid quiet. Carry on, Major.

Wednesday, November 14. Farm Video, Take Four:

"Hello, I'm the syndicate manager for the Country Life Farm stallions."

Cut. Cut. Cut. Say it with some enthusiasm. Or say 'cheese.' Do something to pump some life into it. We gotta get this intro right. Okay? Ready? Roll.

(I keep my sanity by repeating: One hundred and fifty hours equals fifteen minutes, one hundred and fifty hours equals fifteen minutes.)

Thursday, November 15. Today we rewrote several clauses in our Stallion Service Contract. In boldface type we declared that our Breeding Season extends from February 15th through July 10th.

Often in early June, a party who purchased an outside season will get cold feet about a late cover date. Despite contracting for the entire breeding season, such a breeder might call it quits before the stallion manager pleads his case on behalf of the owner of the breeding right. Granted, some mares should be excused, but now our contracts make it a bilateral understanding, not a unilateral decision on the part of the mare owner.

Friday, November 16. Northern Dancer died today. You could hear the book close, the thick heavy buckram-bound book on the Northern Dancer story. No longer is the aberration a reality: That the greatest sire in the world did not stand in the vast Bluegrass, but instead was in tiny Maryland. He was Canada's first, Maryland's second, the world's after that.

I am struck by the coincidence of studying the origins of the breed at the moment perhaps the finest breeding specimen of the twentieth century has died. The similarities, the ironies, the uncertainties surrounding the greatest stallions in history carry the lesson that no one *really* knows where the next great one will come from. Northern Dancer went unsold as a yearling. The Godolphin Arabian was a teaser. Eclipse's sire Marske was "so bereft of external attractions that he was positively despised until he became the sire of Eclipse" (in the words of Dr. W.J.S. McKay in his 1933 book *The Evolution of the Endurance, Speed, and Staying Power of the Racehorse*).

What is there to guide a young man who has staked his family's future on speculation in stallions? I only know the basics. Stick with sire lines. Go with a winner on the race track. *Absolutely* do not accept less than the best in the dam's line. Then pray.

Saturday, November 17. The groom at the trainer's barn: "Oh, he splintered everything. Gone to put him down. They just took him over the hill in the horse ambulance. You'll find him down past the machinery shop."

The horse ambulance chugged back up the hill past me, and a groom stepped out of the cab with only a halter in his hand. As I walked down the long hill to the shop area, the clouds blew away. Orange light from a November sun bounced off new welds on track harrows in various states of repair. It picked out the grain in plywood sheets behind the machinery shop. It gathered in a shadow beneath the fallen body of Altastar, a two-year-old bay colt by Allen's Prospect. It was as if he was simply asleep on the side of the road, until I saw the Ace bandage wrapped around his right front cannon bone, the blood on his pastern, the blood on his hoof. I wanted to throw a blanket over him, but I had none.

I walked back up to the track. A Chevy Suburban drove by with the placing judges inside. A giant plastic fox, the smiling mascot of a local car dealer, was attached to the roof of the Suburban. I jumped in the truck and rode to the gap. I spotted Landing At Dawn walking over for the ninth. I got out and walked with her groom. I foaled Landing At Dawn.

At the race track, everything goes round and round and round—the horses on the track, the lives of the people they affect, and that damn stupid fox.

Sunday, November 18. Bonnie Raitt is on the radio when the kid and I enter the study at sunrise. She's singing:

"You were 'playin' on the horses,' playin' on a guitar, too. You were wearing wings, wearing wings. Now I'm still at the races, with my ticket stubs, and my blues…"

The trouble with me is that I've taught myself to listen, and I'm helpless to block it out sometimes.

Monday, November 19. "It's part of the bargain, the good with the bad," Maj. Stryker said. "Horses who try hard sometimes get hurt. Do you think it was any different in the old days? Why, the great jockey of the eighteen-hundreds, Jimmy McLaughlin, measured his horses by their heart. He said: 'People talk of Hanover, but if Hanover got his head down he was gone, and all you might do couldn't make him win. But Kingston—no matter what happened—would fight it out while he had a leg under him!'

"Unless you understand that the Darley Arabian—from whose loins emanate the great majority of modern-day Thoroughbreds—could trace his ancestry back through a thousand years of deliberate matings arranged by Bedouins to provide the fastest mounts for raiding other tribes, for escaping at a gallop for twenty miles or more, then you don't understand that the racehorse gives his all because *it is in his blood to do so*.

"It's not your fault. That is the point of this morning's lesson."

Tuesday, November 20. An expert is a guy from out of town, wearing a suit and carrying a briefcase. We don't qualify, but there was our video up on the screen at the Sports Palace, Michael's voice-over urging you to "put the Maryland Million in your plans when you are booking your mares."

I guess I've attended too many horse seminars where the expert speaker takes every opportunity to promote his own product. It drives me crazy. We weren't trying to plug the home stallions in front of a captive audience. We were just showing how we capitalize on the Maryland Million, at a luncheon devoted to the future of the event. I think Michael struck just the right note of enthusiasm. Sustained success over a course of years qualifies one as an expert, and we are a long way from that goal.

Wednesday, November 21. The hay that arrived today from Wyoming was so heavy it challenged the human spirit. A high dose of comic relief was necessary to survive such a backbreaking chore. Naturally, the bulk of the farm crew departed yesterday for an extended Thanksgiving. As Richard, Michael, and I dragged bales weighing more than a hundred pounds down the sixty-foot length of the stud barn hayloft, we carried on a slave's litany of idle chatter. We almost broke into a spiritual when the fuse blew on the hay elevator and it appeared we might have to *throw* these bales of rocks from the truck bed to the loft.

As the countdown entered the final ten of four hundred and fifty such bales, I was giddy with relief. I challenged Richard to a "Strongest Man on the Farm" contest. He blushed and said, "Aw, shucks," and I let him hoist the last bale in place. I'm so sore right now if I weren't laughing at being so foolish, I'd be crying.

Thursday, November 22. Alice drove the feed truck for Richard this morning, which relieved me of chores. I stood on the stoop with the kid in my arms and waved as they rolled over the hillsides dispensing a Thanksgiving meal of alfalfa hay to the in-foal mares.

The twentieth anniversary of the Turkey Bowl attracted an international gathering of old high school chums and some new ringers. Dermot from Ireland let a punt hit him in the head (Ireland punters are hard-headed horseplayers); Ali from the Middle East counted One Mississippi, Two Mississippi, the entire time he chased the quarterback. Afterwards, the two teams took a knee for a photo, hiding the beers under their legs so the local press would not condone what TV commercials glorify. We all said a prayer of Thanksgiving that no one would miss dinner as a result of a debilitating injury suffered in pursuit of bygone youth.

Tonight we headed to Norah's for the feast.

Friday, November 23. Citidancer is a bundle of pent-up energy as I show him to a prospective breeder. I recite his tantalizingly brief racing career. This colt was a six-furlong specialist, and nothing excites mare owners as much as speed in a sire—not conformation, not pedigree, not staying ability. Speed sells.

"I have a mare who needs speed," breeders will say. They never say: "I have a mare who was slow."

Maj. Stryker was right. A horse race is like a raid. You have to get out ahead of the guys chasing you. You *need* speed.

Saturday, November 24. "You remember me! My name's Ritchie Trail. I was named for part of the hundred-mile ride near Hot Springs, Virginia, and as my name would suggest, I'm no sprinter.

"Mr. Hadry knows I don't get rolling early. I ran in the Selima Stakes. I finished sixth, but only lost by four lengths. Two weeks ago, he was kicking dirt in the shedrow because he couldn't find a long race for me. He said he'd have to train me right up to the big race, the Maryland Juvenile Filly Championship. If I'd won today, I would've been champion Maryland-bred.

"But today some girl went out way ahead—I could barely see her. I couldn't catch her, but I caught everybody else. Mr. Wilson, he was smiling and seemed very pleased. Mr. Hadry, he still said it would be nice for me to win a race before the year is out. Then I can go south to Mr. Whiteley. I've earned fifty-three thousand dollars and haven't won a race. Mr. Wilson, he said he'd like to have a barnful of fillies like me who can't win a race."

Sunday, November 25. Rocks jut like icebergs above the creek's rushing water. In a good pair of sneakers, you can hopscotch out into the water farther than prudence should permit. A misstep is a cold wet foot half a mile from home. Yet water moving through a forest valley utters a siren call, and at dusk tonight, perched on a triangular rock as if on a saddle, I marveled at nature's physics: Waves flowing back upstream, crystal droplets capturing falling light to form a tiny meteor shower of water hurtling through space.

I recalled a passage written by Kentuckian Wendell Berry in a book arguing against damming the Red River Gorge:

"The stream is mottled with the gold leaves of the beeches. The water has taken on a vegetable taste from the leaves steeping in it. It has become a kind of weak tea, infused with the crown of the forest. I wander some more among the trees, the thought repeating itself in my mind: This is a great Work, this is a great Work."

Monday, November 26. We arranged with a nearby horsewoman to lease some pasture today. Her land is the former acreage where Betty Worthington raised millionaire Jameela, dam of sprint champion Gulch. It is good land. A house trailer is included in the bargain, and Irishman David O'Meara, John's younger brother, will now have a place to call his own. Since he arrived from Tipperary last June, he has shared an attic bedroom in the big house with John, but the young man is twenty-nine years old. He needs his space. Our young horses need their space, too.

John Nerud said the best present for your worst enemy would be to buy him a farm. So great are the headaches from operating one, perhaps I should have my head examined for thinking we can handle two. Yet young horses need to roam, and pastures must be rested. An overhorsed farm will not produce superior animals. Already mares are shipping in to be placed under lights. Raw land nearby is too expensive to purchase, and the cost of developing a farm is frightening to me. Personnel is the key to this plan.

Tuesday, November 27. The day the photographer arrives to take publicity photos of the stallions is a day of high anxiety on any farm. Stories abound of photographers insisting on standing the horse against a beautiful background, despite the fact the stallion has never left the unphotographic stud barn complex. One glimpse of a broodmare five fields away is enough for a stallion to dream of breeding, and no amount of coaxing can restore the stud's calm attention for a professional conformation shot. With Citidancer and his injured leg, safety was a concern as well.

"You'll hate this background when you see the prints," Neena Ewing told me, after I refused to permit the new horse to leave the enclosed show ring. We have only hand-walked him occasionally since his arrival. His cracked sesamoid will not have fused sufficiently until January.

"I'll hate it worse if he gets loose and runs down Old Joppa Road in one-o-nine and change," I said. The colt spent more than a few moments on his two hind legs, rearing in delight. We settled for casual walking shots and a few head shots. Then we put him back in the barn. Neena is right, of course: I will hate the background, but I'll get over it, so long as the horse is still in one piece.

Wednesday, November 28. What a morning! The farm mechanic Steve Brown is explaining how the pickup has just broken down on Route 1, and asking where's the tow chain? Meanwhile, Mom is hooking up the brand-new used fax machine while the *Maryland Horse* office is ringing the phone off the hook. The fax won't answer. Dad is on the other line with an A.A. crony who is threatening to fall off the wagon—Dad talking him back from the edge. You can't hear yourself think.

Alice, Michael, and I cram into a Toyota to get to the accountants by eleven. We are stalled by our own stalled truck, as two state police cruisers surround it with flares and flashing lights. They summon a tow truck to haul this heap three hundred feet back to the farm driveway.

We rush back to call off our own towing crew, which is a good thing because Steve Earl broke the rear window out of the other farm pickup while loading fence boards last week, and Alice backed it into a tack box Sunday and broke a rear light out. With plastic trash bags as a rear window and no brake light, the second truck might look worse than the first truck to the police officers.

The deep cleansing breaths I learned in childbirth classes come back to me. I recall one day last summer, when Neena Ewing aimed her camera at the farm crew, standing in a huddle behind the beat-up old ten-foot mower which had just sheared away another patch of welding, as if staring might fix it.

"Isn't this a great picture?" Neena asked, like a tourist.

"No," I said, the price of a new Bush Hog giving me a headache, "but I'll admit it's a picture any farmer could appreciate."

Thursday, November 29. Richards's Lass, a first-time starter by Allen's Prospect, won at Aqueduct today. The purse for a New York maiden special weight race is ten thousand dollars more than here in Maryland, but if the off-track wagering bill is approved, someday soon our purses might rival those of California and New York. The bill will be recommended at this upcoming session of the General Assembly.

Billy Boniface, who would have made as fine a lobbyist as he has a trainer, heads the Legislative Committee. He knows his way around Annapolis. I've been so busy with the kid and the stallions and the farm that I have not contributed to a cause that benefits the entire sport—from bettors to owners to hotwalkers. I am so fortunate to ply my trade in a business whose leaders are so unselfish with their time.

Friday, November 30. "My son is very frank with me," Charles Lym told me today over the phone from Canada. Mr. Lym is originally from Jamaica, and he speaks in a soft British accent. He has brightened many a dark day for me. Even when Hurricane Gilbert flattened his business in Jamaica, he found a way to laugh. He said this morning:

"My son asks: 'Dad, when are you going to quit with those horses? You are squandering my inheritance.' "

And he laughed, and laughed, and laughed, and his jolly mood stayed with me throughout the day. As I walked home, the sun set on one side of the farm while the full moon rose on the other. The sun was so pink and the moon so white, so close, so early, I was sandwiched in celestial splendor. Autumn is on the run, winter on the rise.

Thoreau had the right idea. He asked: "Why do you flee so soon, sir, to the theaters, lecture rooms, and museums? If you will stay here awhile I will promise you strange sights. You shall walk on water; all these brooks and rivers and ponds shall be your highway. You shall see the whole earth covered a foot or more with purest white crystals, and all the trees will glitter in icy armor."

Like so many farmers, we move towards winter with trepidation for the challenge of cold weather. I can take great warmth from my acquaintances. After all, from Capt. Byerley to Mr. Darley to the Earl of Godolphin to Charlie Lym, the horse business is really a people business, and a remarkable people are they.

MUSIC FROM A FINE YEAR

Saturday, December 1. We once stood a stallion named Divine Comedy. He died on April 12, 1963: Good Friday. He won the Roamer and Saranac Handicaps despite a circulatory problem that baffled trainer Charlie Whittingham, yet Liz Whitney was able to secure insurance on the colt as sound for breeding. He collapsed of a heart attack after covering a mare. Dad phoned insurance agent Ed McGrath.

"The policy states that you were to notify us the moment his condition deteriorated," said Mr. McGrath, unhappy that a six-year-old horse would die so suddenly.

"Well, Ed, his condition has deteriorated all right," Dad explained. "He's dead."

McGrath and Dad often spent Sunday mornings in August in the same Saratoga church, kneading rosary beads. Dad tried to soften the bad news with a little religious humor: "I didn't see black clouds gather when this colt died. I don't think he'll rise from the dead on Sunday."

Dad followed the letter of the policy, and Liz Whitney got her dough. I think Divine Comedy dying on Good Friday just about sums up this wacky business. It has gotten crazier over the years. Nowadays, you have to call insurance folks *before* a horse's condition changes. The policies state: "If the horse is ever sick, injured, or *requires veterinary treatment*, we must be notified immediately. Failure to do so could result in prejudice of claim."

I made a mistake on this point recently. I sent a weanling filly to Dr. Riddle's for routine umbilical hernia surgery. The filly was a year-round boarder, and I acted as agent for her owner, as if she were my own. However, I did not check her insurance status. Most weanlings in this marketplace are uninsured, but our stallions are attracting better stock. I need to keep up. If the filly had died, the insurance company could have denied the claim.

We scolded ourselves in the office and posted a sign on the bulletin board for the future. I smiled with relief when the filly returned home safely, but the comedy would not have been so divine had we been embroiled in an insurance tangle. In that event, I would have prayed for Dad's track record to save me: Insurance on the filly in question, naturally, was in the hands of McGrath.

Sunday, December 2. A two-day horse sale at Timonium in December is a flea market. Twenty out of two hundred offerings are worth a look. Sundays are for broodmares, as farms from New York to South Carolina pare down home herds. Some mares might be diamonds in the rough, but they are expensive to mine: A December purchase requires a winter of feed before a foal is born. The cost of a mare's upkeep keeps bidders sitting down on their hands.

Monday, December 3. Youth sells in this sport. The auction closed two weeks after the Maryland Million, and small breeders with weanlings by Allen's Prospect hurriedly entered their babies. In the pavilion, Kentucky bloodstock agents plumbed the pinhookers market, taking more than crabcakes back home. A lady from Washington bought a weanling for California racing. A man from Illinois bought one for Arlington. A breeder from New York bought another for the Big Apple.

This business requires sustained effort. Yesterday, the broodmare market made me feel as though our efforts were in vain. Today, the weanling market recharged our batteries.

Tuesday, December 4. The farm across Route 1 from us—the site of the proposed mall—goes on the auction block this month. The developer has declared bankruptcy. Two years ago, he paid eight hundred and fifty thousand dollars for the eighty-six-acre tract known as Mt. Soma, the name in reverse being Amos, for the family who farmed the land for six generations—two hundred years of stewardship that ended in 1915. The developer discussed running a sewer line along a tributary of Winters Run, and the County Public Works office liked the idea. The developer took his idea to a local savings and loan. Six months after buying the property, he borrowed two and a half million against it. The sewer line remains a pipe dream; the developer and the savings and loan are down the drain. It was not always thus. A brochure prepared by the Amos family in 1915 lovingly describes the homestead:

"The handsome dwelling is beautifully situated. Swirling air currents of Summer make it a cool spot. In Winter, hills to the north and west protect it from cold. Four crops of alfalfa hay have been cut in one season. Four hundred bushels of tomatoes are grown to the acre. Before the railroads, Mt. Soma was the first stop for cattle drovers from Baltimore who came to rest and feed in the green meadows where stock was fed corn and hay grown on the farm.

"The fruits begin to ripen with the strawberries and continue through to the winter apple, the hickory nuts, the persian and black walnuts. Enough fruit is sold annually to pay the farm taxes and insurance. The birds of Mt. Soma affirm that nature intended it for a home site. No place will one find more bird-nests in Summer, nor hear more songbirds in Winter.

"Why is Mt. Soma offered for sale? The time has come, as it will come to all homesteads, when in the natural course of events it must pass from the family name."

The brochure ends with a photo of Mr. Amos, his hand raised skyward towards the tassels of twelve-foot-high corn stalks. It is a beseeching pose. Mt. Soma stalls for time.

Wednesday, December 5. The horse dentist floated the stallions' teeth yesterday. Tom Rowles did their feet today. Dr. Riddle performed end-of-year insurance exams. Stallion contracts came and went through the office. The orchestra is tuning up. Mother Nature will soon tap her baton to the music stand. She will instruct the sun to lengthen the day. She will see that the stallions are ready to play, and chide the mares into harmony. On February 15th, the performing arts center at the stud barn will open for another season.

Thursday, December 6. Thirteen weanlings and thirteen yearlings left today to live at the leased farm. The yearlings loaded like veteran racehorses. We took clients' weanlings in the next van, then loaded our own babies into Ellen's gooseneck trailer, which has no ramp—the horses simply step up to load. Many horses prefer a gooseneck, because the action of stepping up requires them to lower their heads, whereas a trailer's ramp sends the head upwards, to shy from the trailer's ceiling.

Tonight, the farm is catching her breath, our hilly fields receiving a rest. We will pull soil samples, dress the fields with potash, spread straw to stem winter erosion in areas where traffic was heavy. For once, it is fun losing boarders to another farm.

Friday, December 7. Maj. Stryker prepped us before we drove off for a weekend gathering of Ellen's family.

"You're headed to the Northern Neck of Virginia? Stop at Stratford Hall, the ancestral home of Robert E. Lee, and Wakefield, the birthplace of George Washington. The Lees and the Washingtons did much to encourage the breeding of Thoroughbreds.

"The *American Turf Register* describes Washington's colt Magnolio, a son of Lindsey's Arabian, a gray stallion Washington and Light Horse Harry Lee—Robert E. Lee's father—had seen during their New York campaign in the Revolutionary War. About 1790, Magnolio raced against a colt owned by Thomas Jefferson at the Alexandria Jockey Club—the only time a president of the United States and a future president opposed each other on the race course.

"Oh, my yes—Virginia! How I wish I had the steam in my engine to explore again the treasures of the Old Dominion! Safe trip. Take care of my star pupil."

Saturday, December 8. Now I've seen everything. Today, grown people painted numbers on the blue shells of crabs, then bet which crab would win the race off a picnic table. I found the spectacle pretty amusing, which reminded me that a Carnivalay colt named Pretty Amusing was headed postward in the Maryland Juvenile Championship. I placed a dollar wager on the Number Six crab, then dialed the Laurel Hotline. Both my picks finished fourth.

Sunday, December 9. On the tour of Stratford Hall this afternoon, our hostess commented on the paintings on the walls: "This handsome horse you see is Dotteral, imported from England in 1765 by Philip Ludwill Lee, who often claimed the horse was the swiftest in all of England, excepting Eclipse."

Ellen studied the red turban on the horse's handler, and the minarets in the landscape.

"Doesn't look like a Virginia horse to me," she said under her breath. "Looks more like a Barb, maybe a Turk."

We politely asked our hostess for particulars.

"Oh, I'm terribly sorry. That's not Dotteral. It's the Byerley Turk. Dotteral is in the parlor upstairs. Philip Lee imported many Thoroughbreds before the Revolution. Upon his death in 1775, his widow could not cope with the business of his racehorses and sold them off."

I found a wall plaque at the handsome stallion barn. "Dotteral was born in 1756, height 15 hands three inches, powerful, strong-boned, the sort most esteemed in England for a stallion. His stud fee was Six Pound ($20). Advertised as a great grandson of the Godolphin Arabian."

Philip Lee could not have foreseen that the Godolphin Arabian would be the least important founder of a tail-male line, almost exclusively through his grandson Matchem. At the time, the Godolphin was the most celebrated of all Eastern sires. The only trace of Dotteral is in a list of horses imported to Virginia in *Blooded Horses of Colonial Days*. Dotteral was probably like nine out of ten stallions through the ages: Unable to live up to his owner's expectations. I suspect Philip Lee's widow was glad to be rid of her husband's hobby.

Monday, December 10. "His foals had the opportunity, they had the looks, they were in good hands—occasionally they showed promise—but they simply didn't train on."

Jonathan Sheppard is a Hall of Fame trainer. He has owned a share in Assault Landing for seven years, which has resulted in three crops of foals, the oldest of which are about to complete their four-year-old season—and not one stakes winner in three crops. Sheppard is one shareholder among forty, but he seemed to best capture the assessment of the stallion's offspring: "They simply didn't train on."

I don't know how long Philip Lee held onto poor old Dotteral before he began sending mares to neighboring Fearnaught, known as the Godolphin Arabian of the Old Dominion, but the handwriting—not an oil portrait—is on the wall for our dear stallion.

Tuesday, December 11. When Tom Rowles trims Citidancer's four white feet, crimson rings of bruises—the legacy from a career of rocketing down hard racing surfaces—can be seen growing down through the wall of his hoof.

Against Housebuster in the Jerome, he dueled for six furlongs in 1:08 3/5. I understand where the bruises have come from.

"Horses with dark feet bruise, too," Tom told me. "You just can't see them. When the heels are trimmed down on racehorses to give a longer reach, the laminae of the hoof wall is stretched. As we get a heel growing on this new colt, you won't see any more bruising. Those days are over."

Wednesday, December 12. On my walk to work this morning, I passed John King at the yearlings' water tub, crinkling the hose to loosen ice that formed overnight. His five-foot frame is outfitted in the latest fall fashions: Lime green sunglasses, an orange and blue Baltimore Gas and Electric baseball hat, his windbreaker from the Kappa Alpha house at an undetermined university, and a pair of laceless, insulated Moon Boots. He looked at me as he shook the hose. His smile, like his outfit, went from the ridiculous to the sublime as he stamped the day with his indelible impression.

"Hey! Getting a little frisky out here!"

Thursday, December 13. Dad and Uncle John came into the office and closed the door behind them. I must have looked bewildered.

"We've lost top men before, and we've always managed to replace them," they said, voices overlapping. "It's part of the game. We'll adjust. Okay? Good."

The source of my puzzlement is Peter's retirement: How am I going to replace him? A string of nagging physical injuries—compliments of a lifetime with horses—and the allure of Florida, where his son Gordon is a racing official, compelled Peter to switch his tack to the Gulfstream Park neighborhood. His last day was today. Peter was in charge of herd health, the most important job on the farm. Horses these days receive more vaccinations than you can shake a stick at. As Peter's friend, I wish him the best. As manager of this farm, I must immediately replace him, without missing a single shot.

Friday, December 14. "I think Southern States should pay for the damage," Steve Brown is telling me. "It's not our fault the tire flew off the fertilizer buggy and rammed into a Toyota in the high school parking lot. We got lucky. It just missed a Cadillac."

By the time the tire was replaced, it was raining. The buggy, loaded with four tons of potash, now rests in our hay barn, blocking access. It might be February before the ground is right again for a four-ton vehicle to negotiate these steep hills.

Saturday, December 15. Mt. Soma's main house was arsoned fifteen months ago, and so the farm's only access—the wooden bridge—has been scuttled by local fire officials predicting repeated torment. The farm's four wooden barns have no doubt been eyed by the arsonist. Tonight, I crossed the bridge on a twelve-inch timber. Empty beer cans littered the stream below. On the hill, the house—a mansard-roofed, stone structure with scalloped shingles flanking dormers of its steep facade—sits brooding its fate, timbers open to the sky. It looks like a bombed church. The house is beautiful still. The artwork of a stonemason is not easily undone, and I stood to admire the long, single-stone lintels above the windows, the weaving of random fieldstones.

The traffic on Route 1 made considerable noise, but could not overcome the spell that abandoned farmland exerts on the mind, a spell cast by spirits of people who once worked the land. This is no Stratford Hall, mind you. No sweeping view of the Potomac frames this scene. Yet an innate dignity to the land resists the sounds of cars, the lights from commercial development.

This land deserves better. It is the same Maury loam as Country Life, separated by forty feet of blacktop, and an endless stream of speeding cars. I left Mt. Soma tonight feeling hopeless about its future, but things change in ways no one can forecast. If I do nothing else, I will at least chronicle the changes. It will help young Josh someday to see beneath the surface of whatever stands on Mt. Soma, and understand its history. I hope and pray Country Life can avoid the sad fate of her neighbor.

Sunday, December 16. "The world was not a pretty place when Jesus came into it," the preacher said this morning, "but the Gospel says he arrived in the 'fullness of time.' "

I think he meant conditions were right, but then daydreaming during a sermon leaves me only a cut-and-paste account. I choose the "fullness of time" aspect since the boy in my arms arrived like that, born on his due date, in a year of milestones for this farm. Yet our country is in a recession, and might soon be in a war. It is not a pretty place for many people. Dad lifts his depressed A.A. members by telling of his dream to be in Kentucky someday, on the first Saturday of May, with a baby by one of our sires. I too hope for great things, but this fine year will always be, for me, the fullness of time.

Monday, December 17. "The big picture!" the professors harped on us the first year of law school. "Do you see the big picture yet?"

I hear twigs snap in the woods, glimpse the white tail of a deer in flight. In the fields, a chevron of geese pass low overhead. On the ground, green blades of grass defy frost and regenerate with impunity since hungry horses left. In the office, wheels turn. No end to paperwork. I tell myself the horse business is a sand trap: Stay in the fairway, aim for the green, carry a wedge. Keep your head down when you swing.

The big picture. A most pleasant business. My word, yes, I did my best this nettlesome Monday to keep my eye on the big picture.

Tuesday, December 18. A troop of the Winters Run Preservation Association showed up for the Mt. Soma auction that didn't happen. We called the number posted on the Public Auction sign tacked to the oak tree along Route 1: "Auction canceled," the real-estate agent said. Thanks for telling us. A lady in a red Monte Carlo drove by, saw us gathered, hit the brakes on the wet pavement, spun a hundred and eighty degrees, drove south in the northbound lane, turned around and drove off, touching her hand to the back of her hair. It was surreal.

Ellen left this morning with the boy for Christmas in Virginia. I join them Friday. Tonight, the scent of the baby was captured in the cotton of his jumper on the sofa. The sense of smell brought them back. We have not been separated except for the night of the Breeders' Cup since he was born on October 1st. I miss them terribly. I can not keep my mind on my work. This is a new feeling, to be a father alone in his family's house.

Wednesday, December 19. Dr. Riddle is the Flying Vet. He flies his private plane to surgery sites all along the East Coast. He flew around the farm today. He ultrasounded a two-year-old's tendons, oiled and gave Banamine to a colicky mare straining in her last month of pregnancy, administered EVA shots to the stallions, and filled out fertility insurance forms for the new stallion—all after X-raying the colt's healing sesamoid. He then discussed at length a new worming medicine that is fed daily, then he left me a cassette tape covering immunization of foals.

Roaring away for an evening of surgery on Standardbreds, he looked back over his shoulder and flashed the "Thumbs Up" sign pilots use. I have often marveled at this man's energy. Today, I thought of the spiritual hymn *May the Work I Have Done Speak for Me*. In Dr. Riddle's case, this would be a long song.

Thursday, December 20. I'm not alone, after all. Maj. Stryker is in my study.

"Have you named your yearlings?'' he asked, his face buried in the thick book *British Flat Racing and Breeding*, Charles Richardson's 1923 tome on *Race Courses and the Evolution of the Racehorse*. "You only have until February 1st to do so for free."

No, we haven't, but all babies by Corridor Key will have Key in their names, the Allen's will have Prospect, the Carny's will have Valay. Racehorses should advertise their sires.

"How droll and unimaginative! Yet history is replete with practical names. Eclipse was foaled in the great eclipse of the sun in 1764. His grandfather was a little horse name Squirt, got from a mare named Sister to Old Country Wench. Squirt's father was Bartlet's Childers, sometimes called Bleeding Childers, because he frequently bled at the nose.

"Just bear in mind that only once in a blue moon does a great horse come along. Be prepared. Name all your foals well. Eclipse was more than well-named. He represented a confluence of the forces of nature—the sun and the moon, the strains of centuries of blood. You simply *must* name your best prospect for the blue moon on the rise as we speak. It will appear in the fullness of time, on New Year's Eve."

Friday, December 21. The stationmaster's voice bounced off canyons of loading platforms, as Christmas travelers strained to hear homelands in the echoing atlas of towns: Charlotte, Spartanburg, Gainesville, Birmingham, New Orleans.

I love the train, the southbound route, starting at Washington's Union Station. The engine switches from overhead electric—all power goes off in dark idled cars as we await the *thump* of the diesel engine engaging the coupler. Power is restored, and we emerge from a tunnel to pass within yards of the Capitol, to cross the smoky Potomac in awe of the circular strength of the Jefferson Memorial, through Alexandria, Quantico, Fredericksburg, slowing down through Ashland, home of the Camptown Races, antebellum houses lining the tracks, candles in every window, a stunning display of light before all cars stop in Richmond.

Tonight's journey had nothing to do with horses or syndications or paperwork. I felt immediately revived, as I do every time I recognize the world outside the paneled fences of a horse farm.

Saturday, December 22. The one-man farm crew at Clover Hill is comprised of Bill, who has a singsong southern voice husky from decades of filterless cigarettes. No voice—not his mother's nor his father's—lifts the boy's curiosity as much as Bill's.

"*Where* you goin'? *How* you doin'?" he sings to the boy. "See those *horses*? They look too fat to you? Dr. Newton said they're too fat. I told him *these* horses are always hungry. Hate to see a hungry animal. Just because you're fat doesn't mean you're not hungry now, *does it*, little fellow?"

The first settlers of Virginia came in search of gold. The mother lode was not gold but tobacco, and John Rolfe improved the Indian strain by planting a sweet, rich seed from the West Indies. English and Dutch smokers couldn't get enough of the aromatic weed, and plantations sprung up on the James, the York, the Rappahannock, and the Potomac, blossoming from the export of rich tobacco crops. Before Clover Hill, Bill worked at one such plantation, Tuckahoe, along the banks of the James. He lit another cigarette, and spoke through the smoke to the boy.

"I can't teach you about horses like your Daddy can, but I can sure teach you about cows."

Sunday, December 23. My holiday in Virginia is made more relaxing because of a farm meeting we held Wednesday to outline responsibilities. With Peter retired and the two O'Meara boys back on the Emerald Isle for a Christmas vacation, the brunt of running Country Life has fallen to Michael, Richard, and Steve Earl. Mom, of course, is invaluable in sorting out urgent phone calls, and Dad will keep the home crew smiling: "One of my A.A. cronies came up to me during a football game. He said he was considering suicide. I told him to come back at halftime. He laughed. You gotta keep a sense of humor in this world."

I used to be unable to leave the farm without worrying. It occurred to me tonight that fatherhood has subtly realigned my priorities.

Monday, December 24. Maj. Stryker phoned this morning: "The cat's on the roof, but that's not why I called. Did you get a copy of *The Rough Riders* by Theodore Roosevelt? His original 1902 book has been reprinted. Read it."

I did. The Rough Riders were three regiments of the First United States Volunteer Cavalry: Seven hundred and eighty wild riders, riflemen, and frontiersmen of the four Territories of the Rockies and the Great Plains. Another two hundred and twenty volunteers came from Harvard and Yale, or from the New York City Police Board, or from neither club nor college, but "in whose veins the blood stirred with the same impulse that once sent the Vikings over sea," in Roosevelt's words.

"Half of the regiment's horses bucked, or possessed some other of the amiable weaknesses incident to horse life on the great ranches; but the men were utterly unmoved by any antic a horse might commit. We had a pack-train of a hundred and twenty mules, and close on to twelve hundred horses."

The Rough Riders really didn't ride much in the campaign to drive the Spanish out of Cuba. Most of the horses, and the entire pack train, had to be left in Florida. Roosevelt wrote: "The deficiency in transportation was the worst evil with which we had to contend."

Roosevelt's emphasis on the role of horses proved accurate for another four decades after the Spanish-American War. Two World Wars later, the U.S. Remount was disbanded. *The Rough Riders* is a link in the chain.

Tuesday, December 25. Alice hung pine boughs from the mantelpiece, and I paused to admire her handiwork, and to study the beautiful oil portrait of the chestnut colt Henry of Navarre, Domino's great rival of the eighteen-nineties. August Belmont gave this painting to Grandfather. Although a great racehorse, Henry of Navarre proved a failure at stud. In 1909, he was sent to stand at Belmont's French stud farm. On February 16, 1911, Belmont announced he would give the United States government six Thoroughbred stallions to start a national breeding program, and Henry of Navarre returned to America to become the foundation stallion of the Army Remount Depot in Front Royal, Virginia.

"The Remount sent offspring of Henry of Navarre to the Western states, and the cowboys loved them," Dad told me.

I smiled to think that this state's oldest breeding farm has a portrait of a failed stallion over the mantelpiece. Yet I've walked past Henry of Navarre for years without knowing his role in the Remount, or that he beat Domino three out of the five times they met, which included a dead-heat match race. As a child, I would stand on the bench in front of the fireplace and stare off into Henry Stull's landscape, imagining myself riding through the rolling fields on Henry of Navarre, just like the leggy jockey captured in a casual pose in the black and crimson silks of August Belmont. How wonderful it felt this Christmas night to revisit the landscape of childhood impressions, fortified with the knowledge of an adult.

Wednesday, December 26. Citidancer is inbred two-by-four to Northern Dancer—that is, his sire is by Northern Dancer, and his third dam is by Northern Dancer. As a racehorse, he was a flying billboard for the benefits of inbreeding. As for his capacity to reproduce himself, we won't even know if he is fertile until February. *The British Racehorse* once commented:

"Perhaps the commonest fallacy is that inbreeding must create undesirable characteristics of some kind, such as infertility, loss of vigor, over-excitability and so on. But inbreeding by itself can create nothing. It is not a creative force, but simply a means of sorting out the virtues and faults inherent in a strain or family."

The current emphasis on inbreeding is a matter of history repeating itself. In Eclipse's pedigree, Mr. Richardson's book points out:

"The name of D'Arcy's Yellow Turk occurs four times, D'Arcy's White Turk six times, the Lister Turk five times, the Leedes Arabian twice, and the Fenwick Barb twice. Eclipse was doubtless greatly indebted to the blood of those horses whose names appear so frequently in his pedigree."

And an old horseman once told me: "To say a horse has too much of a great sire's blood in his pedigree is like saying you have too much money. You can never have too much of either in this game."

Thursday, December 27. This afternoon was "one of those days" compacted into four hours: Hay truck at the loft, van at the chute, accountant on the phone, snow in the clouds, feed in the tubs, mares in the stalls, wood in the stove, heat in the office, bills in the mail, message on the fax.

Out on the farm, men huddle against the snow.

"Your father just called from Laurel," Mom said, back in the office. "He said: 'Find him. Tell him Allen just had winner number fourteen.' "

Bread on the water.

Friday, December 28. The morning sun softened the crust of ice before we turned the stallions out. Carnivalay led the way, pushing down to get purchase on the soft grass underneath. The sharp snow grabbed at his heels. He high-stepped slowly to his field, his hocks working overtime, his long tail skimming behind.

Snow had blown into Allen's Prospect's end of the barn. We spread a sandy substance called Barn Snow to provide him traction. Richard spoke sternly: "Walk! Walk! Walk!" and kept the powerful stallion on a straight course to his paddock. No shenanigans would be tolerated this morning.

At my desk, a zillion end-of-year details clamored for attention. As I heard the melting snow drop from the office roof, I felt calmed knowing that the stallions—the reason this farm exists—were safe in their paddocks.

Saturday, December 29. The fog is so thick the fields are indiscernible from the sky—all one white, melting slice of atmosphere. At the leased farm, yearling fillies gallop through the mist. Who will excel as a two-year-old for the coming year?

Late afternoon. I dial Philadelphia Park. Promised Lad needs to win his allowance race to send Allen's Prospect over the four hundred thousand dollar earnings mark. He was beaten a nose, so Allen is about three grand short.

"The difference is illusory," said shareholder Orme Wilson.

When you stand stallions, you want to live where it's always Saturday. That's when you dream. That's when they run for the money.

Sunday, December 30. Hot air thermals swirled around us as the thermometer climbed quickly to seventy degrees, the eight inches of snow melting before our very eyes, green pastures emerging as if in a time-lapsed film. The creek is swollen with clean water, not the violent brown runoff of topsoil as when hard spring rains arrive.

Ellen is humming *Repeat the Sounding Joy,* whose lyrics are of the land and fields, when heaven and earth shall sing.

Monday, December 31. The blue moon rose a rich yellow sphere as we loaded the black stallion for his new home at a farm on the bluffs of the Susquehanna River. I drove behind the van, filled with mixed feelings at moving a horse who once represented such promise for us. He landed here in white bandages the autumn Ellen and I were married. The following spring, Ellen held the leg strap during a breeding session on a short-handed Sunday morning. I arrived moments later. Uncle John said:

"I think she'll be all right. Her head will be sore for a few days, though."

"What?"

"Blackie threw his feet forward while he was up on the mare—you know, to get a better purchase, and his hoof struck Ellen. She'll be all right. Just a knock on the head."

I ran to the house and found her on the sofa, holding an ice pack to her swollen forehead. I wanted to scream at the horse business for forcing me to expose her to danger. I wanted to take her away from all this effort. It was no use. Ellen loves horses. She was being careful. It just happened. That's all. It just happened. That dreadful Sunday morning incident melted into memory as I drove over the hill tonight, as the moon appeared over the deep black water of the Susquehanna.

"He stands for a tie chain like a trained dog," I told his new handler, Mark Herman. "He was raised at Claiborne. Charlie Hadry trained him. Whatever bad habits he has can be blamed on me."

In the woods tonight at home, the moon shone so bright you could discern the shapes of leaves on the forest floor: Pointed tips of the black oak, the scalloped beech, the familiar maple, the thin, racy willow. It is a worthy light to end such a wonderful year.

This year, we greeted a baby boy and a new niece, a Valay Maid and a great breeding season. We saw Eohippus in Colorado and evolution in Saratoga: Go for Wand. Mom and Dad had their fortieth wedding anniversary, and Allen's Prospect homered on Maryland Million Day. The kid saw Northern Dancer, Rollaids was adopted, and a brilliant young colt enrolled in the stud barn. Uncle John and Dad each celebrated their tenth year of one day at a time. The kicks in the head were not in vain. The blind effort of the eighties is now a pipeline of promise for the nineties.

Tonight at this moment, Ellen is upstairs playing a bluegrass album—Bill Monroe strumming *Blue Moon of Kentucky* as the kid claps his heels together in a Johnny Jump-Up spring seat. In the basement study, the radio is all banjoes and the bouncy refrain: "Let's Kiss This Thing Good *By-ee-iiii*!"

We move to the new year with the music of this fine year still playing in our heads.

❖

Year Three

CENTERED DOWN ON A THOUGHT

Tuesday, January 1. Many farming families in our county are members of the Society of Friends, also known as Quakers. Younger sister Alice attends Quaker Meeting.

"We are very respectful of each other's opinions," Alice has told me. "The meeting begins with a quiet meditative period called 'centering down.' Then, individual members may stand and speak their mind. Afterwards, the hall falls silent, and the members reflect on what has been said."

Such a fine term. That is how I feel about the farm on this New Year's Day. The excitement of the past year is history. I now must "center down" on the future, for the busy days of the coming breeding season, when Alice speaks her mind, and Michael his, and John and Ellen and Richard theirs—all the minds of the people who help me run this farm.

In the calm days of January, a period of concentration will settle over the farm, and we'll ask ourselves: "What has experience taught us?" And we'll center down on the lessons we've learned.

Wednesday, January 2. The hill overlooking the stallion paddocks provided a vantage point to assess the team. Carnivalay's white face gazed from a hundred yards away.

"Where's Blackie?" his eyes asked.

It is so easy to be sentimental about horses. Yet, I confess I do not miss the black stallion, Assault Landing, who left on the blue moon of New Year's Eve for a fine new home. Instead, I feel relieved a decision has been made and carried out. The syndicate vote was almost all in favor; only two opposed—as crisp as can be expected in a business peopled by independent minds. In the past nine years, we have syndicated seven stallions, *three* of whom are no longer here. We attempt to practice the spirit of what a shareholder once told me: "I'll buy into your stallion, but if he isn't a good sire, I want *out*. I am not from the 'Buy 'em and bury 'em school.' " Indeed, the language of many syndicate agreements is drafted only to create the syndicate, not to disband it, yet the vast majority of syndicated stallions do not fulfill the shareholders expectations.

Into the empty paddock, we've rotated Corridor Key, whose first foals race this year. His former paddock will soon belong to Citidancer—so young, rearing in his stall, a cracked sesamoid taking him from Olympic fitness to fitful confinement. He clamors for release. Until Dr. Riddle gives the green light after clean X rays, he stays in. When he does come out, a mare will be waiting, a new career begun. He has expectations to fulfill.

Just like a professional sports team, players come and go on this club we call the stallion roster. Well, it's either them or us, I tell myself as I leave the hill, the pitching rotation set for this season. After all, I've got people to take care of, and people are more important than horses. Unproductive stallions do not put food on the many tables of Country Life.

Thursday, January 3. Alice, who imposes order on the periodic chaos of a farm office, is on two weeks vacation. Paperwork is mounting. Self-help office management books insist that you should never touch a piece of paper twice—just keep it moving downstream. Yet on my desk are the volumes of bills and records that are boiled down into quarterly statements for four syndicated stallions. It's a logjam because I can't proceed until I learn the new computer program. At the moment, I can't even access it. Meanwhile, the copier is crushing my two-sided stallion contracts because they are printed on cotton bond instead of slick sheets; every contract I feed through causes the "J" for jam to flash, the hood to rise on this finicky machine.

Three phones are on my desk, cleverly color-coded in white, yellow, and black. Yet only a trained Labrador could discern the correct ring. I answer all three before a lady from AT&T is heard through the white telephone. I interrupt her sales pitch on her three-thousand-dollar system. The ringing line has hunted to the vacant yellow phone. I feel like the harried phone operator Lily Tomlin played on *Laugh-In*.

I calm myself: The phone is a cash register, be glad it is ringing. But I can't find the black phone, which has Call Waiting. It is buried under the quarterly statements. It is *possible* to be on four calls on three phones at one time. Clearly, we need a new phone system, but they are not cheap. I take a swallow of coffee gone cold. If Alice comes back before I get these bills out, I'm in trouble. I'll be prepared, though. If she asks me what I was doing all this time, I'll look her in the eye: "I was on the phone," I'll say.

Friday, January 4. Seven cronies from Kentucky flew in tonight to spend Saturday and Monday in a goose-hunting pit on Maryland's Eastern Shore. As I approached the Bay Bridge at dusk tonight, on pavement formerly the flat fields Finney managed as Holly Beach Stud, I thought about the legislative report predicting the Bay's health in the year 2020. The conclusion might be called the Dead Sea Scroll.

The 2020 Report works hand-in-hand with the "critical areas" legislation. It affects not only farmers of the Eastern Shore, but my own Harford County as well.

Controlled growth is the goal, development limited to one house for each twenty acres. Farmers in Harford County as well as here on the Eastern Shore are up in arms: "Twenty-to-one will put us out of business. We can't borrow against our land if it's devalued like that," they clamor. "This isn't a bill to save the bay. It's a bill to steal the farms from the farmers."

I was twenty-eight when I first fired a shotgun. I hunt once a year on this annual gathering of Kentuckians. If I kill a goose, Dad will throw him on the grill on Super Bowl Sunday. I am a mixed bag of uneasy rationalization tonight—half hunter, half hating the idea. Yet again this year, the sense of adventure, the prospect of seeing Maryland's vanishing wilderness before the 2020 predictions come true, has overcome the strength of my pacifist convictions.

Saturday, January 5. "You, with the glasses, keep your head down when the geese are working," says the guide for the commercial hunting service as he singles me out of the group. We stoop in a stand-up blind in a cornfield the size of BWI airport.

Apparently, geese don't work on Saturday, and the guide's bad jokes have played out before he discovers a mother lode of fun mining my occupation as a supervisor of animal matings. He is unrestrained when the role of a teaser is described. I am not alone in my calling, however. The Kentucky team includes Alex Rankin, who runs a family farm in Oldham County, Kentucky, lyrically named Upson Downs Farm. He foals mares and ships to breeding sheds around Lexington, and he has seen firsthand the unfunny parts of farm life. He described a mare whose uterus prolapsed following labor. I am certain this predicament will be my next major episode in the *Rocky Horror Picture Show* that life in foaling season can become.

Alex winks at me as I bear the brunt of the guide's teasing, sharing our secret knowledge that we both have a more rewarding life than does this fellow who hunts with the sordid enthusiasm of a mercenary.

Sunday, January 6. Old-timers will tell you that as soon as you move a stallion, he'll come up with a good horse. Mention "Japan" to a well-bred stud, and he'll immediately set about siring a stakes winner. I imparted this odd logic to the gaggle of goose hunters today at Laurel, and they raced to the windows to bet on Assault Landing's offspring—except for Craig Grant, a former shareholder. He was not going to spend another nickel on anything to do with the black horse.

The boys in camouflage coats and hunting caps jumped with glee as Snork cruised home a five-length winner, and rushed off like Elmer Fudd's family to cash generous Exacta tickets—except for Craig. He sat glumly in his seat, mumbling: "That's *twice* that stallion has cost me money—once when I bought that share, now again today when I didn't bet his runner."

Monday, January 7. No geese flew in the snow that fell heavily all day, and today's hunt disintegrated past the bad jokes and into a game of watching the guides shoot the heads off the decoys at fifty paces. Upon my return home tonight, Maj. Stryker scolded me for such idle pursuits.

"A gentleman," he told me, "never shoots at still prey nor angles for a sluggish fish. Leave such practice to the ignoble pot-hunter. And to have left your son champing at the bit for more lessons on the origins of the Turf!"

He's teething, Major.

"Oh. He *did* seem almost rabid about John Stuart Skinner, founder and editor of the *American Turf Register*."

Who?

"John Stuart Skinner—a name which should not pass away without some embalming token of grateful recollection. The pioneer of the American agricultural press, the champion of those who nobly toiled at the plough, the loom, or the anvil!"

Tuesday, January 8. "John Stuart Skinner," the major picked up where he left off, "was the father of American Turf Journalism—born in 1788 on a vast plantation where the Patuxent River meets the Chesapeake Bay, right across the water from where you went goose hunting.

"To provide American readers with ideas and records on agriculture, he founded the *American Farmer*, whose contributing writers included Thomas Jefferson. In 1829, Mr. Skinner published the first number of the *American Turf Register*, and in August of 1835, he sold the *Register* for the enormous sum of ten thousand dollars. In 1848, he launched *The Plough, The Loom, and The Anvil*.

"He was in the midst of planning, in his words, 'a flying visit to the good old Eastern Shore of Maryland—the land of good hominy, good oysters, good ducks, good mutton, and good men,' when in 1851 he slipped down the steps of his office and fell to the cellar. He never regained consciousness. Ripe in years and honors was snatched this true-hearted citizen."

Until today, I had never heard of John Stuart Skinner, a man with the foresight to compile the pedigrees upon which is founded the American influence on the Thoroughbred breed. The things I don't know about this game would fill a library.

Wednesday, January 9. For the "Fifty Years Ago" column in the *Maryland Horse*, Marge Dance today showed me the entry written on January 9, 1941, by her father, Humphrey Finney:

"There were no casualties, except among the foxes, as there were two kills. Drew covert back of house, ran over open country, with nice worm and post and rail fences. Killed on the Patuxent River in full view of the entire field. Mr. Riggs presented the mask to Mrs. Parish and the brush to Mrs. Gimbel. Three of the hounds broke through the ice and could not get out. Mr. Janon Fisher Jr. broke through and found the water over his head. Mr. Waugh Glascock, fearing for the hounds and Mr. Fisher, went to their rescue and also went in up to his neck, but (all) were rescued and none the worse for wear. The next day they all seemed in the best of health and spirits."

This evening, I opened the book on the Skinner family again, and found a passage written by J. S. Skinner's son, Frederick: "My father took such pains with my venatic as my literary education. I could shoot, skate, swim and ride to hounds before I was out of my teens."

I majored in English, but cannot recall having seen the word "venatic," meaning "of or pertaining to hunting." Why is it that once you start studying a subject, it stands out from the background, whether in Finney's diary or Skinner's tutelege of his son? This month has been of hunting and horses and Turf writers, and of the icy waters of a January in Maryland.

Thursday, January 10. The letter in Grandfather's files was written in 1943 to Dr. Bardwell, the veterinarian in Lexington with whom he boarded his mares. In closing, he confided: "You are so fortunate your sons are not old enough to serve." All three of Grandfather's sons were in the service. This farm was sapping his health, already weakened by a stroke in 1939.

The current generation at Country Life was born on the cusp, too young for Vietnam, too old for duty in the Middle East. Yet Richard's son-in-law Ryan Cool is serving as a tank mechanic in Saudi Arabia. Richard's daughter Roxanne and five-month-old granddaughter Ashley are now living here on the farm until Ryan's return. The situation in the Middle East is very much on our minds these days. Grandfather's letters help me keep it in perspective.

Friday, January 11. The best-kept secret in Maryland is the existence of a first-class Turf library open to the public: the Selima Room. Named for the great race mare of the Colonial era, the Selima Room is a repository of such famous volumes as Leland Stanford's *The Horse in Motion* (published in 1882, featuring the silhouetted images of a galloping horse produced from the camera of Muybridge); Lady Wentworth's *Thoroughbred Racing Stock*; the Paul Mellon collection of *British Sporting and Animal Drawings*; Frederick Remington's paintings. Shelves and shelves of books are devoted to sporting art, to American, English, and French stud books, as well to current information: *Daily Racing Forms*, microfiche broodmare records, trade periodicals, racing manuals, sale catalogues.

On my first visit to the Selima Room last month, I was overwhelmed that such a treasure of books existed so close to home—a mile from the Bowie training center, fifteen minutes from Laurel, a half hour to either Baltimore or Washington. I told friends about it. "Never heard of it," they said. At the Bowie branch of the Prince Georges County public library system, the Selima Room should be a required field day for every student of the history of the Turf.

Saturday, January 12. Keeneland is no place for novices—unlike Timonium, where Laddie Dance will call you by name, tease you into bidding another hundred dollars. A few years back, we discovered how quickly the Kentucky auctioneers march mares through the Keeneland ring. John Stuart acted as our agent. He paused to advise us very carefully: "Leave your reserve with the auctioneer. But don't get greedy. She will never be worth more than she is tonight."

A reserve should be an odd number, so brother Michael and I set a figure of eleven thousand dollars.

"Who'll give me twenty thousand for this handsome mare?" the auctioneer boomed. No one would. "Who'll give me ten?" I was confused. Ten? I don't want to sell her for ten. I raised my hand. The board said ten thousand so fast I was not certain it was my bid. "Who'll give me fifteen, fifteen, fifteen?" All I could think was the reserve was eleven. I raised one finger, Timonium-style, but they don't take one thousand bids on Tuesday at the November sale. The board flashed fifteen. "Mike, I just *raised* my own bid by four thousand more than our reserve. What do I do now?" Before Michael could answer, the national marketplace spoke. Somebody had raised two fingers. "Now I've got seventeen. I'm gonna sell her at seventeen. Sold, Paulie, out back."

Johnny congratulated us. "Sure is exciting watching you Marylanders play the big top."

Tonight, we're back in Lexington, hoping to buy a mare for a client this time. I remind myself: Keeneland is not for novices.

Sunday, January 13. Horse folks can almost always recall their first visit to Lexington, the first time they drove down Ironworks Pike, or Paris Pike, or Newtown Pike, immersed in the largest concentration of magnificent farms anywhere in the world. The stone walls of Castleton, the gates of Elmendorf, the sweeping entrance to Spendthrift—these are powerful images.

Michael Cataneo lost Landing At Dawn for sixteen thousand dollars in a claiming race at Laurel ten days ago. He intended to breed her to Citidancer this spring. I told Mike that for twice the amount she was claimed for, he could buy a major-league broodmare prospect at Keeneland. There is a war on the horizon; we are in a recession. Step up to the plate; take a swing.

In our rental car, I pulled up to the closed iron gates of Brookside Farm in Versailles. "This is Mr. Paulson's farm, Mike. He raced Allen's Prospect." At that moment, the pressure plate underneath us triggered the gates, and they swung back to reveal stately barns with chandeliers in cupolas off in the vast distance. I was as surprised as Mike, who seemed to levitate off the passenger seat, so strong was the image of the barns on the horizon.

Yes sir, Lexington is a powerful place, almost illusory. My job, though, is to stay focused on the goal of buying the right mare at the right price for Mike to breed to Citidancer. I might have to sit on him to keep him from bidding. In the emotional ego charge of an auction, you can get caught up in images.

Monday, January 14. We looked and looked and looked, and no mare in our price range gave me that tinge of excitement. As we walked out of the barn area at 6 P.M., I paused to marvel at the dedication of the people who prepare and care for these animals. It was cold and dark at dusk, just as it will be before dawn tomorrow, when these same folks return for another day of cleaning stalls and picking hooves and grooming mares. There are many unsung heroes in this game.

"How'd you do today?" Alan Hutchison, a friend from the Windfields days in Maryland, asked as we left.

"Didn't fire a shot," I said.

"Well, kept your powder dry for tomorrow."

Tuesday, January 15. "How come she didn't race?" we asked her handler.

"She cracked a hind sesamoid in a paddock accident right before she was to be sold at the Keeneland summer yearling sale. Here are the X rays."

Mike, this is a summer sale filly. We didn't come here to buy a pedigree we could get in Maryland.

But she's by Fappiano. She could cost a mint. They don't make any more Fappiano fillies.

Mike, she suits Citidancer. It's the Northern Dancer—Mr. Prospector cross. She's only three. You can breed foals from her for years. She's good-boned and straight enough, and she'll fill out to be a beautiful broodmare.

But she's late in the sale. What if we don't get her?

We'll come back Wednesday.

But we leave Wednesday.

Not until five.

I don't like to get to planes late.

She's the one.

Wednesday, January 16. When Michael P. Cataneo signed the ticket for Spellcast at thirty-two thousand dollars last night, he couldn't wait until the summary sheets were posted on the wall of the walking ring, next to the vanning information desk. He scanned the blue print-outs for his name. I had wanted to sign Country Life Farm, as agent, but did not want to deprive Mike of the fleeting satisfaction of being listed in the box score. He was euphoric. I resisted the temptation to mimic Steve Martin: "The new phone books are here!"

Then suddenly, we both came down to earth, literally and figuratively, and I will never forget the moment the plane taxied to a stop at BWI tonight at eight thirty-five. The pilot said: "As you pass through the terminal, you may want to stop at a TV. Things have started to heat up in the Persian Gulf."

Thursday, January 17. Almost two hundred years ago, an advertisement for the stallion Young Diomed appeared in the *Maryland Gazette*. In the issue dated May 1, 1806, Benjamin Ogle, Jr., spelled out the terms of the breeding contract:

"To cover mare, $15, and one dollar to the groom. Or, a note of $20, payable the 1st of January, 1807, must be sent with each mare, or they will not be received; said note may be discharged by payment of $15 on or before the 1st of November next. The (breeding) season will commence the 1st of April and end the 20th of July."

Essentially, this ad called for a no-guarantee season at fifteen dollars, or a live foal contract at thirty-three percent more. Young Diomed must have stood in a bull market, for this ad appeared before his sire, Diomed, the first English Derby winner, had achieved fame as the sire of Sir Archy (foaled in 1805 and the first great native-American Thoroughbred sire).

We have a stallion whose book is full. Shareholders will only sell their seasons on a no-guarantee basis. I have uncovered an important historical precedent that will make the purchase of a no-guarantee season more palatable to mare owners.

Friday, January 18. After buying Spellcast the other night, we were so excited I neglected to have her vetted immediately.

"Be a good idea to get a vet to stick his arm in there," John Stuart glibly advised that night. "Make sure she has two healthy ovaries. You should also get her oiled before she ships to Maryland. Might be an excitable mare. Don't want her colicking on you after eighteen hours on the van to Maryland."

Although my pride in my agency role suffered slightly, I was not too stubborn to take good advice. Dr. Roger Murphy said: "This right ovary feels like she's just gone through a heat cycle," and my mind jumped at the idea of crossing this stout filly with the fine Arabic presence of Citidancer.

Saturday, January 19. Maj. Stryker was grounded today.

"How can you keep me here when Pretty Amusing is running in the stakes named for me?" he huffed. It was a difficult question. Carnivalay's son Pretty Amusing would be second choice in the Goss L. Stryker Stakes for Maryland-breds. I appealed to his sense of family.

"Because your star pupil gets christened tomorrow, and your study is a mess," I said. "Books everywhere. Papers scattered about. Family will be pouring in here tomorrow. For the sake of domestic harmony, let's just clean up today. Dad will call if he has good news. Okay? Okay."

Sunday, January 20. "It's a religious war to them," Rev. Jack Cooper began his sermon. "It all started after the death of Mohammed, the prophet of Islam, in 632. He left behind a militant following. Exactly a hundred years after his death, the Saracens—as the Moslems were called—invaded Europe by coming across the Barbary Coast into Spain, and up into France."

This was not your typical Sunday sermon: I was paying attention. My recent fascination with the Darley Arabian, Byerley Turk, and Godolphin Barb had taught me that the warriors of the Middle East had been breeding a superior horse centuries before formalized horse racing began in England. I knew the Saracens must have been riding across the plains of Spain on ancestors of the Darley, the Turk, the Barb.

"At the Battle of Tours in France in 732," Rev. Cooper continued, "Charles Martel—'Charles the Hammer'—repulsed the Saracens. Charles had metal shields and lances and armor, and Christian civilization was saved from Mohammedan supremacy."

In my mind's eye, I saw a Saracen falling off his stallion in defeat, the horse captured and soon interbred with native mares of France and Spain. A thousand years later, the Darley Arabian came a more direct route from the high plateaus of Arabia: He was delivered by Bedouins to the Syrian city of Aleppo, a great commercial center on the main caravan route across Syria to Baghdad. Thomas Darley signed a Bill of Lading describing his Arabian stallion, and the horse left by boat for London.

This is how horses should be introduced to a foreign land—not as spoils of war, but as items of commerce. Yet war is again upon the land, a religious war, both sides invoking the name of God. As Rev. Cooper splashed holy water on young Josh's forehead, my mind was racing like wind across sand, relating thoughts of war, of horses, of sons of families who believe fervently in their gods. It was all too new to me to make sense of it. I felt dizzy. Back in the pew for the sermon, I calmed down, and so did the guest of honor.

"I see my little baptized friend has fallen asleep," Rev. Cooper gently interrupted his sermon, his kind eyes smiling at the infant. Then he turned, knocked the red wine off the shelf, muttered something under his breath that sounded unprintable, and suddenly the world was right again—a place where priests are human and babies sleep and life goes on.

Monday, January 21. Over the heavy Dutch doors of the foaling stall, the round faces of young Josh and Ashley peered curiously from the warmth of thick winter baby suits. They stared at the heat lamps illuminating the nativity scene. The first foal of the year lay steaming in the straw, attempting to rise, whinnying loudly, slapping front legs together. Ellen and Roxanne held the baby children high in their arms. Richard smiled at his granddaughter, and I laughed with my son, and the first foal of the year seemed right at home in his new world.

Tuesday, January 22. Last night's colt played in his field happily all day. This evening, however, he appeared to be having trouble urinating. Adhesions on his prepuce could be the culprit. A young colt doesn't relish manual extraction of his penis, but if the source of discomfort is not discovered, continued straining could mean a ruptured bladder. The scent of a foal's first manure is now on our sleeves, a reminder of the wrestling match. The problem was not his penis. We'll just have to watch him carefully. The constant monitoring of the foal crop has begun with Number One.

Wednesday, January 23. "My father-in-law is the chaplain for the Blue Angels, out of Pensacola, Florida," Brad Murray told me today over the phone from Midway, Kentucky. "My wife's two-year-old colt by Corridor Key is a handsome dark gray. We named him Naviator, kind of like an aviator in the navy. We'll send this colt to the Fair Grounds in February. I've made him Maryland Million eligible. I'll keep you posted on his progress."

Brad's voice imparted optimism that Naviator would jet to fame, and before any two-year-olds by Corridor Key have even started, we are filled with anticipation. Glowing reports from distant lands kept spirits buoyant throughout the day.

Thursday, January 24. Belair Stud, which until 1955 was the oldest continuing Thoroughbred farm in America, was not in my hometown of Bel Air. Rather, it was located about fifteen miles from Annapolis, in what is now Bowie, Maryland. The magnificent plantation estate, later referred to as the Woodward Mansion, was built in 1740, and took dominion over more than two thousand acres. Belair was owned by Governor Samuel Ogle, who imported Maryland's foundation stallion and mare: Spark and Queen Mab. (Spark was sired by Aleppo, son of the Darley Arabian; Queen Mab came from the Royal Stud at Hampton. Ogle owned the best-bred stallion and mare in the Colonies.)

After Ogle's death in 1752, his brother-in-law, Col. Benjamin Tasker, Jr., took possession of Belair. One of the finest horsemen of the Colonial era, Tasker imported Selima, a daughter of the Godolphin Barb. Tasker sent Selima out in his silks on December 5, 1752, against four rivals for a stake of ten thousand dollars—a princely sum. Selima finished in first place, and the rich race signaled the beginning of great sporting rivalries between horsemen north and south of the Potomac.

Belair passed to James T. Woodward in 1899, and in 1910, William Woodward inherited it. Woodward kept his stallions and mares in Kentucky, but shipped his foals to Maryland at weaning time. Grooms walked the weanlings the three miles from the train station to the welcoming redstone arches of the Belair barn. Triple Crown winners Gallant Fox and Omaha—celebrated racehorses in the white jacket dotted with red that were the silks of Woodward's famous Belair Stud Stable—had walked through the town of Bowie as weanlings. I've seen the train station. I love that mental picture.

After Woodward's son, William, Jr., was killed in 1955, Belair was sold to Levitt and Sons, developers of Levittown on Long Island. The Levitt legacy is not all bad. They did not build their monotonous houses on the front lawn of the mansion, or in the courtyard of the stables. Instead, they deeded these islands in a sea of houses to Bowie. The city's heritage commission has recently embarked on a renovation of the mansion; the stables are preserved inside a hurricane fence.

This update on the famous nursery is not simply the latest chapter in the history of a Maryland landmark, for this farm is part of the heritage of everyone in America who raises Thoroughbreds.

Friday, January 25. Michael showed the stallions to several Pennsylvania horsemen. Alice was on the phone to a client in Puerto Rico. Mom called Kentucky to say a mare had arrived. Ellen photographed and tagged the new boarder. Richard smiled with appreciation at the box of cookies, bubble gum, and Chapstick to be mailed to his son-in-law in the Gulf. Dad went racing at Laurel and brought home a booking from a Virginia breeder. A lady from Florida requested an insurance form for a no-guarantee season.

I stayed at home and read about stallions in Colonial days. My real job today was baby-sitting the kid, watching him scuttle across the floor in a rolling walker. Staring down at him, I noticed the pulse in the as-yet-unclosed fontanel in the center of his head: his soft spot. He was very excited about being mobile. I picked him up. His head against my cheek seemed to hum. The phone rang: "Hello, I'm calling about the ad for a baby-sitter," the caller said.

I thought of how smoothly the farm was functioning today without me. For a moment, I considered submitting my own applicaton for the baby-sitting job.

Saturday, January 26. "Mike Cataneo's homebred won forty thousand in purses, another eight thousand in Maryland breeder and owner awards, then was claimed for sixteen thousand," I was explaining to Tom Roha and his wife, Ronnie, on their visit today. "She represents about sixty-five thousand dollars to the ledger—before expenses, of course. Just as importantly, she put Mike to the winner's circle, after no luck with his first two racehorses. This business rewards perseverance—usually. Your day will come."

The Rohas entered this business last February, the afternoon Lucky Lady Lauren and Valay Maid won stakes at two different tracks. Carny fillies were hot, and Tom bought a two-year-old filly by Carny named Faithful Tradition. Since then, she has swum more miles than Johnny Weissmuller, been pin-fired and blistered, hand-walked and sand-paddocked. Now she awaits the long road back from a strained tendon. With luck, she *might* race by August. In the meantime, the broodmare they bought, Turn to Money, is in foal to Carnivalay and due February 10th—almost a year to the day after they entered the business. The next time they visit Country Life, the Rohas will hopefully be proud parents of a white-faced foal speeding around the paddock. The promise of a homebred sustained Mike Cataneo through tough early years. I expect it will do the same for the Rohas.

Sunday, January 27. "We've got a filly in the second race, Ellen. I'll go to ten-thirty Mass with you and the baby, but I have to leave at eleven-fifteen. It takes an hour to get to Laurel. Dad said he'd pick me up in the parking lot. I'll be quiet. The preacher won't even hear me leave, I promise."

Rev. Cooper must have a direct line to the Lord, because at eleven o'clock, he waded out into the aisle of the small church, turned back to face the altar, and knelt down right beside me. He then began a Litany for Septuagesima Sunday that lasted twenty minutes. Dad is a good sport about everything *except* getting to the races late. I prayed to God the Father, Rev. Cooper the Father, and Dad the Father, and when the priest wound down, I stood up and made it out of the church without disturbing a soul.

Later, smiling, Ellen said: "I told everybody you went to the races. They didn't care. They were too busy passing Josh around after church."

Monday, January 28. The recent cold snap froze the bottle of Regu-Mate stored in the medicine cabinet in the broodmare barn.

"Don't use it when it thaws," Dr. Bowman warned. "Freezing and thawing might have damaged the molecular structure of the chemical, rendering it ineffective."

You thaw it. You shake it up. You *think* it's all right. But it might not be. And you only discover that it might not be when you casually ask a vet a question you think is silly. At two hundred dollars a bottle, we can't use the Regu-Mate. I swear. There's simply no end to the worrisome details that go into running a horse farm.

Tuesday, January 29. Not since Crazy Roost have I experienced such abnormal waking hours. Crazy Roost was a patchwork of red plumage crowing from fence rails at 3 A.M. or 3 P.M., summoned by an errant inner clock. His daylight broadcasts were often silenced by a well-aimed rock, which increased his nocturnal sense of impugnity. I am reminded of the sleepless nights of Crazy Roost's reign, as Richard leads heavy mares in each hand past the office window at seven-thirty, their breath misty in the morning cold. I am half asleep from a night of pacifying a baby boy. He is teething. I call him Drools Verne. Ellen calls him Prince Farming. He calls us several times a night. At his four-month checkup today, he was vaccinated with short stabs in the fat of both thighs. He cried so hard his tongue turned purple.

Today is Ellen's birthday. I hope Crazy Roost, Jr., sleeps through the night. That would be a fine present to his mother. Yes, a very fine present.

Wednesday, January 30. "Looks pretty normal," Dr. Riddle explains as he passes the ultrasound probe over Faithful Tradition's tendon. "The fibers of her superficial flexor tendon were damaged. The monitor shows small black holes where ultrasound waves bounce off granulated tissue, which is less dense than the undamaged fibers oriented in dense parallel lines on the screen. A tendon is like a piece of string: If you pull it slowly, it won't break. If you yank on it, the fibers tear. That's how horses strain tendons, suspensories, check ligaments, curbs.

"This filly can go back into light training: Six weeks of slow jogging on a firm surface. Shoe her with a flat shoe, no toe grabs, a high angle to her foot, a short toe. Exercise is the best thing. A tendon is not static. It heals and is remodeled according to the stresses required of it. Paint the tendon each evening with an iodine solution called Balls, mixed with DMSO. The odds are seventy-five percent that she'll stay sound."

Thursday, January 31. In pairs the mares arrive. From Virginia comes Castalie and her daughter Gramercy—old friends from years past. Castalie is the mother of Ritchie Trail, foaled here, the favorite perhaps in the Gay Matelda Stakes at Laurel on Saturday. From Kentucky comes Arctic Valley. The wind whips her heels as she prances past, stocking feet still snow-white from sales shampoo at Keeneland. High on her rump, a tiny square of hair remains matted from the glue of the hip number. No pal for her yet. Wait until tomorrow. Don't fret. Today was a day to introduce myself as the proprietor of this horse hotel. I work the crowd memorizing new faces. Their names are mouthfuls: Kneadthepasta, Honorary Doctorate.

"Either I'm getting smaller or these mares are getting bigger," Dr. Bowman is remarking on his first day at the vet chute since last July.

"It's neither," John explains in his lilting Irish accent, proud of his off-season maintenance. "We've just added a foot of stone dust where the mares stand. We'll have it raked down to the clay again by June, never fear."

I examine a small, healing cut on Kneadthepasta's shoulder. Steve Brown walked the field three times and found nothing. Still, the board bill states: "Vet examined wound. Penicillin three days. Topical ointment. Healing nicely." I remind myself to handwrite a memo to accompany the bill, and search for the right words. "Abrasion will heal without blemish," I decide.

Near the vet chute, Ellen records Dr. Bowman's comments on barren mares. *Palp*: Heavy uterus, no sign of heat. *Spec*: Cervix inflamed, red-flag this mare. *Palp*: Just ovulated. *Culture. Treat. Suture.*

In the office, Alice and Mom brainstorm names for eight two-year-olds, trying to save the fifty-dollar fee assessed by The Jockey Club for names submitted after February 1st. They flip through atlases, and let their fingers do the walking through the *Registered Thoroughbred Names* book. I bet them a crab-cake we get all eight names, knowing I'll lose, but straining to hear the new names, a distant clash of cymbals that might one day ring from track announcers: "It's Ordnance by two...and Portage Path surges to the front...on the rail is Block Island..."

The wind is howling through the pines at quitting time, and I walk home through the field of late-foaling mares. My orange jacket startles a high-strung mare, and she dances through the herd—with her tail straight up, like a gaited horse. High above, riding an arctic blast, a flock of Canada geese speed for the waters of the Bay, which I know they can see from their vantage. Hardly lifting a wing, they race through the sky, centered down on a thought, and are gone.

Now so too is January.

A CUCKOO'S NEST

Friday, February 1. Richard lifted oak logs onto the gas-powered log splitter, and all across the farm, over the drone of the engine, you could hear the wood rip. The oak was left over after we timbered ten acres a few years ago. Thus seasoned, it will soon warm woodstoves scattered in various farm lodgings. I watched Richard working like a beaver, hoisting logs twice as round as a man's waist. This was an old oak tree. The tree might have seen Indians and hunters. It certainly had seen loggers and farmers. Its concentric rings transected the history of this land.

February is a month when smoke curls from woodstove chimneys that poke skyward like stacks on ocean liners, a month no more than a fraction of a ring in the diameter of an oak tree. Despite the vexatious weather February often visits on us, she can't keep another spring from coming.

Saturday, February 2. Philip answered the phone when I called back home from Virginia. I was bursting with curiosity to discover how Ritchie Trail fared in the stakes at Laurel. Michael's son, at two and half years of age, enjoys an extended greeting.

"Philip, is your dad home?"

"Hi!"

"Phils, let me talk to your dad, please."

"Hi!"

"Hi, Philip. Where's your dad?"

"With Poppa Jody. Witchie won."

"What?"

"Witchie Twail won."

From the mouth of babes, there you have it—the result of today's tenth race at Laurel.

Sunday, February 3. Beech leaves of the understory clapped in polite applause for the glorious seventy-degree day. The old gray mare danced a jig on the quiet paths of Virginia's Goochland County before settling into a vigorous walk, the dry leather of the saddle squeaking in rhythm.

By the lake, a marsh bird chattered across the water, and twenty-six Canada geese waddled off the bank for the safety of deep water. We encountered a chained gate in a thicket. The gray mare marched into a stand of multiflora rose in response to my leg command. To open and close a gate from horseback requires a good horse, not necessarily a good rider. She moved back as the gate swung open, drew alongside the gate as I reached for it, stood patiently as I redid the chain from the other side. Back at the barn, I rubbed dry straw where the saddle pad had matted her coat, and rewarded her with oats. When I turned her loose, I was surprised to see lather between her hind legs as she cantered away. Was it that warm a day in February?

Country Life is a farm where horses are a business. Clover Hill is a farm where horses are for the absolute pleasure only a ride through the countryside can provide.

Monday, February 4. Many of the old-time breeders fastidiously maintain a split-page pedigree binder, to view at a glance inbreeding or outcross strengths of a prospective mating, to recall precise comments about the resulting foal. The top half of the binder displays the five-cross pedigree of various stallions. The bottom half of the split pages is for the broodmare band.

Mr. Janney could flip the pages back to the nineteen-thirties, to his three foundation mares. Orme Wilson, in whose crimson silks Ritchie Trail races, carries his binder with him whenever he inspects our stallions. Hal C. B. Clagett, perennially a leading Maryland breeder, ponders matings in the warmth of his Weston Farm study, frayed edges of his black binder encasing dog-eared pages, the length of gestation penciled in for every foal.

I have watched these men as they consult their binders. Their faces wince when they recall youthful mistakes. They smile when good horses leap from the page. The binder is their almanac, their diary, their legacy to those who come after them.

Tuesday, February 5. The test mare's name is Quality Gal. Her right knee is sprung, and she is built more like Olive Oil than Betty Boop. Yet she is Citidancer's first love, and I find no fault with her. She lives up to her name as the young colt bites her hocks and withers, strikes at her flanks, dances against her side with a blind instinct that knows only the urge, not yet the satisfaction.

Citidancer is a steaming cauldron of hormones on this warm sunny morning. We have a firm but gentle hold on him with the stallion shank. He is impatient and unfulfilled, and lashes out with his hind feet. Veins on his neck stand out. We are not concerned with disciplining him. We do not care that his penis could use a good washing. There will be time for that later. There are many signals going off in his head. He is being permitted to have his way with Quality Gal, and he isn't certain how to go about it.

He bumps his chest into her rump. A light goes off in his head. He bumps her again, harder, and drops to his knees to bite her legs. I pray she won't kick. We are unable to restrain her with a leg strap because of the old racing injury to her knee. She looks back at him. "What are you doing?" she seems to say, but she does not kick. He sniffs her. She is receptive. Drops of preejaculate trickle from the colt's penis, and he slaps himself against his belly. He wants to mount her. He rears up against her rump, but does not straddle her with his front legs, and so he slides forward off her flank. He is very mad now. I think he might wheel and kick at her. "That's enough for now," I determine, and John tries to lead the colt back to the barn. He refuses, backs up. John eases the tension on the chain shank over the colt's nose. He stops. He wants to try again. "He's too hot, John. Put him away. Let him smell the mare on himself for a few hours. We'll try him again this afternoon."

John walked him for an hour to cool him down, then put him away. The colt wears a blue bandage figure-eighted over his injured leg for support. I watch for signs of lameness. He does not appear to be sore. His eyes are bright as he walks past me. This is all very new to him, but he is a very smart colt. Everyone who ever has been around him has told me so. He is inbred to Northern Dancer, a good-breeding horse. It is important to be patient with a smart horse.

Wednesday, February 6. This test breeding process is very much on my mind. I woke up this morning and clipped my fingernails short. You can cut a horse quickly when he is jumping around, when you are trying to insert him into the mare. I do not want to wear plastic gloves when I handle the colt's penis.

At this afternoon's session, we want him to perform while he is fresh, before he gets steamed up. The fresher the better. We put a rope shank on his off side, to guide him like a double-tied horse right in behind the mare. He has never been led from the off side. From birth, horses are not led on the off side. He fights the tension on the right side of his halter, backs up like an angry bull, his red face and perfect blaze lining us up right in his sights. Yes, it is like holding a very light-footed bull. He marches back towards us, almost puffing his chest forward. We aim him for the mare's rump. He jumps her, stays up, ejaculates, soaks a bit, gets down—quiet as a lamb. He goes back in his stall like, "Wow, what hit me!" A very satisfying moment for all concerned.

Back in the office—to the crew behind the controls—I announce: *"Lift-off. We have lift-off."*

Thursday, February 7. Castalie's arrival papers included a scribbled notation: "Vaginal discharge treated. Please keep an eye on."

There is no such thing as a *minor* infection in the vaginal tract of a pregnant mare. We lifted her tail up. A thin stream of mucous trailed from her vagina. The mare's due date is March 16th. An old mare, she is five weeks away from producing her twelfth foal. Dr. Bowman was summoned. Palpating the foal through the mare's rectum, Dr. Bowman felt that the foal's reflexes were slow.

"Most foals will recoil if you grasp their face," he said. "This foal did not. He's alive, though. That's encouraging."

With the ultrasound, Dr. Bowman pointed out flecks of floating debris in the uterus.

"She's an old mare. She's entitled to some wear and tear. It's difficult to assess how much of this debris is related to the infection, or simply to her age."

He inserted a speculum and a cervical swab. When he withdrew the speculum, it was covered in a mucousy discharge.

"Not good," he said. "Not good." Then he took blood from the mare, to run chemistry tests determining levels of fetal protein. "I am going to start her on antibiotics. I'll call you with the culture results."

Castalie has produced two stakes winners and is quite possibly the best mare in Citidancer's first book. This is an almost incidental observation now. Of greater concern is the foal inside her.

Friday, February 8. You almost want to flip your baseball cap backwards and get tied on, because the ride started today. At dawn, two Sallee vans waited expectantly at the loading chute, their cargo of thirteen mares (and one foal) from Kentucky ready to register for a four-month stay.

"We're here a week early—it was February 14th last year," the Sallee driver said. "I remember 'cause my son wrapped his car around a tree that day."

Everything seems earlier this season. A half-dozen foals are on the ground—a few more than usual. Opened the breeding season earlier, too. Dr. Bowman sent the first *real* mare to Citidancer, a handsome gray dowager named My Friend Irma. Citidancer approached her cautiously.

"Where's my gal Quality?" he asked, hesitant to step in behind a horse of a different color.

"She's gone on, Citi. Changed her mind about love. Wouldn't let Jazzman touch her this morning."

Saturday, February 9. What a morning. Hay truck here. John and David gone to the sales with five mares. Two new guys don't even bother to phone before not showing up. We turn out the Kentucky mares—after walking the fields for hazards. The most expensive mare in the group stops under a tree whose branches are ten feet off the ground. She rears up and puts her head in the limbs. She could have put her eye out—perhaps the only unforeseeable way to hurt herself in a one-acre paddock. I summon Steve Brown and a chain saw to trim the branches. Dr. Riddle is right: Horses try to commit suicide, all day long, every day.

Allen's Prospect breeds his first mare of the year, using a muscle he hasn't flexed in seven months. He gets hard, goes up on the mare, goes limp. He gets down, waits, tries again. We get him in the mare, and he expels a load of stored semen, but it is not a good ejaculation. We wait around for a few minutes. He is rekindled, comes back with purpose. This time we get a good cover.

The maiden mare who foaled last night will not permit her baby to nurse. Constant attention is required to hold the mare, sometimes with a twitch on her nose, before she will allow the foal to nurse. Peyton takes blood, does a colostrum check: two hundred. *Way* too low. We hurry to the freezer, slowly thaw sixteen ounces of colostrum pulled last year. It is as sticky as maple syrup. Peyton tubes it into the twelve-hour-old foal. We are ahead of the game. Another six hours and we would need expensive plasma therapy instead of simple colostrum.

I go into Richard's house by the foaling barn and make a cup of coffee that a spoon will stand up in. The first Saturday of the breeding season. We are in full swing, thank you.

Sunday, February 10. On our Sunday walk down the creek today, we discovered recent beaver activity: Two eighteen-inch-diameter hickory trees felled perfectly into the stream, stumps gnawed to the heartwood. Beavers are quick workers. Numerous poplar seedlings have simply disappeared, a circle of stumps the only evidence. The seedlings will surface downstream, where a huge beech tree left its creekside mooring and now serves as a seineing net for flotsam carried by winter rains. In the days before trappers cleaned out the beaver population, rivers such as Winters Run often were diverted by massive beaver dams. Flooded flatland gave birth to aquatic populations that fed birds and enriched the diversity of area species.

As we surveyed the cottage beaver industry springing up on Winters Run, a blue heron taxied down the rocky streambed and lifted off to soar out of the woods. No doubt he would approve of increased fishing habitat. Beavers can transform a rushing stream into a giant pond. We usually object strenuously to any force that might change Winters Run. The development plans by the beavers, however, has given us pause, for I love the sound of white water rushing over the rocky gorge. I kind of hope the beavers get washed downstream so someone else can decide their fate.

Monday, February 11. A good horse carries a lot of weight: Everyone wants to climb aboard and take credit for his success. At a seminar last week at Laurel, I climbed atop our young stallions and explained it is more than simply breeding fifty mares their first year and praying.

You have to find the right prospect, sift through race records and pedigrees for the proper mix of speed and blood; negotiate a deal; market the young horse; assemble folks who breed to race; weather the early uncertain days before his fertility is established; tease, palpate, breed, ultrasound, recheck; suffer through early fetals deaths and attrition at foaling time; buy his weanlings at auction, raise them right, pay their Maryland Million fees; send the yearlings to training centers; protect his shares from the market before his fruit has ripened; hand-pick a half-dozen trainers. You do this, and more, for *four* years before the first two-year-old breaks from the gate. Then you pray—pray that back there on Day One, you backed a winner.

Tuesday, February 12. Castalie left today for Dr. Bowman's. In the event she enters labor prior to her due date, at least a veterinarian will be present. We contemplated sending her to New Bolton Center, which has an excellent foaling program for problem mares, but Orme Wilson participates in the Maryland-bred program. He did not want a Pennsylvania-bred. I am keeping my fingers crossed. I want to get through February without any casualties. I am reminded of this month *last* year, when a foal put his foot through his mother's rectum, when another mare delivered a foal with deformed hind legs. *Two* years ago this month, Rollaids made his debut, after a few hours of trauma in the birth canal.

Lately, the phone line has crackled with similar stories. In Virginia, where cattle farms abound, are many fields planted in fescue grass, where gestating mares graze on toxic endophytes; their red-sacked foals arrive unannounced and suffocate in placentas thickened by the endophytes. A new Pennsylvania breeder is baptized by fire as his maiden mare aborts at ten months. Here in Maryland, Dr. Bowman pulled a foal out in pieces from a mare last week: "It was revolting. He had decomposed. There was no warning. The mare never even went off her feed."

You drive by a horse farm and see foals bouncing about, and you might think, "What a wonderful time of year!" But don't overlook what goes on behind the scenes.

Wednesday, February 13. "Have you or any member of your immediate family ever owned or operated a handbook or any bookmaking establishment, or been associated with bookmakers? Yes [ ]. No [ ]. If Yes, give particulars."

This question, on page one of the four-page owner licensing application, is no more ridiculous than asking a fox if he has been sneaking around the hen house. What bookmaker is going to answer this question truthfully? Or give particulars?

"Yes, me and Uncle Tenoose make book. We hang around with bookmakers, too. Here is our social security number and address. The password is Swordfish."

If you own five percent or more of a racehorse in Maryland, you must be licensed. I spent four hours this morning on applications for twelve newcomers to Thoroughbred racing. These are not "Partnership Friendly" forms, in an era when small racing partnerships are so commonplace. Poorly drafted, confusing, repetitious, the Maryland application form is intended to prevent bookmakers and felons from obtaining licenses, but instead it infuriates a well-intentioned newcomer to the sport, who has better things to do with his time, such as paying attention to his *real* business so he can afford this game.

I should have known better. Multistate licensing forms and uniform FBI fingerprint cards are available by writing the Racing Commissioners International, c/o the Kentucky Horse Park in Lexington. But nobody uses this service. We all just jump through the old hoops, which vary from state to state. I told the new people in our racing partnership: "Just grin and bear it. You get a parking sticker and two passes to the track as a reward."

Thursday, February 14. A Help Wanted ad is likely to rain responses on you. We advertised for a new night watchman a few weeks ago, and although we hired a young gal, the phone keeps ringing. The recession has created hardship and unemployment for many folks. We are so busy during breeding season that we often lose sight of the real world outside the farm gates. I am so very grateful our stallions got hot before the recession hit.

Friday, February 15. "You ever been stung by a dead bee?" Walter Brennan asks Humphrey Bogart over and over in the film version of Hemingway's *To Have and Have Not.* That is the question I keep asking myself about the developer whose bankruptcy petitions have frozen the Mt. Soma property. Today, we sent a mass mailer to the Winters Run Preservation Association, urging attendance at a reception for county council members at the Liriodendron mansion in Bel Air. A popular site for luncheons and wedding receptions, Liriodendron is an example of a possible future use of the Mt. Soma mansion. We need to plant that seed in the minds of the newly elected council.

Meanwhile, a Florida developer has been sniffing around Mt. Soma. In addressing the Winters Run group today, I felt a little like Walter Brennan, hopping from one foot to the other in paranoid anxiety, warning that even a dead bee can sting you.

Saturday, February 16. Coldest day of winter today. Ice froze a foot thick in plastic water tubs in the fields. Richard broke through it with a hammer, but he was no match for the bitter wind whipping out of the north. Water is a very big concern in this weather. Colic from bowel impactions is a risk in cold weather—when exercise is restricted, when field waterers are frozen, when pounds of dry hay are hungrily devoured. We don't have heated waterers in the smaller fields, so we brought those horses in early and topped off their water buckets before heading to the stallion barn.

As we bred two maiden mares, hair from their coats blew off where the stallions' legs had clung for purchase. The mares shivered. The lighting program had tricked them into believing warm days of spring were coming. They were shedding winter coats instead of putting them on, unable to resist hormones triggered by photo sensors in their eyes. A great deal of effort and expense has been lavished on these mares since December 1st, when they went under lights each evening instead of roaming with unfettered access to dark run-in sheds. After the mares were bred, we gave them HCG to encourage them to ovulate tonight. In the small hours of the morning, a conception will occur inside these mares. Two weeks into next year, they will produce valuable January foals for their breeders.

Meanwhile, the wind howled Mother Nature's objections.

Sunday, February 17. Finally, late this afternoon, the baby surrendered to sleep. Ellen followed his cue. The little house on the prairie was quiet. Then the phone rang on Ellen's bedside table. I think I said hello.

"You sound rather soprano-like," Marty Friedman said. "Did I wake you?"

Marty loves the opera. On car rides back from the Meadowlands, he often joins Pavarotti in a sing-along that brightens the bleak New Jersey nightscape.

"Been a long weekend, Marty. Can I call you later?"

I set the phone down absently. Then I heard a voice: "If you'd like to make a call, please hang up and try again." I rushed to the phone before the beep beep beep alarm could awaken Ellen. Marty is calling to tell me his mare foaled last night—that's my guess. I want to call back and congratulate him, but working in the cold weather just has sapped us all.

Monday, February 18. My heart pounds before I summon courage to speak at a public gathering, but I can always calm myself by invoking the democratic spirit of the Founding Fathers: Organize your thoughts. Speak your mind. Be brief.

"You said more sites like Liriodendron should be acquired and preserved," I said when a local historian finished his slide presentation to the county council members. "I would like to suggest Mt. Soma. It is less than two miles from Bel Air, and the mansion—although recently arsoned—has a magnificent vista over-looking Winters Run and the hills of Country Life."

"My word, yes, Mt. Soma would be ideal for adaptive reuse," the historian declared enthusiastically, adjusting his bow tie in reflection. "It's unfortunate that commercial development has so encroached upon it, but the site still has great potential as parkland."

The county has no money to fertilize this parkland idea. Yet hopes rise and fall. In private conversation afterwards, the historian—with the detachment of one who often has fought for causes and lost—remarked that the work of the arsonists might have spelled an end to the mansion: "Without a roof, the floor timbers will rot. When the floors rot, the walls will lose their bracing and collapse."

I contemplated the idea of throwing tarpaulins over the shell of the mansion. To buy a few months here and there, perhaps that would make the difference. Perhaps someday the county council might sit in front of the fireplaces of the Mt. Soma mansion and watch slides of how even an arsonist could not usurp the work of nineteenth-century stonemasons.

Tuesday, February 19. The crew warns each other of dangerous mares. Many horses simply *hate* veterinarians. Recently, while holding a horse for Dr. Rachel Blakey on a neighboring farm, my childhood friend Tom Polk took a hoof to the jaw, requiring a hundred stitches. His teeth are intact, we understand. When Polkie worked here, he was kicked in the stomach by a mare named Tooth Fairy. I am glad he did not meet the Tooth Fairy last week.

We are six, seven, eight people—thrown together in an outdoor classroom where lessons are learned at a dear price. We are each other's teachers, each other's guardians:

"Careful, she 'bout kicked the chute down yesterday."

"Doc said breed her, but she doesn't agree. Awfully early to get a stallion hurt. *Always* too early to get a person hurt."

"Don't twitch her. Look at the tag on her halter: It says 'No Twitch!' "

It is an underappreciated aspect of this business that so many people take physical risks *every day* on the farm, on the race track. The only real protection we have is each other.

Wednesday, February 20. In his report *Venereal Diseases of Mares*, Dr. D. J. Simpson of Newmarket, England, stressed that "control of infection is by detection and treatment *before mating*." This is precisely the reason all maiden and barren ship-in mares should have clean cultures prior to breeding. We insist on clean cultures, and it paid off this week. Dr. Bowman pulled a routine culture on a barren ship-in mare showing her first heat. He phoned us immediately with the results.

"Positive for Klebsiella," he said. "Do *not* breed her!"

I rushed to the textbooks: "*Klebsiella pneumoniae* is one of the most dangerous organisms...characterized by inflammation of the cervix and accompanied by a purulent discharge from the vagina...readily transmitted through the act of breeding...can migrate to the testes of the stallion...very persistent and very difficult to eradicate."

This breeding season could have been over before it started had not Dr. Bowman discovered the bacteria, had not Ellen done her paperwork to reveal the need for a fresh culture. Grateful for their vigilance, I am aware that we just dodged a bullet. I shake my head and mutter: There is always something, there is always *something*.

Thursday, February 21. "This filly's so tough she'll knock the taste out of your mouth," the head man at a Florida training center told me today, about Allegedly's two-year-old, a member of Corridor Key's first crop. "Not like some fillies, soon as you sit on them, they fall apart like a three-dollar suitcase. No sir. This filly's owned by a fellow who's closer to his money than nine is to ten. When he sprung for her, I knew she had to be good. I won't breeze her until her knees close, but I'm telling you, she's all class."

This fellow's opinion, however salty, sounded sweet to me.

Friday, February 22. The risk of breeding the wrong mare to a stallion is a constant fear to a syndicate manager. It has happened at every breeding farm, including this one. We've addressed this problem with an updated Breeding Shed Form (requiring clean cultures) and with Stallion Service Contracts mandatory even for shareholders. A breed-and-return mare must arrive with completed paperwork, or the van driver signs a Breeding Shed Form and becomes responsible for her. A mare boarded here is double-checked with the office records, then we sign her Shed Form.

Of course, we already had bred a dozen mares before I printed the form out of the word processor. From today on, however, the new rules apply.

Saturday, February 23. "All members of the Marchers and Shouters Society must be on hand," Orme Wilson cajoled the cheerleaders for Ritchie Trail, entered in the Jameela Stakes for Maryland-bred three-year-old fillies at Laurel today. "Charlie says she will run the best race of her life today."

So persuasively summoned, we hurried through afternoon chores and arrived near post time. Front-running Wide Country put some wide country between herself and Ritchie Trail, and I was hoping the last shall be first when

Ritchie collared Wide Country just yards from the wire, but we lost by a diminishing head. His voice hoarse from leading the Marchers and Shouters during the long stretch duel, Orme accepted defeat graciously.

"She loves a distance," he gasped. "We'll aim for the Pimlico Oaks the end of March. The stakes I would really love to win, though, is the Alabama."

The way Orme said it, for a moment there I had visions of Ritchie Trail standing breathlessly in the splash of lime they call the winner's circle at Saratoga after she captures this summer's Alabama.

Sunday, February 24. All the world dances to a different beat on a Sunday. This winter—until today—I had not heard the distinctive song of migrating tundra swans, a Sunday chorus from on high, a hundred soprano voices at once, singing "woo woo woo woo!" Weekdays are for the atonal music of pressing details, of making the money to meet payrolls, of running a family-owned service station. The song of the swans today was a dreamy musical reprise, a reminder of why we chose to be farmers: For the luxurious freedom of being outdoors.

Monday, February 25. Richard roared away on the Case tractor to harrow fertilizer into the soil. Perfect timing, for snow is in the forecast—snow, to land on top of the granules, to dissolve them slowly, no runoff, no waste. Rich grabbed me at dusk, so contrite I thought he might cry: "The wind caught Carny's gate and the chain harrow got hung. Now the gate's twisted. Can Steve Brown fix it?"

Richard had been on the tractor since 8 A.M., stopping only long enough to switch his baseball hat for a blue woolen cap, the type Jack Nicholson wore in *One Flew Over the Cuckoo's Nest*. He had been ten hours on a humming diesel, a popping frozen manure across one hundred acres of pastures, raising a winter dust storm, prepping the ground for spring. I clapped him on the back, and a small cloud of dust lifted off his jacket, like the character Pigpen in the *Charlie Brown* comic strip. Brown fields smooth as brush strokes rolled away in all directions where he had harrowed. I stared at the landscape, at the four-panel fences disappearing over the round earth, at the stone driveways beside the young pines that carry the eye to one-hundred-foot spruce trees enveloping the big house. I admired the white farm buildings huddling in a cozy curtilage under silhouettes of ancient English walnut and Norway maple. Pretty as a country club before a reception, I thought.

"Rich, don't worry. We'll fix the gate first thing tomorrow," I said. "Now git home and git ya' some dinner. Ya' hear?"

Tuesday, February 26. Richard is "Chief" to this cuckoo's nest of a farm crew. He anchors the team of Steve Brown, our hippie handyman; John "I Got the Gate" King (alias Shorty); Prescott, whose age of twenty-eight outnumbers his consecutive days sober by ten; Cheryl, the "Radar" of this M.A.S.H. unit; Steve Earl, reliable but as temperamental as any stallion; John O'Meara, whose calming Irish voice ("Steady, harse, whoa.") brings stallions down to earth; Ellen, the resident equine paramedic, a figure in the early morning fog, papoose on her back, gliding through the Indian camp, making good medicine.

For the five months to come, this cuckoo's nest is an asylum from the outside world: It is breeding season.

Wednesday, February 27. What we do for a living is rather a specialized absurdity, an inside joke, easier understood if experienced in animal husbandry and wifery. The two fellows from Woodstock Farm who brought Native One for Carnivalay to breed understood our predicament today. At the very moment Carnivalay wished to ejaculate, Native One stepped forward. Carnivalay tried his best to follow her off his stone-dust breeding ramp, but she went from having the rump of a fifteen-hand mare to her actual size of sixteen and a half hands. Carny simply could not reach her at the magic moment.

"I think he bred her," John said.

"Sure wasn't much of a cover," we replied. We had been too busy holding the mare to have a firm opinion. John held Carnivalay near the mare, hoping the stallion would return, would drop back down. The stud further confused the issue by being indecisive.

"Mind if we put the mare in a stall and try him again after we breed Citidancer?" I asked. The Woodstock boys nodded.

"Might save us a return trip," they said.

Citidancer is getting his act down, but he is still very playful; he has not been turned out in a paddock yet. This stage will pass once he plays in a field instead of on a mare's back. He went up and down three times before we could insert him into the mare. Then he bred like a pro. The Woodstock boys nodded patiently again: "Can't rush a young horse," they declared.

We retrieved Native One from her stall and set her in front of Carnivalay's ramp again. If a fifteen-hundred-pound mare wants to walk one step, five men against her chest cannot prevent her. Carnivalay clung to her sides. He fell out of her with the end of his penis engorged to the size of a trumpet. "Bred her again!" John said. It wasn't the best cover, but we had reached the point of diminishing return, literally.

"It only takes one," we wisecracked to the Woodstock boys, who smiled, loaded the mare onto the van, and drove off for Chesapeake City, Maryland.

Everybody with a mare in heat wants a great cover, but it just doesn't happen every time.

Thursday, February 28. For the crew who work inside, the dry heat of a woodstove can become stifling. Alice opened the front and back doors of the office today, and the smell of spring revived us. On my desk are the month's board bills, item by item, twenty-eight days of sundry entries that a fellow who gets a bill for five or six hundred dollars ought to know: Culture, uterine flush, ten days of progesterone, one shot of Prostin, halter repair, insurance exam, foaling fee, IGG levels, colostrum, palp, breed, ultrasound, in foal, recheck, no twins, balance forward, amount due.

Alice's data entry efforts were splendid. After years of running a commercial farm, though, I am still overwhelmed at the labor required to send a mare home fat and in foal—and it is only February. What trifling details these chores seem tonight. I want to put them out of my mind, and stand under the stars and the full white moon rising over the watershed. I want to stop these round hills from spinning towards morning for just a moment, so that I can appreciate that I am a young man lucky enough to carry on a family tradition. On another farm, though, in another family, I might have been called to have fought in the war that ended today in the Middle East.

We put February behind us on this high note, and move forward into March with a great sense of relief. Spring is imminent.

THE OGRE

Friday, March 1. Mother Nature in March can be an ogre—a dreaded giant shaking us as if we are merely one of those glass paperweights with fake snow in it. When she shakes the glass ball, it's a roisterous winter scene. When she lets it settle, tranquility reigns over tiny people on a postcard farm. She decides whether we are scrambling over the countryside one minute, or resting on a bale of straw near sleeping horses the next. The force of Nature will send newborn foals in waves that stop and start up again, and mares to cycle in bunches, hustling us up the hill to the stud barn in sun or snow. Then the calendar will suddenly go blank, as Mother Nature rests.

How I feel about March depends upon where we are on the shake.

Saturday, March 2. "One swallow does not make a summer," wrote Aldo Leopold in *A Sand County Almanac*, from his Wisconsin farm in the nineteen-forties, "but one skein of geese, cleaving the murk of a March thaw, is the spring."

We sat on the stoop at sunset with the boy at six months, and together we watched the airshow: Geese in a dozen flocks of a hundred each, carrying low clouds on their backs north towards Canada, spaced across the sky within a quarter-mile, their honking arriving on the wind before the spectacle of so many birds in flight.

"This is life in the flyway, son. How do you like it?"

Sunday, March 3. All quiet on the racing front. Valay Maid, out since the Breeders' Cup, preps for her return. Ritchie Trail hones her game to face Wide Country again in the Pimlico Oaks on March 30th. Shine On Sarah soon will debut for twelve new folks to racing.

The Jockey Club returned half the names for the two-year-olds. Block Island and Prudence Park got through; Ordnance and Fini Key (by Corridor Key, the last foal out of Auntie Freeze) were denied. The Amazon Queen will be doing business as Reprimand (out of Scold, of course). Edgar Allen Pro, although a mystery, is now a registered name. We sustain ourselves on their promise. This feels natural somehow, as if we were accustomed to seasons of racing, when trains migrated north from the Carolinas, stopping at Havre de Grace and Pimlico, at Garden State, then on to Belmont in June, Saratoga in August.

Maryland's list of pioneering contributions to racing—the first Jockey Club, the first state-bred program, the first million-dollar sire showcase—also includes the first modern winter racing, at Bowie in 1958. I was four years old. I have never *not* known year-round racing. My genes, however, sense how Grandfather felt each spring, anticipating the debuts of Ladkin's offspring, the return of Ariel's veterans. Racing was a season back then, not an every-day-of-the-year grind.

Monday, March 4. Ellen records ten various vaccinations, including shots administered in a series, for a circulating population well above a hundred horses through the course of a breeding season. The status quo changes every time a van pulls in. Add teasing charts, breeding records, and veterinary reports to these daily records, and a snowball can become an avalanche. Arrival papers from well-intentioned mom-and-pop operations that carry the notation, "She's had *everything*," are not enough. When were the shots given? Was the series completed? Were prefoaling shots given?

Dr. Riddle gently preaches perseverance to us: "Gotta get it right. That's all there is to it. You gotta get it right." Overwhelmed by the job at first, Ellen has grown into it. It is, most emphatically, a task you just gotta get right on a horse farm.

Tuesday, March 5. Several barren and maiden mares stubbornly obey nature's mandate not to cycle until the days are *in fact* longer, not simulated through artificial lighting. The mares were sent here to produce follicles, to ovulate, to send the egg to the sperm, to begin the miracle of conception, when cells—each containing sixty-four chromosomes, the genetic blueprint of sire and dam—become an embryo, sex determined the moment sperm penetrates the egg, to result in a colt or a filly as fast as the wind or pitifully slow, with a blaze or no white, offset in the knee or perfectly correct, lop-eared or blessed with a fine head.

We'll take our chances, girls, just give us the opportunity. Please, for the sake of the owner who keeps calling, asking: "Why isn't my mare bred yet? She's been there since December 1st! What are you guys doing?"

Calm down. It's early. Remember the standard response to reproductive whims: "Welcome to the broodmare business." When the board bill is eclipsing the value of the mare, sometimes that is all I can utter in defense.

Wednesday, March 6. The first person to the broodmare barn in the morning steps into a cavernous bank barn, neat as a pin after the night watch girl has picked piles of manure out of stalls and swept up, and before the day crew has arrived to tease, turn out, muck, bed, water, and feed—the calm before the storm of activity.

The first person hears mares munching an alfalfa breakfast, notices stall screens flush with the floor, where air moves at ground level, where foals sleep with heads in the pure draft. The first person is betrayed by a whinny, which in turn alerts other mares, and before the spell is broken, the person backs out quietly, to look up at dawn, where other voices call—high flocks of tundra swans, sunlight refracting off white wings. Unfortunately for me, I am seldom the first person to the barn.

Thursday, March 7. A quietly-touted Allen's Prospect three-year-old, Prospectapade, makes his first start tomorrow, in maiden claiming for thirty-five thousand. That's about the cost of producing a racehorse. For our racing partnership of new folks, we deliberated submitting a claim. The condition book contains the rules of racing, but says nothing about taking horses from friends. I discussed this point with Dad.

"I'd call the trainer," Dad said.

"Well, I'd be kind of burned up if you claimed him after I'd told you how good he's been working," the trainer told me.

Dad was right.

Friday, March 8. Prospectapade led for three furlongs, tired, got bumped, finished last, beaten twenty-three lengths. I looked at Dad. He looked at me. We breathed a sigh of relief. Tomorrow, the racing partnership folks are all headed out to Bob Manfuso's Chanceland Farm to see Shine On Sarah work.

"Glad you don't have to tell them you claimed a horse that finished dead last," Dad said with a compassionate smile. The unwritten rules of this game are best learned from trusted advisers, whose instincts have been honed from experience.

Saturday, March 9. A typical Saturday in the breeding season.

At 7 A.M., Richard delivers Jazzman to the broodmare barn. It takes ninety minutes to tease and turn out some sixty mares, after filling in teasing charts, preparing vet lists, looking at foals for signs of trouble. By eight-thirty, we head for the morning breeding session; we finish jumping maidens and twitching untwitchable mares by ten. The vet comes at eleven. For two hours, he palps, treats, specs, and ultrasounds perhaps twenty mares.

Folks not holding mares are forking manure produced continuously since 4 P.M. the previous afternoon. Dr. Rupert Herd said mares defecate almost seven times in a twenty-four-hour period. I respectfully submit he's a little short on his figures. Mares save the majority of their business for the privacy of their stalls, and seven piles is nothing to them.

The crew breaks for lunch, then returns to battle manure. (I've read of organic sheep fertilizer called Baa Baa Doo, and am considering similar marketing.) Then stalls are bedded, with hay, water, and feed supplied. The afternoon breeding session begins at three. By four, we are bringing mares back into stalls, retracing our morning steps. If this crew—for whom I have such respect—can say, "See you in the morning" by five-thirty, we've done well. We are responsible for any mare who chooses to foal in prime time. The night girl comes on at ten.

Sunday, March 10. At noon, the cat is asleep on the sofa as two clients describe their experience with a commercial farm. I haven't been in my office all morning.

"They called to say she colicked. We drove up that Sunday to see her. When we arrived, everyone had gone to the sales."

I try to listen, but I'm having a bad Sunday. A key player is AWOL. I had to enlist Richard on his day off to hold mares for breeding. Rich works sixteen hours a day during the week. I hated to ask him. I am angry, but I try to keep cool. The clients want to breed their mare here. I am vaguely aware of a bad smell in the room. I glance suspiciously at the sleeping cat.

"We found our mare all right. She was *dead!*"

The clients stop speaking. Their eyes fix on the floor. They seem overwhelmed. I want to defend the unnamed farm, to suggest an explanation. I am also trying to figure out these people. I imagine myself on their hit list if their other mare decides to die *here*. I follow their stare to the floor. Then I understand why they are so distraught. Stingo, the lethargic Beagle, has left his calling card at the foot of my desk. He must have been locked in the office overnight. I explode in laughter. "I'm sorry about your dead mare, but I have to clean this dog dirt up."

"We were wondering if you'd noticed," they said. "We weren't sure what kind of operation you ran if you didn't even notice this in your own office. By the way, what's the stud fee on the gray stallion?"

Monday, March 11. The conversation as I *hoped* it would go:

"I did stuff like that when I was eighteen."

"Well, I couldn't very well have driven."

"You could have gotten up early and been back by nine, in time for breeding!"

"Sorry."

"The mares don't know it's Sunday. Missing one person from the breeding crew is a big deal."

"I said I'm sorry."

"Maybe Wednesday should be your day off, not Saturday."

"It *won't* happen again, believe me."

"I believe you."

Instead, there was no conversation. I was so mad at him I was afraid to speak. "I'm disappointed," was all I could say. I can't trust my tongue when liquor benches one of our players.

Tuesday, March 12. Any farm with a pond should install a PVC-pipe fire hydrant. Period. Last night, in a wet dry run, the Fallston and Kingsville Fire

Departments maneuvered pump trucks down steep hills to the pond, where rookies and veterans alike took turns playing two hundred and fifty gallons per minute upon the pastures.

"Now that we know this hydrant is here, we'll put it on our maps," the captain said. Alice walked the firefighters through the huge bank barn, built circa 1885. Their mouths dropped open imagining the fire load of seven thousand bales of hay and straw. The weight of combustible material per square foot of floor space in a wooden barn is almost off the charts. The firefighters then climbed three flights of rickety stairwells that would chimney the fire.

The hydrant would aid in watering down the buildings next to a burning one. I smiled when a rookie who'd seen the pond said: "Know what I'd do if this barn caught on fire? I'd get the stock out. Yep, I'd get the stock out—quick. Hey? Mind if I bring my boy fishing here Saturday?"

Wednesday, March 13. Peyton's vet report read like an emergency room file: "Princess Mistletoe's foal: weak, lethargic, unable to rise unassisted, umbilicus dripping blood, nursing poorly. Tentative Rx: 1) neonatal septicemia 2) neonatal isoerythrolysis 3) blood loss anemia. Treatment 80 mg gentocin 1.5 million units (6cc) Penicillin IV, 1 liter .9 nacl with 25% dextrose. Drew blood for CBC 2 profile/blood culture."

I called the foal's owner.

"This colt would be wolf bait in the wild, but Peyton is determined. I think this foal will survive."

Thursday, March 14. Sleet fell all morning, and we hustled to bring foals in off damp ground. Dr. Bowman showed up at lunch. No time for soup today. We worked straight through—bred all the mares he advised. Finished up late. Long day.

At quitting time, the house swallowed me into a deep chair, and young Josh in his walker strode purposefully—mouth agape with the drool from his first tooth—towards Pappy. From the stereo, Talking Heads sang:

> *"Hoooome!*
> *That's where I want to be.*
> *Pick me up and turn me round.*
> *Hoooo-ooo—mmmme,*
> *that's where I belong,*
> *Did I find you or you find me?"*

This fatherhood business is a ride down a sliding board into waiting arms.

Friday, March 15. If I had opened my eyes at chosen moments on the Amtrak ride from Baltimore to Providence, Rhode Island, I would have seen only the stately sycamores lining the boulevard outside the Philadelphia Zoo, or the crew teams from Penn sculling down the Schuylkill River, past the boathouses, under the bridges. I would not have seen the first of hundreds of discarded car tires along the tracks that somehow signal the entry to a world so different from the country life.

If I could have chosen when to look, I would have seen only the swans stopping to feed in Long Island Sound, within a train car's distance from the tracks. I would not have seen hulls of housing projects in the Bronx, or junkyards in Connecticut cordoned off with concertina wire (why such a pretty name for such deadly wire?), cars stacked five high, hoods open, as if disemboweled by huge urban birds of prey. A man wouldn't climb up five flights of crashed cars, would he? Could he? I couldn't.

No one can close his eyes riding the rails on the Northeast Corridor, immersed in Urban Blight 101. Any farmer on the train says a silent prayer of thanks for the opportunity into which he was born.

Saturday, March 16. Tom Baylis drove past Brown University and recalled his own college days: "I went to the University of Rhode Island. My senior year, I sold my overcoat for cash. Winter came back, and I about froze."

Ten years ago, Mr. Baylis sent me my first mare: Small Star. It was twenty degrees below zero that night. The brakes on the tractor-trailer froze, and the mare spent the night at the State Police Barracks on Route 1, only four hundred yards from Country Life. She could have used Mr. Baylis' overcoat. Small Star was no La Troienne, and we unsentimentally sold her for eight hundred dollars last December at Timonium. Mr. Baylis asked about her. "Somebody good bought her," I told him. "She has a good home."

Mr. Baylis loves his horses, winners or losers. He won five straight with Small Star's daughter, Travelling Star. She earned almost sixty thousand racing at Suffolk Downs. That's a lot of win pictures. Alice phoned from the farm this morning. Travelling Star foaled a filly by Allen's Prospect last night.

"Does she have any markings?" Mr. Baylis asked Alice.

"Yes, she has a small star."

Mr. Baylis is eighty-one years old. In December, he had angioplast surgery. He will not invest in offerings that take years to mature. But of this hours-old foal with the small star? He intends to stick around and race that baby in his Bridgham Farm colors. Folks like Mr. Baylis are the heart of the horse business.

Sunday, March 17. You don't simply buy a stallion prospect: You buy his family. I say this because Citidancer's pedigree got a big boost today, just as Carnivalay's pedigree did when Go for Wand became a champion. Citidancer's dam, Willamae, is a half-sister to Tong Po, who this afternoon won the rich Federico Tesio Stakes. Tong Po's owners are thinking of the classics now.

Following a strong female line is nothing new. The Bedouins routinely traced a foal's family back fifteen generations, and did not stress the sire line. William Woodward, breeder of two Triple Crown winners, wrote in 1925:

"I am one of those who believe female lines are more important than the male, for the simple reason that public performance and public judgment select the stallions, and but relatively few horses are used in the stud, whereas, for the most part, all mares of any merit whatever are put into the stud. It is the duty of the individual breeder to select his mares with the greatest care and judgment."

In other words, don't follow an empty wagon. Nothing falls off it.

Monday, March 18. It was rap music time at the stud barn. Citidancer was *stylin'* today—four white feet, low-fiving and high-fiving, when he climbed on Carny's stone-dust ramp and bred Magical Powers as if he was *Allen's Prospect, Jr.*

"That's a wrap, boys, that's a wrap," I said after Citi's one-jump performance.

Tuesday, March 19. A horse's total genetic makeup, what he carries in each cell, is his *genotype*. His outward appearance, the visible and measurable characteristics, is his *phenotype*. Peter Pegg, who measures the engineering elements of stallions and mares, commented:

"Structurally, Northern Dancer does not fit as many mares as Mr. Prospector, as he is slightly smaller and very powerfully built. However, we have found that horses of this phenotype are often able to very successfully breed larger mares of the same phenotype. It is interesting to note that practically every son of his that has been successful at stud has had the same phenotype as him, even though they have been of different sizes."

Mike Cataneo's mare Spellcast (by Fappiano, by Mr. Prospector) is in foal to Citidancer (by Dixieland Band, by Northern Dancer). Peter's numbers, taken after the mating, were not favorable. Yet he wrote:

"Don't worry about the poor computation with Citidancer and Spellcast, as they are of similar phenotypes even though they differ greatly in size."

We are not yet a third of the way through the current breeding season, and already I am looking forward to next spring, when a genotype emerges from the womb as a phenotype.

Wednesday, March 20. *Finally,* we have begun turning Citidancer out for a full day of play, instead of brief periods of hand-walking. His sesamoid has healed sufficiently. He's pasture-sound.

St. Patrick's Day was his first day out. I reminded myself that he was a Saratoga yearling, probably not turned out regularly since early spring of that year. Four days of exercise has improved his muscle tone, boldened his attitude. He drops down immediately when he sees a mare. Still, we take it one day at a time in this evolution of turning a young racehorse into a mature stallion. Patience, green grass, exercise, sunshine, and a little psychology are the necessary ingredients.

Thursday, March 21. Schoolmate Denver Dan Slattery writes:

"Won't be long before things start heating up, with little 'J.P. Three, slithering around the stalls, miraculously avoiding calamity the way Sweet Pea or Mr. McGoo used to crawl/step into thin air at the construction site only to find a new girder swinging obligingly into place. Little Luke loves the brown bear you guys sent him, and I often let that bear deliver little monologues about life in Maryland, the massive country breakfasts, comfy pillows as big as your thumbnail, celebrity basketball hoop battles where aging white men jump 4-5, no, 2-3 inches—guys with glasses who talk alike running around squawking at everybody.

"That's a strange little bear. Hey! How about this parenting, huh?"

Friday, March 22. Slip of the Tongue hates vet chutes, teasing chutes, loading chutes, flings herself backwards if you *try* to twitch her, strikes with front feet if you *try* to lip-chain her, only shows heat to other mares, then turns and kicks them. Jazzman hates her, that's how mean she is. But ten days of progesterone and one shot of Prostin have brought Slippie into estrus, if you can call it that.

Dr. Bowman knows her well. He performed a deep suture operation on Slippie in the off season, stitching the floor of her rectum to repair a tiny tear that permitted coliform bacteria to seep into her vagina, contaminating her reproductive tract. This was her second such operation, after her last foal caused a rectal-vaginal tear.

We gave Slippie two cc's of Ace to tranquilize her. She permitted Rich to put the chain shank gently over her lip, then Steve Earl took her leg as Jazzman found his courage. "Give her her leg," I said as Jazzman rested on her back. She immediately kicked him in the shins of his hind legs. He squealed in pain and looked back at me. I swear at that moment he vowed never again to trust me.

We brought Dew Burns out. "Keep her leg this time." Dew never met a mare he didn't like. He jumped her. She acquiesced. The window of opportunity was open a crack. I made the decision to proceed. I stuck K-Y Jelly directly into her vagina and squeezed out half a tube. I felt as if I was oiling a catcher's mitt. I let the jelly get warm, holding my hands against her vagina.

I know Allen's Prospect. He would breed a tractor if it had a halter on, but he assumes we would have the tractor ready to breed. He trusts us. John brought Allen out. The stud mounted Slippie, felt crowded by the two deep-suture operations, and he refused to breed her. "Face her uphill, Rich. Change the angle." We squeezed the other half of the tube of K-Y Jelly into her, and rubbed the jelly on Allen's penis. We had one chance to get this mare in foal, and this was it. Uphill, Allen changed his tune and bred her great. We made a quick exit from behind her hind legs. Later, as I shuffled pink phone messages at my desk, I saw my hands shaking. It only confirmed what I already knew. My concentration was shot. I was as spent as the stallion.

Saturday, March 23. The ogre is shaking us this morning, pouring rain on us. Prescott, new to the crew, fell off the wagon last night, is now wandering the streets. Jason, head of the teen team, is telling us:

"My sister has a hundred cancer tumors—a year to live. Insurance won't cover it. She was on the local news last night. Mom and Dad are going crazy. I'm keeping my sanity, though."

I tell Jason about Dad, about our dark years of his drinking. Jason bites his lip, keeping his sanity under control. I keep talking. I tell him things happen we don't understand. We'll figure it out someday. Meanwhile, Richard bursts into the office. He is steaming mad. Seems Steve Earl got the truck stuck in Richard's immaculately harrowed fields. He might as well have torn up the greens at Augusta, Rich is so mad.

David calls from the satellite farm. A colt we should have lost on a claim for thirty-five thousand dollars (but they spelled his name wrong on the claim form) is bleeding from being gelded. Dr. Riddle has had to tranquilize and repack him. The ogre is watching it rain on us. A foal absolutely will not walk up the hill, steps on my feet with hard pointy toes. "Baby him along," I tell myself. "Coax him with your last ounce of patience, then grab the little creep by his tail and manhandle him to the field." He kicks me in the thigh when I turn him loose.

I want to put out an all-points-bulletin for Prescott. If I could find him, one side of me would grab him by the shirt and yell: "Don't you understand? You are *powerless* over alcohol. You're breaking your girlfriend's heart, and I'm an inch away from firing you. You have a *disease*. You should thank God there is a cure—A.A. Ask Jason about diseases. His sister doesn't have a cure. Whaddya think of that, Mister?" The other side of me says, "What's the use?" I hate this side of me, this side that wants to say he's hopeless: Nobody is hopeless.

Comic relief is provided by a windswept foal, born the other night with legs like quotation marks. He bucks when he's turned loose, almost falls down, but catches himself and scoots off like the winner of the equine Special Olympics.

Dad tells me about a groom on cocaine, how the colt the fellow rubs tested positive—maybe off the traces of cocaine on a tongue-tie that had been in the groom's pocket. The owner lost the purse. The trainer was set down through no fault of his own. The Feds raided the trainer's car and his tack room and turned everything inside out—and he doesn't even hardly drink a beer. It's a crazy world, with more substance abuse than anybody admits—here on the farms, there at the race tracks.

Now it is dusk, and every horse is in a warm stall, bedded deep, hay in the corner. I open the car door, turn the radio on. A smart-aleck deejay is playing *Blue Sky* by the Allman Brothers. At the house, I look back towards the fields, shrouded in fog, empty of horses, the rain over. The roar of Winters Run, swollen with brown topsoil, carries away the day.

There might not be much money in horse farming, but damn if it doesn't make you rich with the thrill of being alive.

Sunday, March 24. The boy at six months reaches for colorful objects, loves orange and blue—the farm colors. On the floor of my study, he slaps at orange books, lifts them to his face to chew their corners. Favorites are *Tesio, Breeding the Racehorse*, and the *Life and Letters of General Robert E. Lee*. He flips these two great horsemen cavalierly over his head, leans forward to focus on the bald eagle against the blue sky of *Endangered & Threatened Wildlife of the Chesapeake Bay Region*. He skims past the foreword by Russell E. Train, president of the World Wildlife Fund, partner with Orme Wilson in Ritchie Trail's half-sister Gramercy, due to foal tonight from a cover in Allen's Prospect's book.

Out of the glass doors of the basement, he can see Winters Run, a hundred feet down a hillside of beech and poplar, in the hundred-year flood plain no building can endanger. If he'd been watching from his car seat this morning as we drove across the dike on the pond, he might have glimpsed four Great Blue Herons, who startled and lumbered in succession—a rookery in motion—back to the riffles of Winters Run.

The boy lifts the orange sunset cover of *A Sand County Almanac* and holds Aldo Leopold to his forehead as if reading by osmosis. Then he tosses the paperback and greedily reaches for Richard Scarry's *Busiest People Ever*. I want to read him this wonderful children's book, but I need to check on Gramercy.

Monday, March 25. All the horses bucked and careened off into fields turning emerald green now from the weekend of rain. Today is our reward. Daffodils are in bloom as the sun rises against purple clouds of weather headed in the other direction. Gramercy foaled on her due date. There is some logic, some control, some order to life in breeding season.

Tuesday, March 26. "Sure, I've seen plenty of 'em choking," trainer Marty Fallon said when he happened to call tonight during the controlled crisis of Gramercy's clogged esophagus. "Do everything you can for her."

"What can I do?" I asked.

"Nothing you can do. Just do everything you can."

That's the way Marty Fallon talks. That's the way I feel: Doing everything, but doing nothing. Grazing placidly with her day-old foal at 3 P.M., Gramercy one hour later was foaming at the mouth and nostrils, producing enough saliva to fill a feed tub. I've never seen a horse choking, and I thought at first that she might be allergic to spring onions or fresh grass.

"No, she's choking," Dr. Riddle advised by phone. "Give her ten cc's of Banamine as a muscle relaxant and three cc's of Ace as a tranquilizer. Put her in a stall with no bedding—just water. Often the saliva juices will break down the fibers in the obstruction. If she is not better by nine o'clock, bring her to the clinic."

Peyton arrived at nine to check on a foal. He examined Gramercy. She had stopped foaming and was rooting about the stall for something else to lodge in her throat.

"I think she cured herself," I ventured.

"Let's pass a stomach tube, make sure," Peyton said. The lining of her nostrils was inflamed. Her nose began to bleed as we passed the tube. She sneezed, and red dots blew all over the old white sweat pants I wore. I felt like one of the Marx Brothers feigning the measles, blowing ink through a colander. Peyton tried the other nostril. Success.

The mare seemed okay at ten-thirty. We can't let her eat until tomorrow, however. The Pimlico Oaks is this Saturday. I'm glad I don't have to call Orme tonight, tell him a sister to his Oaks filly choked to death on a piece of grass.

Wednesday, March 27. Don Capper, an equine nutritionist, visited today with Brenda Holloway of Bel Air Farm Supply.

"Do you still have any of that fancy Wyoming alfalfa?" he asked. "If so, don't wait to feed it. The first ninety days of a foal's life are the most important in the mineralization of bones. His mother can only give him what you give her. Feed the best hay *now*."

Thursday, March 28. Silhouetted high in an old walnut tree by the pond, a huge bird sat motionless. He might as well have been a limb. Then he spread his wings to reveal white mottling on his chest, and I caught a glimpse of white on his head. Was this a *bald eagle*? Nah, probably just a hawk. Hmmm...too big for a hawk. I sat still. The bird lifted off and flew straight away, the strong, heavy wings of a raptor disappearing over the hill. This was no turkey buzzard. Since DDT was banned twenty years ago, the Chesapeake Bay has witnessed an increase in the eagle population. Ten years ago, ninety-three active nests were sighted. Today, I think I saw an immature bald eagle in a tree at the pond.

Friday, March 29. "Yes, she's impacted. I think oil and oral fluids will work. However, if she doesn't improve, could we send her to the vet clinic, where they can run constant I.V. fluids in her? Good."

First she ships here in heat at seventeen days past foaling. Her days aren't right. Vets say breed, but she throws front feet, almost mauls Richard.

"Transport neurosis?" Dr. Bowman suggests.

Whatever. Her days are telling us something. Then she develops an impaction. I hand the case over to Michael. Ellen and I drive young Josh to Richmond for Easter with the grandpartents. We hit D.C. traffic. The radio plays *Road to Hell*. We arrive at Clover Hill with our own case of transport neurosis.

Saturday, March 30. The coals go out. I set the steaks down on the porch. That way the dogs won't eat them off the carving board. I get the coals lit as the phone rings. Hey, maybe that's Mike! Maybe Ritchie Trail won the Oaks!

I rush inside. Benjamin, the old wolfhound, pushes past me to the porch, swallows a tenderloin whole. I pull my sacroiliac out trying to kick him. That wasn't Mike on the phone, Ellen tells me.

Guess we didn't win the Oaks, I say to myself, poking a screwdriver down through frozen burgers.

Sunday, March 31. Ten deer, four herons, countless doves, three geese, one beaver, and a cast of a thousand spring peepers were featured this afternoon at sunset, as Ellen drove Abby the Appaloosa in the Amish courting buggy across the fields of Clover Hill. The boy and I were along for the ride. Easter Sunday in the country, old-fashioned style.

The ogre might be shaking Country Life right now. I don't know. I'm not expected to call home to tell the crew how to do their job. They're the best. I'm supposed to be relaxing, filling up a reservoir of patience to cope with April.

SOMETHING ABOUT ANIMALS

Monday, April 1. "You don't *know*?" Mom asked when I phoned home from Virginia this morning.

A long pause. My mind clicked down the casualty roster. What don't I know? A family member? No, they would've called. Ritchie Trail break down? What *don't* I know?

"Orme Wilson died Saturday morning. Your father and Michael were waiting for him at Pimlico before the Oaks. Charlie Hadry told them."

Another long pause.

"I can't believe this, Mom. He was sharp as a tack. He played tennis. I spoke to him Thursday when the entries came out."

"The obituary was in Sunday's Washington *Post*."

I set the receiver down. A part of me wanted someone to shout "April Fools' Day!" Another part of me let the news sink in. The last time I spoke to him, he corrected me about a term I borrowed from him: "It's the Marching and *Chowder* Society, not Shouters. It's Chowder!"

Later, I thought back to a day in law school when I quibbled with a law professor about the exact meaning of a legal phrase. I said: "I'm sorry, but I don't see the difference." He said: "Don't ever apologize for being *precise*."

To me, Orme was like that law school professor—precise, in the most gentlemanly way. And with his precision came a dependability that all who knew him relied upon. I have lost a friend as well as a mentor.

Tuesday, April 2. No I-95 for us yesterday. We tootled up old Route 301, through Bowling Green, Virginia, the American home of Diomed, winner of the first Epsom Derby in 1780. At the age of twenty-one, his popularity with English breeders at a low ebb, Diomed was sold for two hundred and fifty dollars to Col. John Hoomes of Bowling Green. Col. Hoomes, the leading importer of English stallions, wrote to fellow breeder Col. John Tayloe:

"I wish you could see Diomed. I really think him the finest horse I ever saw."

Nothing enhances a horse as much as owning him. Hoomes' own agent quickly warned Tayloe:

"Diomed was a *tried* and *proved* bad foal-getter. Mr. Weatherby recommends you strongly avoid putting any mares to him."

Weatherby compiled the first stud book, yet John Tayloe paid no heed to the agent's advice. In 1805, the year Diomed became impotent at the age of twenty-eight, Col. Tayloe bred a Diomed colt from a mare he bought through Weatherby. That foal, named Sir Archy, also bred mares until he was twenty-eight, and he became the first great American-born stallion—the Bold Ruler, the Northern Dancer, of the early eighteen-hundreds.

Such historical footnotes put the boy to sleep faster than a moving car, and thus we avoided transport neurosis on the ride home.

Wednesday, April 3. "How'd it go yesterday?" Dad asked.

"I felt like I was a kid applying for a job."

"What'd he say?"

"He said: 'We're dark tomorrow. Come back Thursday. I'll talk to my stall man.' "

"A lot of trainers are hurting for horses. There should be some stalls open."

"I know, but the meet's going on."

"Does he know that a lot of our new owners live in Baltimore? They don't want to drive to Bowie to see their horses work."

"I mentioned that."

"We've paid our dues."

"Yeah, but it's *his* game. He's the racing secretary. He holds the cards. All I could do was try to make him see that shipping in to race is like playing an away game. We *need* to be on the home field."

Thursday, April 4. For the memorial service today in Upperville, Virginia, folks flew in from the many worlds in which Orme Wilson lived. The preacher was overwhelmed. Twice, he asked us to sing *Eternal Father Strong to Save*, the refrain of "For Those in Peril on the Sea" ringing off the stone walls. I don't think it was intentional to sing the hymn twice. Had Orme been in the pews, instead of in the flag-draped coffin near the altar, he might have raised his eyebrows coyly from his hymnal, and smiled at the person next to him—a schoolboy politely overlooking a teacher's gaffe.

After the service, I waited by my car, letting Nanny the dog stretch her legs. Orme boards a mare with us named Norland Nanny. He always asked about the two Nannys—the horse *and* the dog, being precise in his very thoughtful way.

Friday, April 5. Things get in your head and they just stay there. Yesterday in a church in Upperville, I stood in the back, behind Charlie Hadry, trainer of Ritchie Trail, Gramercy, and Norland Nanny. He wore a sports coat reserved for stakes days. I think he wore it when he saddled Private Terms and Finder's Choice in the Preakness Stakes for Mr. Janney. The losses of Mr. Janney and now Mr. Wilson are very personal to Charlie. Those gallant sportsmen respected ability in other men. They are now Charlie's absent friends. He stood stone-faced and died a little himself yesterday.

Tonight, Richard and I delivered a foal, easy as you please, out of the mare Wedding Dress, who Charlie trained for Mr. Janney. You shouldn't let friends go easily, I thought, watching a foal take its first breath of life. You should hold onto them in a healthy way—always aware, like lines reaching back through a foal's pedigree, of the subtle ways they continue to influence your life.

Saturday, April 6. "The foal you asked me to examine?" Peyton began, very seriously. "She just had a grand mal seizure."

"A what?"

"A grand mal seizure. Lips smacking, drooling, flailing limbs, violent involuntary muscle movements. These seizures can be caused by lack of oxygen during birth. If the placenta became unattached from the uterus, oxygen could have been cut off. It's like an epileptic fit."

"Rich and I foaled her last night at eight o'clock. It was a very easy foaling. I was eating fish sticks by eight-thirty."

"Well, I think we should start the foal on DMSO to reduce swelling of the brain."

I called Stuart S. Janney III.

"It's an electrical thing in the brain, Stuart," I said. "This foal just short-circuited. I'll call you tomorrow with an update."

I hate the smell of DMSO. I hate to touch DMSO. My heart goes out to Wedding Dress' foal, who tonight is recumbent in a deeply bedded stall, exhaling fumes of DMSO coursing through her system.

Sunday, April 7. Ellen would like to plant a garden where we piled rocks during excavation of the house site.

"Will you please make me a stone wall in the yard?" she asked.

On Thursday, I had driven past the resurrected stone walls of the old Brookmeade Farm on Route 50 near Upperville. "Nothing to it, Ellen." A few hours later, I quit, half a wall in disarray in the yard. Even with an extra hour on the clock, I was unable to finish. I thought about Steve Earl's comment about the extended days, "It's Daylight *Slave*-ings," as I pulled down a book whose title I could not refuse: *Stone Work, Reflections On Serious Play & Other Aspects of Country Life*, by John Jerome:

"Stone work does not demand a lot of thought. Some other part of your attention starts taking care of selection and alignment of stones, of the mechanical principles of moving weight, and your mind is set free as a colt turned out to pasture. Personally, I find that using the other part of my attention—the part that knows how to stack stones—is a powerful restorative."

Jerome must pursue a sedentary occupation. I am just the opposite of restored. The things he doesn't know about country life.

Monday, April 8. To Jazzman and a tape recorder in the fields this morning: "Check Cleverness. I think she's coming around the mountain. Call owner about blood clot we saw on ultrasound of his mare. Change routine with his mare. She could lighten up with that foal pulling on her. Get her some of that Wyoming alfalfa. Talk to the Lucases about April Showers. Can we let her cycle naturally? Check Princess Mistletoe, she's blinking. Jazzy, the flies came out yesterday, the trees came out today. Toes Knows goes home tomorrow. Let's get her feet done. The new foal has a small umbilical hernia— watch."

I think I know how a teacher feels, immersed in the idiosyncrasies of individual students.

Tuesday, April 9. "I was a carpenter, making good money, but I wasn't happy," Tom Rowles has told me of his decision to become a farrier. "I came home one day and told Pat: 'I want to be a blacksmith.' She said: 'You're the man of this house. You've got to be happy in your work. If you think you can support two boys on a blacksmith's wages, go do it.' So I worked nights as a janitor at Union Hospital while I went to farrier school. I've never regretted it. I love being with horses."

Richard holds the foals for Tom the first time their feet are trimmed, at two months of age. It is a backbreaking labor of love, patience pouring out of the man holding spring-loaded feet. This is Tom Rowles at his best, his hands stuck like glue to young legs that resist restraint. A knee hits him in the cheek. He does not let go. The foal drops like a bull to his knees. Tom bends with him. Man and beast reach an understanding. Tom gently rasps the edges of the foal's hoof, sets the rasp in the straw, holds the foot to emphasize just who won this little war, then eases the hoof down. The foal sighs in resignation. Richard eases his grip.

Wednesday, April 10. Mom and Dad hold this family together. All five children, and all the in-laws, have worked together to further the farm's horses, sometimes at the expense of the family relationship. Brothers and sisters working side-by-side, under stress, after a long day, may not want to share a communal occasion: a birthday, a cookout. What is there to talk about? Work? No thanks. Not hungry. See you tomorrow.

But no, you bury the topic of work. This is your family. Today is your mother's birthday. We are people who have interests other than horses. Norah is back in college at the age of thirty-seven, in pursuit of a degree denied her during the difficult years of Dad's alcoholism. Alice has traveled around the world. Andrew is a dry wit. Michael puts a spin on perspective.

With grandchildren in laps and underfoot, Mom unwrapped the boom box purchased after Alice assembled a syndicate of siblings and spouses. The first song the radio ever played was the *1812 Overture*. It was silly and fun to hear the refrain of trumpets, to sing along: "Boom Boom. Da-Da-Da-Da-Da-Da-Da Dot-Dot. Boom Boom."

Happy Birthday, Mom.

Thursday, April 11. Peter Steinhart, author of the "Essay" column in *Audubon* magazine, believes:

"Children love to name animals—and explore the differences between them. My two-year-old daughter shouts with glee when she sees a horse. Animals are part of her capacity for joy. Animals are part of our minds, a part of the loom upon which we spin out thoughts. We are bullish and bearish, hawkish and dovelike, busy as beavers, timid as rabbits, sly as foxes."

In his book *Thinking Animals*, Paul Shepard, a professor of natural philosophy, writes:

"Animals are among the first inhabitants of the mind's eye."

I thought about the animals in young Josh's life: Nanny and Boxy, Chessy the cat, the mares he stares at, silhouetted at sunset outside the glass doors of the living room. He has teddy bears and rocking horses, books with Babar the elephant and Zephir the monkey, ducks on his coveralls, bunnies on his bib. Naturally, the book *Thinking Animals* was written by a fellow named for the keeper of sheep, who wrote:

"From the standpoint of the developing brain, an assembly, say, of horselike animals—donkeys, tarpans, zebras, asses, and other groups of their odd-toed relatives—is as essential as blood."

I'm tempted to say "Hogwash! Horsefeathers!" to Shepard's observations, but the boy keeps looking out the window, watching the mares graze. He is transfixed. There is something about animals. There is more here than meets the eye.

Friday, April 12. On the average, we foal fifty mares a year. Well, I should say, forty-nine mares, because no matter how vigilant we are, we miss one—every year, it seems. The circumstance is generally the same: The mare does not bag up and enters labor well before her due date. She is not in the foaling barn. She is in the main barn, and waits until the night watch girl leaves at six o'clock. Richard gets up at five-thirty, does fifty push-ups, then checks the mares in the foaling barn. At six-thirty, the boys are in the main barn, loading feed onto the truck. They glance in at the mares.

Between seven and seven-thirty, a mare can sneak a foal into our world. This morning, a wet baby stood and whinnied at me.

Saturday, April 13. The careers of most stallions can be graded on a curve. First year at stud, the novelty attracts a good book. Second year, some glitter wears off. Third year, a syndicate of breeders must carry him. Fourth year? Well, his babies haven't run, let's wait and see. (Sharpies like to breed to fourth-year stallions. They either hit or miss quick.) If the first crop runs well at two, the stud will breed his best book of mares in his *fifth* season.

White-faced foals by Carnivalay, more than fifty in all, will dart across pastures from Florida to New York this spring—the fifth-season curve—their dams sent here *last spring* because of first-crop stakes winners Valay Maid, Lucky Lady Lauren, and Groscar. When this spring's foals race two years from now, Carnivalay will be twelve years old: in his breeding prime. No one who has ever stood a stallion has not been keenly aware of the curve.

Sunday, April 14. At 7 A.M., a breeder called to discuss the morning ship-in mare to Citidancer.

"She has a very soft follicle, but she's not showing heat," she said. "I could tranquilize her, but I'm not certain that would keep her feet on the ground."

No stallion, but particularly a first-year stallion, should be exposed to injury unnecessarily. One mare won't make him, but one mare could *break* him. A kick in the testicles in April would be disastrous. He could be out for the season. Only a small first crop would represent him at the races. Worse still, he could be mortally injured.

"How about this?" said the breeder, sensing my apprehension. "I'll let her go, then short-cycle her. We won't lose much time, and maybe she'll come in heat better."

I thanked her. As I turned Citidancer out, I told him: "It's a day of rest, son. Enjoy it."

Monday, April 15. Michael's son Philip, aged two and a half, sits at the lunch table with the farm crew, tearing the crust off his tuna fish sandwich—just one of the boys. I wonder what he thinks of the crew's chatter.

"Did you hear about the van that flipped over? The boy driving hit a mailbox, veered left, then right, and flipped the van over on its side. One mare scrambled to her feet, the other mare couldn't move. The emergency squad cut the roof off with a torch, and they walked the horses out."

Philip listens, along for the ride. Peyton comes in.

"What's Peyton do?" Michael asks his son.

"Animal dok tor."

Peyton smiles. The kid struck a chord in the vet. Mildred, who raised us, sits on her stool by the stove, and loves what she sees. Tonight at dusk, fog blankets the land, spring birds trill noisily. The cold rain that fell today is on the move. Tomorrow will be seventy-eight degrees. We are tacking through breezes blowing through the breeding season. Becalmed this evening, I feel just like Philip, absently munching Mildred's lunch, taking in the stories—just one of the boys, along for the ride.

Tuesday, April 16. I was wrong about Slip of the Tongue. She gave us another chance to get her in foal. This time, when she blew, she blew sky-high, brown haunches lifting off the ground, a leap to have made a Lipizzaner proud. A five-man crew and a fifteen-hundred-pound stallion ran for cover. Richard stood over Slippie, who came down on her flanks, a heave of hot air whooshing out. We *all* blew a sigh of relief, as Richard said: "I just had to give her her head."

One thing about Slippie: You always know what kind of mood she's in.

Wednesday, April 17. The first thunderclap of the spring was heard this evening, sending the dogs under the beds and the horses up the hills, and the air turned cold, and the sky filled with white clouds that parted for the sun, as two geese flew straight into the gap where God lives and were absorbed into the heavens.

Thursday, April 18. The breeding season turned itself up a notch today.

Peyton says the new foal's heart is beating very rapidly, gums are yellow—like a foal with jaundice. Blood settled out: twenty percent red, eighty percent white. Mother's milk is killing him. We hold the foal off the mare, and Peyton gets blood-harvesting equipment. Jazzman doesn't volunteer to be a donor, but complies. Peyton and Jazzman save this foal's life.

From Virginia, the patriarch of a breeding family arrives with his brood. We are all in Carnivalay's paddock, John holding the stallion, when a fellow across the street starts up a chain saw. Carnivalay careens towards us. The patriarch slips, falls face down as I swing the gate shut on his back. Over crabcakes later, the old sportsman smiles: "I've been treated worse. I once had a trainer who treated me like a mushroom—kept me in the dark and fed me manure."

Broadway Gal arrives, drunk on Dormosedan—stronger than Ace, stronger than Rompum. She can move only enough to tattoo both hind feet against the teaser's stomach. We called her owner: "You'll have to find another mare for this season. We refuse."

Back in the office, we see the *Sports Illustrated* article: "Fading Fast, the Sport of Thoroughbred Racing." At quitting time, we treat a foal for scours. It is the smell of bacteria heated up that finally gets to us.

"Let's go *out* to dinner." The boy sits in a high chair and stares as other children shriek. I forget all about the day. What wonderful therapy this boy is to his father.

Friday, April 19. Dr. Bowman preaches that the breeding season doesn't *really* begin until May 1st, when green grass and sunshine send mares through crisp heat cycles, when fertility becomes natural, not coerced, when even the most spiteful of mares will consider conceiving.

Well, then, eleven *practice* mares just left for Kentucky, their coats rubbed to a shine by Ellen and Cheryl, white flannel on halters, sixty-day-old embryos in their wombs, five days into a precautionary ten-day Regu-Mate routine to offset shipping stress, mineral oil swishing in their bellies en route to the intestines to prevent impaction, twenty cc's of combiotics in their bloodstreams, feet trimmed, manes pulled, fetlocks clipped, tails brushed.

Bowman's right. Like the weather, we're just warming up.

Saturday, April 20. Umbilical abscesses win the Disease of the Month Award. In the past, when foals would succumb to various ailments, we performed very few postmortem examinations. I bet if we had posted more foals, we would have discovered complications involving the umbilical area. But you frequently can't see such problems. Lots of foals have hernias. You have to get down close to discover an abscess, to touch a patent urachus dribbling urine, to feel the heat of an infection.

We have two foals here who aren't doing right. Blood work shows infection. Ultrasound reveals fluid-filled masses—the inflamed remnants of umbilical vessels. Peyton has lent me the literature from Urachal and Umbilical Disease 101. While he preps the foals for surgery, I prep the owners. The surgery is seven hundred and fifty dollars. Add a couple hundred for board and vanning to the clinic. You're looking at a thirteen-hundred-dollar belly button.

Sunday, April 21. At the end of a day like today, you're ready to snap, to push the kid aside, to turn your back on your wife, to wrap a pillow around your head and just force yourself into a fitful sleep. Another lost Sunday—damn dreary rain covering you and your foals all day. The last foal you handled left the scent of scours on your hands *again*. The two new foals you intended to turn out for one hour stayed out for two in the rain, because you couldn't get their stalls ready. Deep inside, you worry about your livelihood. You keep thinking about the *Sports Illustrated* story.

Tonight, Kevin Kellar visited, to drop off a pass for next Saturday's Hunt Cup. Kevin manages the stallions at Worthington Farm, Lord Gaylord's home. We went to school together. He worked here when I left for Kentucky. Without his help, maybe I wouldn't even have a farm to complain about. He didn't know it, but he put today in perspective for me. It really is not such a bad thing—the rain and all, the short crew, the manure piled up in the aisleway. Tomorrow will be better.

Now the baby is sitting on the bed in blue pajamas. He is smiling at me. It is all worthwhile. Today was just the hard part of the bargain.

Monday, April 22. "The foal's doing fine," Peyton says, brandishing the extended belly button, nine inches of networked necrosis surgically removed. The umbilical artery is lined with infected yellow tissue. Peyton draws a crude map outlining nearby organs. I get the picture, thank you. I set the nauseating atlas on top of the cabinet in the broodmare barn. It will be worthwhile for the crew to understand the mechanism that caused this foal to be so sickly.

Tuesday, April 23. Don Reynolds is retired from the Baltimore Gas and Electric Company. Jim Glenn runs a food-brokerage business four miles from Pimlico. Bill Honore worked for the National Park Service ("Did you know that the White House is a park?"). John Adams is a political fund raiser. Betty Miller split one unit five ways and calls her partners "The Golden Girls." Ken Schnell uses his West Point education for Westinghouse.

These folks and a half dozen others comprise the Country Life Discovery Partnership. The name is a *double entendre*. Grandfather raced champion Discovery when the colt was a two-year-old in 1933. As the proprietor of Sagamore Farm, twenty-one-year-old Alfred Vanderbilt needed a flag waver, and for twenty-five thousand dollars, the silks on Discovery changed. The chestnut colt became one of the greatest handicap horses of all time.

To "discover" is to find some existent thing previously unknown to you. Two years ago, we began turning over stones for new people anxious to discover the sport of racing. Today at Pimlico, the effort bore fruit. Shine On Sarah, a three-year-old filly, became the first starter for the Discovery Partnership. "I own her foot," said one Golden Girl. "I own her ear," said another. "Look, those are our silks!" said a third as the jockeys appeared.

Sarah broke last. Turning for home, the orange and blue silks appeared to be gaining. The terrace dining room rocked with the Golden Girls' chorus: *"Sarah! Come on Sarah! Come on, girl!"* Sarah passed tired horses, couldn't catch the leaders, finished third, earned sixteen hundred dollars, paid three dollars and twenty cents to show. We were rich! We hit the board in our first start! The Golden Girls rushed to cash their tickets. Don Reynolds lit up like BG&E had plugged him in:

"This is the most exciting thing I've ever done. I've always wanted to own a racehorse. Sarah's *my* horse, *our* horse. This is great."

Wednesday, April 24. It was a scherzo of a storm, performed in triple time—first wind, then rain, then hail. It ended as fast as it had begun.

"For once, the weatherman was right," Dad said, rolling down his car window. " 'Severe Storm Watch until 2:15.' It's 2:14. Guess that does it for today's mowing." And he puttered off to town to rescue A.A. cohorts who might use a storm as an excuse to drink.

Pebbles of hail could have chilled an impromptu mint julep in fact, but Aunt Jin was on our minds, not Uncle Bourbon. Aunt Jin, winner of the Monmouth Oaks years ago, arrived early for her three o'clock appointment with Carnivalay. Aunt Jin was in the same crop as Ruffian, Mr. Janney's great filly who, like today's storm, ended so quickly. Could it be possible that Ruffian would now be a *nineteen-year-old* broodmare? Mr. Janney owned a share in Carnivalay. He supported his stallions. He might very well have bred Ruffian to Go for Wand's half-brother. I caught my mind fancifully wandering.

Enough of such shenanigans. Focus on the matter at hand.

"That's a cover," I told the van driver, and we loaded Aunt Jin back on the truck for her ride home to Virginia.

Thursday, April 25. Blue Lass is a grand old mare approaching her due date. Last year in Kentucky, she hemorrhaged after foaling. She was touch-and-go for several days. Now in foal to Carnivalay, the mare is comfortable in her surroundings; we do not want to send her to a vet clinic to foal. I think if she blows a uterine artery this time, she's a goner, whether here or at a clinic. Peyton dropped off several liters of lactated Ringer's solution at the foaling barn, to run I.V. fluids in her if she hemorrhages again.

Everybody's on their toes for this one.

Friday, April 26. Spent this morning on the road, visiting the Manor Equine Clinic, where Wedding Dress' foal tugged halfheartedly at her dam's bulging bag. This was our second foal of the month to have umbilical surgery. Dr. Jim Juzwiak said: "How would you feel if you were opened stem-to-stern?"

Nearby, at Ray Mikkonen's tidy little training center, an Allen's Prospect two-year-old nursed pin-fired shins, and two Corridor Keys galloped. Ray, his wife Anja, and their son Leo worked like beavers to get through the morning. Every stall was full. Ray explained: "Trainers cannot bring their horses along slowly, like in the old days. If they do not run their horses often, they'll lose their stalls. We get their horses ready here."

When Ray vans mares to breed to our stallions, he does not dawdle. In his choppy Finnish accent, he told me: "When you are poor, you got to boogie." So I boogied on home.

Saturday, April 27. "Please clear the course. The horses are *under starter's orders*."

These three words emphasized by the announcer today are also the name of a famous painting by Sir Alfred Munnings. The Maryland Hunt Cup would be a scene to fascinate Munnings, bewitched as he was with the anticipation and the aftermath of a race. From where we stood at the seventeenth fence on the backstretch—if the four-mile course over the most beautiful terrain in Maryland could be reduced to mere race track parlance, for no other backstretch is a field of clover, shaded by budding dogwoods and virgin stands of Atlantic hardwood—the start of the race was as magnificent as Munnings' imagination, fourteen silked riders circling their horses, like a chain swinging, edging crabwise, poised on the limit of control until they were almost in a straight line.

Afterwards, in the courtyard of the receiving barn, blacks and bays and dapples and grays—"all the pretty little horses," as the children's song goes—walked in circles, and stood with ears alert, as if listening for the huntsman's horn, the hounds' cry. Munnings wrote: "I am watching shadows cast on the

short grass, and the look of the sky and maybe the shape symmetry and the lighting on a horse—a scene for the artist."

The smell of linament hung in the warm air as grooms tidied tack boxes and loaded up, to set off for the next stop, an aristocratic circus on the move through the springtime countryside, readying for the moment when the horses are again under starter's orders.

Sunday, April 28. On my Sunday nature walks along the banks of Winters Run, I attempt a stab at geology, pocketing blue stones with red stripes that might have been formed at a time hot enough to melt two rocks together. John McPhee's book *In Suspect Terrain* takes local geology only as far south as the Delaware Water Gap north of us. Suspect terrain is the term for land whose origin is uncertain. Was it heaved up through plate tectonics? Or left behind by retreating glaciers?

For the two-year-old gray filly who "is so good she'll knock the taste out of your mouth," we submitted the name Suspect Terrain. Her father is Corridor Key, a name that has nothing to do with being locked out of your room. Corridor Key is by Danzig, and the port of Danzig was the key in the Baltic corridor terrain that the Germans needed to jump to Russia in World War II. He is a thoughtfully named colt. The mother of Corridor Key's daughter is Allegedly. Hence: Suspect Terrain.

Carlos Garcia will train the filly. I told him her name. He looked sky-wards, held his palms out, and in his inimitable Argentine accent, laughed: "Suspect To Rain? What kind of name is that?"

Never mind. I'll explain later.

Monday, April 29. Edgar Allen Pro has outgrown the three-furlong training track at the satellite farm. This two-year-old freight train barrels around the tight turns. The racing secretary at Pimlico gave us two stalls earlier this month, and tiny Zola Valay promptly ran like tiny Olympic star Zola Budd in her first start on the home field. Missed by a neck, but earned a check.

"Do you know the difference between 'Ready-to-Run' and 'Running?' " the secretary asked us. He was saying he can't fill races with "Ready-to-Run" horses. He's not interested in two-year-olds. I don't envy him his job. Filling a race card almost every day of the year must be a grind.

Said David, who manages the satellite farm with crisp punctuality: "Eddie's ready to do more than I can do here. It's time for him to *move on.*" Edgar Allen Pro is on the go to Pimlico.

Tuesday, April 30. Amazed by objects within his reach, the boy on the eve of his eighth month sidled his walker up to the coffee table. He puts everything in his mouth, but before devouring the tidy little program from the Breeders Awards Dinner, he held the pamphlet in front of his face, and stared at the silhouetted image of a horse standing on a waving flag of Maryland. Animals. There's something about animals.

I thought how symmetric the program had become: At the dinner last Friday night, Uncle John accepted an ice bucket for freshman sire Allen's Prospect, the presenter kindly noting that Country Life in 1946 had bred Raise You, the dam of Raise a Native, the grandsire of Allen's Prospect. Moments later, Dad accepted Carnivalay's brass stall doorplate with Stallion of the Year engraved on it. The old boys don't keep ice buckets as handy as they did when they ran the farm; the doorplate will hang proudly outside Carny's stall. The dinner was history for me until the young boy revived the memory. This is full-circle material, from Raise You to raising him.

A TOUGH MONTH

Wednesday, May 1. The two Canada geese nesting on Winters Run suddenly appeared on the pond at dusk, flanked by seven infants—a flock of fist-sized yellow goslings stuck like glue to swimming parents. A few days ago, I found broken shells scattered on the Winters Run island, and despaired that the eggs had been attacked by predators. But no, the eggs had simply hatched! How stupid of me!

Goslings on the pond, foals in the field, a baby boy on my back—youngsters everywhere. It's springtime on the farm.

Thursday, May 2. The ship-in mare stood for the teaser. Yet when the stallion nudged her flank, she reared straight up, as high as a horse can go, and tried to strike, walking backwards on her hind legs, pawing the air, her torso twisting. The moment her front feet landed back on the ground, she lunged, and bolted away from us, away from her nine-day-old foal. Her left hock buckled, then her right hock. She did not, *could not*, control her hindquarters. She stopped at the side of the barn and trembled in pain.

"I think this mare broke her back," I said to the men. "I think she broke her back when she reared in the air."

That was nine o'clock this morning. For the rest of the day, we watched her die. Oh, there were numerous important consultations with veterinarians, who ran ball-point pens down her spinal column and stuck her hind feet with needles to determine sensation; with the mare's owner ("I'm sorry, he's in a meeting. I'll have him call."); with her boarding farm, who wanted to be apprised of every development; and, of course, with insurance agents. Various voices said:

"Could be compression of the spinal column. Might subside."

"Miracles happen. Don't put her down."

"Try anti-inflammatories."

"Let me know if her condition deteriorates!"

After a while, this quartet of voices was a sideshow. Years from now, the only thing I will remember is that we watched an animal die. Dr. Lewis Thomas authored essays for *The New England Journal of Medicine*. In "Death in the Open," he wrote:

"Animals have an instinct for performing death alone. You do not expect to see dead animals in the open. It is always a queer shock, part a sudden upwelling of grief, part unaccountable amazement."

The mare would crawl on her front legs, then fall flat in exhaustion, only to then lift her head suddenly, breaking the surface of the field like a sea monster. Her foal would stand over her, and bend down to nurse. At 4 P.M., the mare struggled one final time. The foal nursed hungrily when the mare's bowels relaxed, and we took him away to a waiting nurse mare. We covered the foal in peppermint spirits, and rubbed peppermint into the nurse mare's nostrils. In the distance, we heard a chain saw. I knew that would be Peyton, removing the mare's spinal column for the insurance autopsy. I thought about Dr. Thomas' conclusion:

"All the life of the earth dies, all of the time, in the same volume as the new life that dazzles us each morning, each spring. If it were not for the constant renewal and replacement going on before your eyes, the whole place would turn to stone and sand under your feet."

The nurse mare had first been bred as a two-year-old: Her name is Young Romance. Yesterday, she lost her own foal. Tonight, she adopted an orphan.

Friday, May 3. "It's just a fungicide!" the young man hollered at me after I hollered at him to turn his tree-spraying machine off until I could get the stallions out of the nearby barn. "It just kills bagworms. It won't hurt nothing. I'm twenty-five years old, been doing this five years, and I'm fine."

At the very moment I contemplated this man's health when he is sixty years old, the Valley Protein truck pulled in to reassemble the dead mare, and the nurse mare and the orphan foal were carefully being led to a tiny paddock to become further acquainted.

"Well, why'd you pick such a windy day?" I asked him. I hated the tone of annoyance in my voice, but sometimes things just get to you.

Saturday, May 4. Not a good week. Wedding Dress' foal never recovered from umbilical surgery and was put down last night. Today, Peyton palped the mare at the clinic: "Breed tomorrow." So Wedding Dress returned home quickly this evening. The farm was calm, everyone inside, watching the Kentucky Derby. Through the living room windows, I could hear Wedding Dress galloping in the orchard field with the problem mares of May—the single girls, the not-quite-ready-for-conception gals. I checked on her. Milk was streaming from her bag. The exercise will benefit her, I thought. Tomorrow we start all over, breed her again. Constant is the renewal and replacement of life.

Sunday, May 5. "There are guys out there who are thirty!" a young voice in the stands exclaimed. She was almost ten years off. A handful of forty-year-olds were the oldest fellows represented at our annual high school alumni lacrosse game. In college, we participated in a study that measured your mood after exercise. The study proved that depression could be offset by hormones released during exercise. I have never forgotten that lesson. I ran up and down the field and happily put the worrisome week behind me.

Monday, May 6. It is 5:28 P.M. A tornado watch is in effect. It is ominously dark. Buds of maple leaves drop like rain from the trees as the wind whips the limbs. The song from *The Wizard of Oz* pops into your mind: "The house began to shake." The temperature has dropped twenty degrees in fifteen minutes. The farm braces herself; her tenants huddle. When the rain comes, it falls so hard you can watch the creek rise.

Ellen and the baby are in the basement. We are tiny creatures, powerless over the whims of Mother Nature, who today sent withering humidity on the run, replaced in one furious blow of air.

Tuesday, May 7. The day broke cool, and new life arrived in the aftermath of the storm. Filouette and Lisa Hackett, two mares who spent their gestations on Regu-Mate, delivered healthy foals last night. Filouette is the dam of Restless Con, who sprinted away with the Haskell Handicap last summer. Lisa Hackett is the dam of stakes winner Bal Du Bois. Their foals were eagerly awaited.

Dr. Bowman immediately began preparing the mares to get back in foal. As he lavaged their reproductive tracts with antibiotics, I was struck by how quickly we must change gears. We cannot rest on the satisfaction of these live foals; the clock is ticking on conception for next year.

Then I stopped myself: Filouette is owned by a lady of great character, whose joy at seeing the video of her hours-old foal will be immeasurable; Lisa Hackett's owner believes that mares often prosper if given a year off. These owners, their mares, and today's newborns, are exceptional.

Wednesday, May 8. In the fireplace in the basement study, wet oak logs hiss against the efforts of kindling. Jim Juzwiak has just called from the vet clinic.

"We *did* have to put that foal down," he said of Arctic's colt, the other umbilical surgery patient from last month. Cheryl had called me to the barn this morning: Arctic's colt was throwing himself down, rolling against the wall. When I arrived, he was belly up, his feet doubled back, like a puppy waiting for a scratch. Nothing cute about him, however. We called for a van, shipped him speedily back to the vet clinic for his second surgery. I told Jim: "No heroics."

"We opened him back up," Jim said tonight. "He had extensive adhesions from the umbilical abscess. He also had a twist in the small intestine. His prognosis was poor. We put him down right there on the table."

This tiny basement office is a refuge for the wild life of a farm manager. The whistling of sap from the oak logs is punctuated by the call of a goose headed down the creek at dusk. I am detached from the scene—simply the scorekeeper, through this diary, through this record of my private landscape. The score tonight is three dead horses in one week—first a mare with a broken back, then the two umbilical cases: Wedding Dress' filly, Arctic's colt. I keep saying to myself, "This empty feeling will pass."

Thursday, May 9. Three foals were born last night—renewal and replacement. Blue Lass, the old mare who hemorrhaged last year, delivered a chestnut colt into the waiting hands of John and Richard at seven-thirty. She rested, exhausted, until eight-thirty. Dr. Bowman told us of a mare at his farm dying of uterine artery hemorrhage: "You could *hear* the blood escaping through the hole inside her. It sounded like water through a garden hose." I trained my ear to her flanks, heard only strong gurgles from her bowels, the way plumbing sounds in an old house. The night watch girl took the mare's pulse every hour, and pressed the mare's gums for capillary refill time.

A foal's first walk to the small turn-out paddock is always gratifying to us. When Bess and her baby were led out this morning, I felt like standing and applauding in admiration as they passed.

Friday, May 10. Our annual Preakness Party is less than a week away now. The party provides the impetus for improvements, particularly to the main house. Last year, Steve Brown repaired the cracking plaster ceilings in the living room. It was unsettling while he was working, with Stull's portrait of Henry of Navarre missing from the mantelpiece. Soon Henry returned to take dominion, like the jar in Wallace Stevens' poem, over all it surveyed.

This year, no one will even notice Steve's major project. All this week, scaffolding has circled the house, as Steve has been replacing the old sagging galvanized rain gutters with aluminum gutters. We should videotape these projects, and submit the footage to the PBS series *This Old House.*

Saturday, May 11. Richard's son-in-law Ryan Cool emerged from the Persian Gulf war without a scratch, then broke his arm horsin' around with the guys at the Army base. He came home today, and the yellow ribbon on the farm sign on Old Joppa Road can rest now.

An exciting renewal of the Pimlico Special came and went, the breeding season carried on, but the only thing folks really talked about this evening was Ryan—in a cast, pushing his nine-month-old daughter Ashley in a baby carriage over the hills of the farm, making up for lost time.

Sunday, May 12. Arctic's owner is a young man. I want him to breed mares with us for the next hundred years, but he is unhappy at present. Not only has his foal died, but we have been unable to get a cover on his mare, despite her follicles. She behaves worse than a maiden. This owner has paid a significant board bill for an empty mare, and a slew of vet bills for a dead foal. His season to Carnivalay has not left the stallion's loins.

When Arctic was on her thirty-day heat, she would not stand for the teaser to jump her. Her front feet were lethal weapons. Tranquilizers didn't help. We missed that heat. So we gave her Prostin on Wednesday, and tried again this morning. Same story. She reared forward, pitched her front feet, threw a fit. Her ovaries said yes, her brain said no. We tranquilized her again. No change.

Anticipating my phone call to the owner, I asked Alice to videotape our efforts. We let the mare have her leg. She reared. We took her leg. She reared higher. We tranquilized her: Three cc's of Ace, one cc of Rompum. No difference. We switched teasers—brought out Jazzman instead of Dew. She *almost* let Jazzman get on her back. I told the crew: "I've seen enough."

Tonight, I watched the instant replay. It ended when the mare threw her feet at Steve Earl and lunged out from underneath Jazzman. Alice's voice on the tape exclaimed: "Oh, my God!" That said it all. I watched myself put Jazzman back in the barn. This is the first time I have ever felt it necessary to videotape a mare's behavior. Tomorrow morning, we will try again. I hope the follicle holds.

Monday, May 13. Drugged or not, Arctic wanted nothing to do with reproduction again today.

"Was she this bad for the Sunday crew?" the Monday crew asked.

Ellen again videotaped the mare trying to land on people, as if we were filming a documentary.

"Did you get that?" I asked, when Arctic sent the twitch flying into the air. "Yes."

I mailed the production to Arctic's owner.

Tuesday, May 14. A *real* TV crew showed up today, announcer Brad Ganson and cameraman Jack Miller of Channel Two News in Baltimore in search of a horse-related Preakness feature. They were supposed to be here yesterday, when everyone was outfitted in farm baseball hats and polo shirts. Today, we looked like the Beverly Hillbillies, in jeans and shorts. Trying to beat the ninety-five-degree heat, Cheryl appeared in a tank top: "I can't go on TV like this! Why didn't you tell me these guys were coming?"

For this country's Bicentennial, Brad rode his Quarter Horse mare from Maryland to Maine, fell asleep in the saddle on a side road. He woke up when the mare walked up the Interstate–80 ramp. "That was not the route I wanted to take." A toll booth attendant permitted him to walk the mare across the bridge over the Hudson River. Brad knows plenty about horses. For two six-minute segments on the Thursday and Friday evening news, Brad and Jack filmed for eight hours. They captured Jazzman teasing, stallions breeding, Peyton with his ultrasound, a new foal, yearlings galloping, two-year-olds in training—a Thoroughbred's progression from the womb to the races.

Wednesday, May 15. Every dog gets one free bite, and Prescott had bitten us once already. He definitely picked the wrong week to fall off the wagon for the *second* time. We had been depositing his paycheck directly into the bank, but he found enough funds to get drunk, and so his girlfriend said: "That's it!" She stuffed him onto a train bound for his hometown in upstate New York.

Now we are a man short, in the busiest week of the year. Dad has always stressed that an alcoholic will only get straight when he has reached his bottom, and the bottom is different for all people. I am so vexed at Prescott, if he suddenly appeared, I'd reach his bottom with the toe of my work boots.

Thursday, May 16. Forty-eight hours ago, the wraparound porches on the big house were like cottages not yet opened for summer. Scaffolding still rimmed the house, privet hedges were unruly, the lawn unraked. There appeared to be no one in charge, and no one to spare from the paramount task of breeding mares—the business of the farm. This crew has been to the playoffs before, however, and without fuss, rakes appeared, flowers were cut, floors mopped, storm windows replaced with screens. Suddenly it was springtime. At precisely 7 P.M. on the most glorious cool evening of May, the rhododendron reached peak bloom, draping pink petals over guests on the front porch, on folks spilling out into the yard.

It has been exactly thirty years since Mom started this party, when the press corps surrounding Carry Back's quest for the Triple Crown found libation here, where Carry Back was conceived. The bar tab has lightened in the intervening years, but there is still an intoxicating romance to a grand old farm dressed up on a spring evening.

Friday, May 17. At 5 P.M., minutes after delivering deadly bolts of lightning to high school lacrosse players sixty miles away, the storm struck Baltimore. The Black-Eyed Susan Stakes was contested after the rains stopped. The track was a sea of mud, and Ritchie Trail trailed the field in the most uncharacteristic race of her career. To the north, in the direction of home, purple clouds rained lightning on the countryside. When we returned home, Mom was standing outside the office, a long look on her face.

"Castalie was struck by lightning," she said. Under an ancient English walnut tree lay the Buckpasser mare Castalie, the dam of stakes winner Ritchie Trail, the dam of stakes winner Gramercy, the dam of a two-month-old foal suddenly orphaned. Castalie was twenty years old, the best producer in Citidancer's first book. She arrived in February with a uterine infection. Dr. Bowman treated her, and she delivered a live foal. Just this morning, we bred the old gal to Citidancer. This evening, she's dead. I stared at Castalie and wondered if the semen were still alive in her.

"The boys checked all the other mares," Mom said. "Everyone's all right. The foal is in the barn."

We were speechless.

Saturday, May 18. A pride of new faces to the Thoroughbred game arrived early to board the rental van for Pimlico and the Preakness. They were bouncing basketballs outside the office, or taking pictures of foals, or scanning the *Form* spread out in the yard.

Then, for the second time this month, the Valley Protein truck pulled in. The basketballs stopped. The *Form* was abandoned. The new people stared in grim fascination as Castalie was winched up into the bed of the dead-animal truck. "It is always a queer shock," Dr. Thomas' words echoed in my mind. "Part a sudden upwelling of grief, part unaccountable amazement." Ellen appeared, brandishing powdered milk, reading the instructions for feeding an orphan foal.

"Five gallons a day?" she asked. "Foals drink five gallons a day?"

The plastic buckets bore the label: Mare's Milk Replacement. I thought: ' "Renewal and Replacement.' This process takes many forms."

Sunday, May 19. Nature does not pause in her reproductive agenda. No sooner had the celebrity basketball game between the guests and the home team ended at dusk than Follies Star foaled a beautiful filly, and Fuel to Burn produced a strapping colt. In the hope that the two recovering mares would be kind to an orphan, we turned Castalie's two-month-old filly out with the newborns.

Orphan Annie cantered with her tail held high, prancing like a giant question mark across the pasture. Where is my mom? Are you my mom? Raised hind legs were enough to warn the orphan away from the old mares. "Please don't corner her," I warned the mares. "Please just give her company." The exercise seemed to do Orphan Annie a world of good, as she drank several quarts of milk from the shallow bucket we hung on her stall screen tonight.

Monday, May 20. Drove Johnny Weninger to BWI. I got stopped by the metal detector. Emptied pockets. Beeped. Took off glasses, removed belt. Still beeped. "The boots," J. W. said. "Steel shanks in the boots." Took boots off. No beep. Go ahead, sir. Watched J. W. fly away. Last guest of the weekend. Alone now. First time in months. Alone at an airport. No phones. No foals. No friends. Just jets.

Don't wanna fly anywhere. Not done yet. Unfinished business. Breeding season. July, soon enough. Drove home. Music loud. Pulled in lane. Crew still breeding at 7 P.M. Thanks for working all weekend, guys. Thanks for still working.

Tuesday, May 21. This morning, a client watched as I stalked a thirty-day-old foal who arrived here without ever having been handled. When I put my hand on his halter, he bolted. I was afraid to let go—might never catch him again. He carried me down the steep hill at a gallop. Suddenly, his legs tangled and he fell, sending me flying over his head. I came to rest on my knees, like that famous photo of Y. A. Tittle after being tackled. My glasses were gone. So was the foal. I knelt there, blind, and said the serenity prayer. I wondered how many folks at other commercial farms were victims of unled foals.

"I'm glad a client was watching," said Ellen. "I thought you might get up and kill that foal."

Wednesday, May 22. "The trainer went to the cops, said the dealer had come right into his shedrow, stared him down. The DEA put three undercover agents to work as hot-walkers. A week later, a paddy wagon carried away nineteen drug users. That night, two bullet holes were fired through the front door of the trainer's home. Drug dealing. It's big business on the backstretch."

Thursday, May 23. Normal daytime temperature is seventy-seven degrees. It was ninety-seven today. Despite record heat blanketing us, we cannot stray from farm policies. Every mare must be jumped by a teaser before being bred. Every new arrival must be properly checked in—vaccinations noted, lip tattoo recorded, side and front photos taken, orange name tag and number affixed. (Every new arrival is a potential Typhoid Mary, for in this heat, bacteria and viruses flourish.)

Departing mares must have manes pulled, feet trimmed, baths given, flannel placed on halters, and board the vans with accurate departure information for their home farms. Two hundred mares will come and go this spring, and almost every mare is owned by a person who will have more money invested in board, vanning, vet charges, and stud fees than the mare would bring at auction. In my mind's eye, I see these owners walking to their barns in the cool of an evening to greet their returning mares. We have only that one moment when they first see the condition of their mares to keep them breeding with us next year.

Friday, May 24. Fat and in foal. Send mares home fat and in foal and you will never have a complaint. I recited this credo today, as we strained to lift hundred-pound bales of alfalfa, irrigated from artesian wells in Artesia, New Mexico. Hay of this quality will offset the effect of the drought we are experiencing. Bluegrass in the pastures has already browned out. Foals need milk; alfalfa hay increases milk production in mares. Fewer sick foals, more fat mares—a two-for-one benefit to feeding good hay. Only the chattering of the truck driver diverted our attention.

"Watch for rattlers in that hay," the slender fellow said. He reminded me of the actor Kris Kristofferson. He ran his fingers absently through his beard, rolled his blue jeans up to show us his boots, which climbed all the way to his knees. "Folks say rattlers bite your ankles, but they don't. They bite you in the calf. Softer there. These boots have seen a few rattlers."

The driver didn't offer to help us. He sat on a cooler of Cokes and raved about good-looking girls in Arizona. The entire crew stopped when he said he was working on his *fifth* wife. That's when I noticed an unpleasant smell in a nearby bale. I flipped it over, and the fetid tail of a dead snake hung limply between the wire strands.

"Yep, that's a diamondback. Female. See? No rattles."

Saturday, May 25. "Whatever you do, don't accidentally inject yourself," Dr. Bowman warned as he withdrew the blue box of tranquilizer from his vet truck. "This stuff'll kill you." I must have looked suicidal about the prospect of breeding Arctic, because he reconsidered. "Here, I'll give it to her." He handed me the empty box. I read the label: "Dormosedan. For use as a sedative and analgesic to facilitate minor surgical and diagnostic procedures in mature horses and yearlings."

Dr. Bowman injected Arctic in the vein with a mere two cc's. When we led Arctic from her stall, her hooves dragged across the blacktop shedrow. When Jazzman teased her flanks, this mare's repulsion fought to the surface of her cloudy consciousness: She kicked the chute.

"Don't dawdle," advised Dr. Bowman. "Jump her with the teaser. She can't fight now."

We did as we were told. Jazzy jumped her. She stood.

"Now bring out the stallion."

Before she knew it, Arctic had been bred. We quickly gave her HCG, to encourage her follicle to ovulate. As Richard dragged Arctic past me, I whispered encouragement to all the little sperm cells the stallion had just deposited into her uterus: "Please find that egg soon. I don't want to do this again in eighteen days."

Sunday, May 26. I sang:

> *All the Federales say,*
> *"Could have had him any day"*
> *They just turned and walked away,*
> *out of kindness I suppose.*

Willie Nelson's voice in tribute to the Mexican bandit Pancho Villa played in my hot head this afternoon as we struggled through the Memorial Day weekend with a slim crew. Two days ago, a colt by the stallion Pancho Villa blasted one of our Federales, Steve Earl, with both hind feet. Steve went reeling across the lobby as Dr. Riddle administered solace: "You just got the wind knocked out of you. Easy now. Just take it easy."

I knew from experience that Steve would be out the entire weekend. First Prescott on a drunk, now Steve from a kick. We are still chasing elusive follicles, but we've now lost two men from the posse. The day ended when we bred a mare by Policeman named Stop Thief.

Monday, May 27. And the heat goes on. A van from Bright View Farm in New Jersey arrived at 2:30 P.M., half an hour early for the three o'clock breeding. This was the key to finishing early today. I feared the van would hit holiday traffic on the Jersey Turnpike, or on the Delaware Memorial Bridge, and be delayed for hours.

"Nah, I left early," the driver said. "There was an air show at the military base near us. Had to get the mare and foal loaded before the Blue Angels got revved up overhead."

His mare was soaked in sweat. We stood her in the cool grass and let her foal nurse. This is very stressful weather on horses being transported. We bred, she left. We turned our young foals back out at four o'clock. It was still ninety-four degrees.

On Saturday, a newborn overheated after half an hour outside. Said Peyton: "These young fellows don't have much of a thermostat. Foals have a higher ratio of surface area to body weight. They attract the heat." Peyton wanted to know if the pond had a shallow area. No. "Well then let's turn on the hose. Run cool water on his jugular, hose his genital area. That's where the major veins are." The foal cooled down. He had been one hundred and five degrees. Now he is one hundred two. We can live with that. We hope he can.

Tuesday, May 28. A huge vet list today, but no breedable follicles. This is frustrating, but not surprising. With some of the problem mares, we must wait for one soft follicle. That's all we ask. A green light from the vet; a good cover from the stallion; a bit of luck from Mother Nature—and one soft follicle. The opportunity.

Wednesday, May 29. Dad celebrated his eleventh anniversary of sobriety this week. At the A.A. meeting, he confided:

"I'll go to *any* lengths to help someone get treatment. I prefer it, though, if the treatment center is just a few lengths from Pimlico."

In another era, booze and horses went hand-in-hand, or hand-to-mouth. The annual Virginia Yearling Tour, for instance, set a high-water mark. Wrote Snowden Carter in the July 1966 *Maryland Horse*:

"The tour began at Nydrie Stud. The bar was open. The heat was intense. The caravan moved to Morven Stud. Mr. Stone served box lunches by the lake. Humphrey Finney put on a Chinese mask and was pulled in a rickshaw by his host. Glenmore Farm was the next stop. Then came Keswick Stud. Dazed by pedigree, dust, heat, and bartender, the Tour group disbanded at Keswick. Then there was the party at the Boar's Head. Morning was too soon to Dr. and Mrs. O'Keefe's Pine Brook. John Finney, stuck with announcing, kept his coat on. He was more than washy. Bowes Bond went in close for a good look. The caravan moved on. Hubert Phipps was the final host for the Tour."

There are more than a few casualties of the times in Raymie Woolfe's accompanying photos. When Dad tells about the old days, he knows whereof he speaks.

Thursday, May 30. A nervous mare walking the fencelines with a newborn can kill the baby in this heat. That almost happened today. At midnight, I coaxed Peyton into one more emergency call for May. We had little difficulty flipping the foal onto his side. Peyton shaved a one-inch square on the foal's jugular and inserted an I.V. needle. He started the foal on Gentocin to prevent infection, and on dextrose and electrolytes to arrest diarrhea, so runny it covered the foal's buttocks as if water. I thought it was urine. No, said Peyton. We milked the mare out and tubed it into the foal.

We just spent eleven months getting a healthy foal from an old mare; we almost lost that foal in one hot afternoon. I climbed into bed at 2 A.M. and watched the moon move across the sky outside my window. This has been an incredibly tough month.

Friday, May 31. What a way to end May! Way to go, Sarah! Way to go, girl!

Everybody but the promoter made it to Pimlico for Shine On Sarah's turf debut. Dr. Riddle had just departed when Dad phoned. Twenty new faces to the sport of racing discovered the thrill of victory, as Shine On Sarah surged to win by a neck. It was a track photographer's dream come true.

"How many prints do you want?" Jerry Frutkoff asked Michael, after the Discovery Partners overflowed out of the winner's circle and onto the track, a graduation photo. The Golden Girls posed for their own photo after the group shot. So did a number of other partners, because when Sarah won, everybody owned *all* of her. No one felt as if they owned just a piece of her. She was everybody's horse, entirely. What tonic a winner is! We sail for June on the wind of a win from Pimlico.

THE THINGS WE CARRY

Saturday, June 1. What will June bring? So much is in motion now, a job started, but not completed. Today, we compiled the "June Hot List," comprised of mares not yet declared in foal. As if ripping pages off a calendar, we will delete mares as pregnancy is ultrasounded, until by month's end, owners of the impregnable will say, "That's enough. See you next February."

In phone calls to boarding farms at fifteen days past breeding, we'll ask: "Did that mare catch? Haven't checked? Please let us know. Thank you." If we lose a heat in June, we lose a foal—a May 27th Northern Dancer, perhaps. Refreshing Canadian air arrived on the heels of Shine On Sarah's win yesterday. For the moment, the heat has abated, the crew is momentarily revived. The cool air is our second win. Our goal this month is to finish strong.

Sunday, June 2. No church again today, as the boy grows up without religion during the breeding season. Attendance has been sporadic since his christening. The Apostles must compete with the twelve Mary Magdalenes on the palp list this morning.

Let's watch *Mass for the Shut-Ins,* I tease young Josh at lunch. He crawls across the porch floor, a hitch in his giddyup, unable to unfold his left leg from underneath him yet—his Civil War crawl, I call it. He pulls himself up the sides of the coffee table, grabs the remote control. The TV springs to life, channels flip, confusion reigns. "No!" I teach him the word. "No!" He looks back at me, grins, continues. I confide to this diary that I hate to work on Sundays, but I love who I work for.

Monday, June 3. Lady Cameron, a mare bred way back on February 15th, has lost her pregnancy. We can't get her back into strong heat, even with Prostin. Her pregnancy had progressed past the magic "thirty-six-day" mark, when endometrial cups form in the uterus, producing equine chorionic

345

gonadotrophin, a hormone that inhibits estrus. So there she stands, in the orchard field, tentative to Jazzman—not quite in heat, not quite out.

"The incidence of early fetal death at modern stud farms is usually ten-to-fifteen percent," writes Dr. Kerstin Darenius. "Sometimes erratic estrus occurs. The conception rate for such mares is poor."

This handsome gray mare has resorbed in other years. Her owner is Jeff Huguely, a very understanding fellow, experienced in broodmare misfortune. But this is an unforgiving economy. This mare is so kind, if he no longer wants her, we can find her a good home as a riding horse. Gray mares are easier than bay mares to give away.

Tuesday, June 4. The elementary school class that visited today was chaperoned as we went about our chores. I tried to recall growing up on this farm: At breeding time, a stallion was led from the stud barn down a long hill to the broodmare barn courtyard. He would dance down the hill, a lion on a leash, only to emerge moments later, docile as a lamb, to walk placidly back to his own barn. I never received a satisfactory explanation for the sudden change in temperament. Today, various children asked me why I led Jazzman on a shank through fields of mares and foals.

"Hmmm, well, let's see, gee…"

Wednesday, June 5. Lost a client's four yearlings today to a competitor who undercut our board price. I understand the economic motives. I also know that the yearlings will not receive the type of care they had here. You can't do it on the cheap. You get what you pay for.

Yet so very few yearlings these days are worth the cost of raising them. A yearling owner must feel like the driver of an old truck, towing these babies up a steep hill to the crest. I think the clutch is going out on many of these trucks.

Thursday, June 6. A good day follows a bad day in this business. That is the dictum of perseverance. Today's ultrasound results were splendid, Dr. Bowman pronouncing success on a half-dozen reproductive reprobates. It is human nature to sometimes regard the problem mares of June with animosity, as if the mares have intentionally tried *not* to conceive. A transfer of pressure occurs, from the owner, to the boarding farm, to the mare—the subject of often fruitless effort and expense.

Steve Earl glided past the office working on his harmonica lessons. The warm sound of the reeds recalled a line from a song by the Cowboy Junkies:

"Everybody knows, good news always sleeps till noon."

Right after lunch, I began phoning owners of suddenly pregnant mares.

Friday, June 7. "She's pooling urine on the floor of her vagina," Dr. Bowman said yesterday, tilting the speculum out of the mare's reproductive tract, urine flowing down the tube to splash on the ground. I wondered how

long stallion semen could survive in such an environment. Not long, I was certain—we have been breeding this mare since February. Dr. Bowman inserted a
rubber catheter into the mare's bladder, to act as an extension of her urethra, so
urine would be voided outside her vagina.

With a catheter inserted, the mare cannot control her bladder, and urine
dribbles out as it is produced. In this hot weather, flies are frenzied around such
a mare. Today, we brought her in early, hosed off her hind legs, applied
Vaseline to prevent urine scalding, and crossed our fingers that such a simple,
yet drastic, measure might result in pregnancy this time around.

Saturday, June 8. They are Single Mares only because they are without
foals. They should be called Foalless Mares. They either didn't conceive last
year, or resorbed, or their foal died this spring, or they are maidens. Yes, they are
called Single Mares, but this shortchanges varied histories.

I know these girls well. They roam the back acreage near my house. They are
more active than mares with foals. They trot off in groups of two or three, or the
whole herd of a dozen mares gallops in a cloud of dust, free in the cool of an
evening. They congregate on the hill, for best exposure to air currents. They stand
head-to-tail, swishing each other's faces, quietly suffering the gnats and
mosquitoes, but mostly the relentless flies—huge biting flies so peculiarly
attracted to equines that they were named *horseflies*. These mares are like ladies
who move into town surrounded in mystery, with rumored pasts, a shared guilt at
being foalless:

"She had no milk so her foal died."

"She kicked a stallion last year and the farm wouldn't try breeding her
again."

"She was barren in Kentucky, and they *think* they've got the best vets in the
world."

"Her foal suffocated in its afterbirth—a fescue foal."

"She had steriods at the track. Her ovaries are the size of walnuts."

"Her owner gave ninety thousand for her, but he can't get her in foal."

Here they come again, at sunset, fluid silhouettes on the move, except for the
race mare with the fused ankle, bobbing her head as she throws her stiff leg forward at a walk. The brass on their halters glints as their hocks brake on the steep
incline. In the mornings when I am on the phone with their owners, I see the
mares only as economic units. In the evenings, I see them as animals, as evocative as a herd of elephants. Being without foals only adds to their solitary mys-
tique.

Sunday, June 9. Back to work on the stone wall today. Recruited teenagers
Jason and Rob for rock detail. They drove the Subaru Brat through the fields, harvesting this year's crop from around gates and waterers. I was Sergeant Carter;
they were Gomer Rock-Pyles. By evening, we had added ten feet of stones to the
wall. This is slow work, but it will outlive me, either in a heap from frost heaves,
or in a straight line, as it appears now from a distance, as if always here, its massive weight anchoring the land.

Monday, June 10. "You'll never get a job in the mailroom," Betty Miller said. "My framed photo of Shine On Sarah arrived in a hundred pieces today. You should know better than to send glass through the mail."

In my zeal to present the Discovery partners with their own winner's circle photo—framed and ready for hanging on office walls—I skimped on proper packaging. I could make excuses, that I was rushing to get the photo *and* an update in the mail ("Sarah might run again June 13th. Mark your calendars. Tell your bosses."), or that the post office closed at noon on Saturday, or that Dr. Bowman was driving in the lane. I only hope I get another chance to send a winner's circle photo.

Tuesday, June 11. Missed her. Doggone it. Missed her. Late yesterday afternoon, Dr. Riddle said: "Breed this mare tonight." She did not concur. Her hooves still had red paint on them this morning from kicking the teasing chute last night. Her attitude, however, had improved a degree. We bred her at 8 A.M. When Dr. Bowman arrived at ten, he ultrasounded the follicle: "Ovulated. Maybe twelve hours ago. Not much hope for conception here."

The mare is owned by trainer Carlos Garcia, who humbly brags that he *made* Carnivalay, having trained stakes winners Valay Maid and Lucky Lady Lauren. He quietly boasts that he will *make* Allen's Prospect: "Last year's Maryland Million is history." He has two-year-olds coming up from Florida by Corridor Key: "We will *make* him, too." The mare I missed was bred twelve hours too late to Citidancer. I am doubly unhappy that her last chance at romance occurred at the expense of Citidancer, who now will have one less foal in his freshman crop at the races.

(We interrupt to report that at 9 P.M., the sound of light rain on the dry leaves was heard.)

Wednesday, June 12.
MEMO TO: Farm Pets.
FROM: Irate Farm Management.
Dogs guilty of excessive barking, particularly after dark, will be reported. Goats who yank their tethers from the ground will be impounded. Cats who reproduce at their own whim will be spayed. We all have to live on this farm together, in peace and harmony. So pass the word: *If your owner does not control you, we will!*

Thursday, June 13. As we teased the mares in the Route 1 field this morning, two herons glided overhead on their breakfast run to Winters Run. Meanwhile, rush-hour traffic crawled towards Baltimore. A lady gazed up at the field of foals. She dawdled a second too long before inching forward. A man behind her in a Saab leaned out his window, shook his fist, and cursed loudly at her. His voice carried up the accoustically correct hillside. I listened to him swear. Some days I like my job more than on other days. Today, I loved it. I waved to the guy in the Saab, but he didn't see me.

Friday, June 14. We once stood a stallion named Mr. Doughnut. His major contribution to the breed occurred when Mike Price recorded Mr.

Doughnut's studdish voice on a cassette tape, thus providing photographer Neena Ewing with the sound of an excited stallion, to use whenever an alert pose for a conformation photo is required. To this day, Mike's hoarse voice still puzzles unsuspecting stud grooms during a Neena-shoot: "Doughnut! Doughnut! Make some noise, Doughnut!" and Mr. Doughnut obligingly hollers.

Mr. Doughnut literally *carried* Mike in behind any mare, whether she wanted a doughnut or not. From one such wild mating came the imaginatively named Lady Doughnut. At Pimlico today, Lady Doughnut's Carnivalay gelding, My Baby and Me, became the first two-year-old winner for a home stallion this year. Old-timers here shared the inside joke: "Doughnut! Doughnut! Make some noise, Doughnut!"

Saturday, June 15. O-for-four! O-for-four! So, you went O-for-four on ultrasounds today. Well, what do you expect? Fifteen-year-old mares bred on foal heats! You should be ashamed. What is the conception rate for foal heat on *young* mares? Fifty percent? Thirty? And you're breeding twenty-two-year-old mares? What's that in human terms? Figuring four years to one, how many pregnant eighty-eight-year-old women do you know?

What about the maiden mare, you ask? She ovulated crisply thirteen days ago! Why isn't she in foal? Did the stallion trick you by not ejaculating? Even old pros can be fooled, unless a dismount sample is examined under a microscope. So now you walk away with that hundred-yard stare in your eyes. Someone asks:

"What's wrong with him?"

"Ah, nothing. It's just June, and he went O-for-four on ultrasounds today."

Sunday, June 16. Early on Sunday mornings, Richard mows his hilly yard, weed eats around his deck. He is back inside when the crew arrives for work. He wipes the stove, the refrigerator, fills ice trays, takes a long shower, and puts on new jeans so blue and stiff all anyone could do in them is watch TV.

He sits under the ceiling fan, watches television. Today, *Super Sports Follies* visited the race track. The voice-over was bad jokes, sounds of cars crashing—mind-numbing blooper music to accompany scenes of racehorses crashing through inside rails, jockeys being tossed, horses straddling outside rails. It was a show to delight imbeciles, and Richard is no imbecile. He watched it only because he hates golf, playing on the other station. He thinks *Super Sports Follies* might film him standing underneath the striking legs of twitched mares, or holding a foal for its umpteenth shot in the buttocks—the foal lifting off all fours, straight up, leaping until reaching the anchor chain of Richard's strong arms. He dreams of his boyhood home in Bluefield, West By God Virginia—the "Air-Conditioned City," where if the temperature hits ninety, the city fathers spring for lemonade. He counts the hours until July 15th, when he takes two weeks in Bluefield. He'll go fishing in the New River.

By midday, a pot roast cools on the counter, as potatoes boil for Sunday dinner. This is how our best man spends his day off: alone, but not lonely.

Monday, June 17. The things we carry: Worker Number One carries his temper, and a retractable measuring tape to get the things he can control perfectly plumb. Number Two carries a ready apology for minor miscues, and a harmonica he bought for eight dollars at the Bluegrass festival. Number Three carries a gentle upbringing that peers into the souls of co-workers. Number Four carries the feeling that it was always going to come down to this. Number Five does not carry fear of failure, only fear of not having tried. Number Six carries a work ethic as strong as his back, and a Barlow knife because nobody ever has a knife when you need one. Number Seven carries the good and the bad in direct polarization, no gray area.

These are the things we carried to work today.

Tuesday, June 18. In the heat and drought of late, foals in stalls at midday are only slightly better off than those outside. To beat the heat, foals sleep in dust by the gates, susceptible to breathing in bacteria and viruses airborne on that dust. The swarming flies in the fields make you question God's purpose behind such a nettlesome insect. This is indeed miserable weather for young foals.

Wednesday, June 19. This hard ground might be the culprit behind a sixty-day-old foal developing symptoms of club feet. The foal's front feet have worn excessively at the toe, causing him to become footsore and stiff in his gait. Tom Rowles is preparing plastic shoes to lengthen the toe, encouraging proper growth. We've curtailed the foal's exercise, and backed off the mother's feed to inhibit the foal's growth until a better foot develops.

None of these practices in themselves are curative, but the combination just might relieve the problem. Naturally, the colt is owned by a newcomer, who only recently remarked how swimmingly his career in this game had progressed.

Thursday, June 20. Carnivalay is eleven years old, but by the time he is twenty, Maryland's place in history as a Thoroughbred-producing state will be in its final chapter. I hate this thought, but it's true. We don't have enough affordable land. A cadre of Thoroughbred farms might exist in the form of drive-in semen banks, where you purchase a Maryland-sired embryo in a breed-and-return; or as maternity hospitals, when that same mare is shipped back across state lines the following year to foal—presuming of course the Maryland Million and Maryland-Bred Fund programs still are viable.

Whether or not the national Thoroughbred industry recovers its health is academic. The demand for land on the Eastern seaboard will eclipse the profit from raising horses in the Boston/Washington corridor. It is already happening too quickly: The number of mares bred in Maryland in the past four years has dropped twenty-five percent, from approximately four thousand to three thousand. The number of stallions standing for ten thousand dollars or more dropped sixty-nine percent (from sixteen to five).

Across Route 1 from our front fields, wheels are in motion for public water and sewer. Savvy land speculators, even in this recession, are jockeying for parcels up and down Route 1, and the county has yet to even vote its approval.

I named my son after my father, in part for the stability a namesake sometimes provides, and I pray that the boy will have more than this diary to remember our farm by.

Friday, June 21. I listened as she spoke:

"I gave up asking *why* a long time ago. Sometimes there isn't a reason. I've been in this business a long time. Things just happen. It is our job to accept, and to go on. I'm sure you did all you could. Don't look back. It won't help."

Mom asked me later, "How did Geri take the news about her foal?"

Better than I did.

Saturday, June 22. Richard and I were laughing so hard we almost dropped the Spot-a-Pot, which we had hoisted on two-by-fours, as if we were Chinese coolies carrying the emperor's throne. Ellen's hoot echoed from the kitchen window. She appeared instantly, with her camera, to capture the ludicrous scene as Richard and I moved the Spot-a-Pot a hundred feet closer to the tent in the front yard for the yearling show crab feast.

Sunday, June 23. "I like a horse with muscle," Hall of Fame trainer Allen Jerkens addressed the crowd at last night's crab feast. "A horse with muscle can stand the stress of racing."

Judges over the fifty-seven-year history of the MHBA yearling show have relied solely on their trained eye. Pedigrees are not available to them. Today, Jerkens' gaze stuck on a granddaughter of Mr. Prospector—a line often stamped with an abundance of muscle. Coincidentally, the filly came from the same farm as the crab feast, and the trophy for the class (sixty-three Maryland-bred fillies were entered) was donated by that farm. Alice giggled as she handed the trophy to Michael, and Ellen closed out the family affair by accepting the julep cup for the Amateur Judging Contest, her picks most closely reflecting Jerkens' top five yearlings. It sure looked like the fix was in, but I think we were just making up for lost time.

You see, the last time we earned a blue ribbon was in the late nineteen-forties, when Max Hirsch pinned Loraine, who became a stakes winner. In recent years, participation in this show has made a yearling eligible for the lion's share of a twenty-five thousand dollar award, based on two-year-old earnings. (Valay Maid won it two years ago.) Personally, I'd rather win the cash award next year than the blue ribbon this year, but farm morale in the aftermath of the show has not been as high in the forty-four years since Max Hirsch and the summer of forty-seven.

Monday, June 24. Finished at eight o'clock tonight, as a slim Monday crew returned after dinner for a twilight breeding session. Pleasant weather in the low eighties eased the overtime. We are all aware that if we miss a mare now, her poor owner will lose thousands of dollars—once you factor in six hundred per month in board for the season lost, the barren year with no foal, and the long holding period until she next produces a foal. We attempt to prevent this loss by finishing the season strong.

Tuesday, June 25. The hot-air balloon shooshed overhead tonight, a vibrant red sphere low against the watershed, against neighboring cornfields, which look much better after the weather of late—not so much an adequate rainfall, just a lessening of throttling heat. Only eighty-five degrees again today, a strong breeze keeping flies off balance.

Every mare we breed now is a repeat offender, winners of the Back Again Purse, recidivists in the vet stock lockup instead of free in the north forty; tough dames, reproductive Hall of Shamers, resorbers and aborters, with hostile uteruses and pelvises pronounced from repeated foal bearing. Yet if we catch one out of two of these mop-up mares, we'll consider it a job well done, and allay mythic fears of missing the next Northern Dancer.

This evening, Mom and Dad came down to see their grandson. They sat on rockers on the balcony porch. Dad tossed ice cubes at me as I mowed the lawn. There are moments as the breeding season draws to a close when you can see the end, and are not desperate to reach it, but stride purposefully towards the goal, in the fitness only such rigorous daily regimen can impose.

Wednesday, June 26. "And Castalie's orphan foal is doing splendidly!" reported Mrs. Orme Wilson, who accompanied her late husband throughout the world during his foreign service career. "Even the young broodmare Norland Nanny is treating her well. You know about Norland Nannies? Why, every one in the British Foreign Service employs Norland Nannies to care for their children. They are the absolute best nannies. I find it so interesting that our mare by that name would be baby-sitting the orphan."

Thursday, June 27. Never, *ever*, judge a book by its cover, especially when it pertains to potential clients. When Mr. B. walked into our office seven years ago, announcing that his mare when mated to Carnivalay would make us all millionaires, I turned him over to Michael, and made my excuses.

"You should hear his story," Mike told me later. "He and his wife go to Atlantic City often. She won *eight hundred thousand dollars* on a slot machine. I told him not to expect such miracles in the horse business, but he loves horses, and always wanted to own one. He also runs a deli and has promised us lunch when he visits."

Mr. B.'s timing has always been exquisite. Today, he arrived with fifty cold cut hoagies, and farm help poured in from all directions. A crowd of twenty-five jammed the front porch for lunch. Mom baked a coconut cake, and I turned thirty-seven to a chorus of: "Now get outa here. Use that excursion ticket. Go play lacrosse with your college chums. We'll cover for you."

. ebbp

Friday, June 28. Grandfather's duties included keeping Samuel D. Riddle's racehorses by Man o' War eligible to events such as the rich Futurity at Belmont Park. Riddle had nominated fifteen Man o' War foals of 1931. On this date in 1933, Riddle wrote:

"Please declare out of the Futurity of 1933 Fighting Mike & Willow King & Gold and Black."

V. E. Schaumberg, racing secretary at Belmont Park, responded to Grandfather's cable:

Dear Mr. Adolphe Pons:
I beg to acknowledge receipt of your communication dated June 28th making declarations out of the Futurity for a/c of Riddle.

Futurities in particular suffered from cash-flow problems during the Great Depression: The Pimlico Futurity, won in 1932 by Swivel in Grandfather's silks, was suspended in 1933 and 1934. The news was not all bad in 1933. That year, Grandfather bought this one-hundred-acre parcel of land for eighteen thousand dollars. The former Rockland Farm became Country Life Farm.

Saturday, June 29. All that remains of Camp Hale, high in the Colorado Rockies near Leadville, are the poured cement foundations of dozens of barracks, and a historical marker identifying this plateau as the home of the 10th

Mountain Division—the famous skiing soldiers whose all-white uniforms were bloodied with the highest fatality rate of any World War II unit.

Nazi fortresses in the Italian Alps could only be conquered by men and mules. I know that Uncle John and Dad, stationed at Fort Robinson in the northwest neck of Nebraska, over a corner of Colorado, shipped U.S. Army Remount mules to Camp Hale, and to Camp Carson as well. Way out here in the middle of God's country, I am connected.

"If you get down to Colorado Springs," Uncle John told me yesterday, "stop at the Broadmoor Hotel. The Remount commandeered the Broadmoor, used its stables to house Remount stallions. It was quite a luxurious facility."

Sunday, June 30. A good book makes a holiday complete. In a Rocky Mountain bookstore, I snared the intriguingly titled *At Play in the Fields of the Lord*, a novel about missionaries and a vanishing Indian tribe. Today, I ran up the sides of a mountain, conditioning my lungs for the four-day lacrosse tournament, huffing through thin air at eight thousand feet, passing purple flowers of mountain lupine—up, up, up towards snowy peaks slicing into the heavens. I know that back home, the crew is still mopping up, pushing through the heat, finishing strong. When I call home, and they ask me what I am doing, I will say:

"I am at play in the fields of the Lord."

A TRUE FARM

Monday, July 1. Back home, the drought continues. Many a recent evening, I have walked the fields, kicking dried manure, dreaming of irrigation. On vacation drives to Ocean City, we have passed huge rolling wheels of pipe that spring to life, slackening the thirst of flatland crops of beans and peas. A Jolly Green Giant must roll these pipes through thousand-acre fields level to the horizon. I want to ask him if I'd be successful irrigating a hilly one-hundred-acre farm.

Here in Colorado, melted snow flows down Gore Creek, just past the side-lines on the lacrosse field. Sixteen teams play continuously on two fields, yet the grass resists the traffic: Moments after the day's last game, sprinklers kick on. The grass stays as green as the irrigated alfalfa hay delivered to Maryland from this arid region, hay for which we pay two hundred and twenty dollars a ton. On this hay, mares bloom, foals prosper, stallions maintain fertility despite breeding season demands. Yet would not irrigated fields of Maryland bluegrass accomplish the same objective, at a fraction of the cost?

From Winters Run, a hundred feet upstream from Country Life, the local water company extracts three million gallons of water per day. What rights does a riparian landowner downstream have to Winters Run, a mini-Colorado River in our own backyard? The doctrine of water apportionment is "first in time, first in right." Country Life predates the water company. Why, even the generous aquifer underneath this farm could supply an irrigation system.

Oh, well. Just a thought. Won yesterday, twelve to two. Lost today, eight to five.

Tuesday, July 2. The Indians who crossed from Asia into the New World shed their nomadic hunting habits when they discovered the secrets of planting and cultivating, of harvesting and storing crops. They settled just south of our lacrosse fields, on the raised mesas of the Rockies. They built pueblos and they made baskets to carry corn and beans, squash and gourds. They had no horses and no livestock. They never developed the wheel and they used no metals. They were peaceful farmers—until the last quarter of the thirteenth century, that is. Twenty-five years of extended drought forced the Indians to abandon mesa pueblos and cliff dwellings, to compete against other Indians for land where rains fell.

Water. From Maryland to Mesa Verde, farming has always been a matter of water.

Wednesday, July 3. From the playing fields, we can see Interstate-70. Instead of tractor-trailers, however, horse trailers crowd the highway. These are not Thoroughbreds rushing towards the last cover of the breeding season, for Colorado mirrors the industry's declining numbers. Over the past six years, the state has seen a thirty-five-percent drop in mares bred, a twenty-percent drop in stallions. No, on the holiday highways of Colorado, Quarter Horses head for dude ranches, and show horses continue the summer circuit.

At a beer blast after today's game, a college teammate asked me:

"What do you do now?"

"I breed horses for a living."

"Oh, I've read where that's a pretty tough business nowadays."

"It has its ups and downs. How 'bout yourself?"

"I *was* with a bank."

We shook our heads for the dashed years in pursuit of a vice presidency. Suddenly the horse business didn't seem all bad.

Thursday, July 4. Denver Dan showed me a graduation photo in the moments before I boarded the plane for home. We all had hair. We all were thin. We all were so untried, so untested. On the flight home, racing over the curve of the earth, the sunset accelerated by the speed of the jet, this trip became more than a passage from the mountains back to the coast. It became a journey through time as well, a notion reinforced by how rapidly the cabin grew dark, sunlight staying on the Great Plains, refusing to follow us back.

I have spent four days at play in the fields of old friends. Careers bounced around in quick quotes after games. Kids were counted. A subconscious clerk kept taking inventory. How are you doing? Good. I'm doing good. I work on a farm, a horse farm. Yeah, it's dry, but it'll rain soon. Either that, or I'll irrigate.

"Please bring your seats and tray tables to their full upright and locked positions. Once again, we wish to thank you for flying with us. Have a safe Fourth of July. Good night."

Friday, July 5. At best, nine out of ten mares will conceive during the season. Every breeder you encounter in July owns that tenth mare—the unpregnant one. Finney wrote that this is the time of year for frayed dispositions on the stallion's part. He might have included mare owners. The brief trip out West refilled my reservoir of patience. Thus rededicated, I took over the phones, permitting Michael, who had manned the hotlines in my absence, to enjoy the cool air-conditioning at Laurel today, where Shine On Sarah surged to the lead, couldn't stand the prosperity, swished her tail, looked around for company, found some, and finished second. The Golden Girls sighed.

Saturday, July 6. It takes five years from the autumn a horse arrives at stud until his two-year-olds run, another two years before his first crop matures as four-year-olds. *Seven years.* Carnivalay arrived here in September, seven years ago. Now, his four-year-olds are maturing. Today, Valay Maid split horses to win the Molly Pitcher Handicap at Monmouth Park. Last week, Lucky Lady Lauren won by five lengths in New York. In between these two headliners are twenty-one winners from twenty-six other starters in the stallion's first crop. They win at Thistledown and Philadelphia Park, Calder and Rockingham. Dad ferrets the charts out of the *Form.* "Look, Dum Crambo won a two-mile grass race at Delaware. Maybe that's his hole card: Two-mile races." Each little win flows like rainwater into the sire's progeny earnings, but each big win, like Valay Maid's race today, is a veritable thunderstorm.

Sunday, July 7. If a stallion doesn't get the job done in his first breeding season, he becomes an equine Sisyphus, rolling a huge stone named "marketability" to the top of a hill, only to have it roll back down into that vast commercial wasteland where nineteen out of twenty stallions end up.

Citidancer is determined not to let that stone roll over him. Although he is only four years old, he has withstood a demanding season. Numbers are everything to a first-year stallion; attrition works fast. To wit: Castalie, covered at 9 A.M., struck by lightning at 5 P.M. the same day; Frankie's Turn, in foal on one jump, suffers a bout of colic a month later and dies; and the poor mare who spooked and broke her back on May 2nd. Three mares in the first book will bear no fruit, no foals to grow into promising two-year-olds. The stone slips. The only protection against attrition is numbers. At 9 A.M. this morning, we crossed our fingers on the rump of a twenty-year-old stakes-producing mare. Citidancer danced. The stone moved back up the hill.

Monday, July 8. Owned by a longtime client, Trillora Gold is contentedly driving us nuts. You see, Trillora arrived in January—in foal but with no breeding date. She had been pasture-bred sometime last summer at a large plantation down South. We hoped the passion had occurred in early summer, but the size of her belly indicates it might have been closer to Labor Day. This is the latest foaling mare anyone here can recall. To the farm crew who check on her several times a night, I reason: "Look, July means nothing to her. She might hold until August. Don't kill yourself waiting on her. Labor Day might mean *Labor* Day. Just keep an eye on her bag."

We have delivered forty-five foals this spring. We await one more, in this endless summer of expectancy.

Tuesday, July 9. It was dark, and I couldn't find a shovel. Then I found one. It was a flat shovel. No good for digging. I found another one. This one had a broken handle. The ground was too hard anyway.

"Alice, can we bury him in the garden? The ground's much softer there."

"Sure, he'd love it in the garden—a place of food. I can talk to him while I'm planting. Here, underneath the sunflowers."

Stingo was named for a character in *Sophie's Choice*, a depressing novel at odds with this canine comedian's countenance. A chestnut Beagle with no interest in hunting, Stingo was the perfect lap dog, until his weight reached forty-two pounds. An ottoman with legs, he became a moveable feast, a meal on wheels, visiting the umpteen dog bowls on this farm on a leisurely circuit.

Alice arrived home on crutches from the Peace Corps nine years ago. As she convalesced in front of the office computer, the brown puppy provided company. He was her dog. Tonight, the vet told Alice that Stingo had suffered congestive heart failure. "We could treat him, but it will recur. He can't get enough oxygen." Stingo became Alice's Choice. We dug a deep hole in the garden. It was very sad to say goodbye, but Alice had made the right choice.

Wednesday, July 10. "Sorry to bother you on vacation."

"Nonsense. No bother. What's the trouble?"

"Your mare was in foal at twenty-two days, but we checked her today at thirty-two days, and she's resorbed."

"Oh, my goodness. I thought Mother Nature was telling us something when she needed four covers to get in foal. Well, if we breed her now, I'd have to give her off next year. Let's just let her go. We'll put her under lights and try again."

Thursday, July 11. Still phoning farms to determine the status of mares bred in late June. Rewarding to hear they are in foal, disappointing to find out otherwise. Oh, well. The crew received a bonus today—those about to embark on vacations to Ocean City or to Bluefield, West Virginia. Thanks for throwing us a bone, Steve Brown said, as he barked like a dog, his special way of showing gratitude.

A breeding farm at the close of the season is very similar to a school whose semester is ending for summer break. Some of us are not quite through exams, but others are washing their cars and wearing flip flops. We all are settling into a cruise mode, a necessary psychological healing from the battles of the past five months. In the pipeline are scores of foals for next spring, and that thought alone provides a substantial measure of job satisfaction.

Friday, July 12. In correspondence to the Discovery partners, I hesitate when describing routine occurrences in the horse racing game. Isn't it too glib to simply remark that Shine On Sarah "bled"? That now she runs on Lasix? What does "bleeding" really mean? Will I frighten a Golden Girl? The following is also germane: Edgar Allen Pro's testicles bothered him when he galloped. So we gelded him. He'll be out of action while he heals. Ellen's Prospect bucked shins. That means she's torn the periosteum covering her cannon bones. She's too sore to train, and so she'll be blistered. Stable Gossip developed puffy ankles from training and from the cement floors in Pimlico's new barns. We've stopped on her. We might pin-fire her.

All this bleeding, blistering, gelding, and firing has a ring to it I've been unable to soften for the Golden Girls. Then I think: Heck, those dames are tougher than me. Give 'em the straight poop. There's good and bad parts to every sport.

Saturday, July 13. At 4 A.M., Richard got a head start on all that traffic headed for Bluefield. Sans our server, we played in the annual Bonita-Country Life volleyball game, a chest-thumping spectacle between a farm that plays every Sunday night and a farm that plays once a year. Without Richard to serve winners, we couldn't even rotate properly—the sure sign of a green volleyball team.

Bonita and Country Life are overlapping family affairs. We use the same veterinarians, same feed men, same labs for cultures. We dream the same dreams, too, and I thought it was appropriate that today, while Bonita graciously spiked the ball down Country Life's throat, a runner sired by Bonita stallion John Alden was beating the unbeatable Wide Country, not to mention our dear Ritchie Trail—in the Pearl Necklace Stakes at Laurel.

Sunday, July 14. We are still cleaning up from last Sunday's violent windstorm, when seventy-five mile-an-hour gusts tore the tops out of one-hundred-and-fifty-year-old Norway maples in the yard at the big house. Sycamore branches were hurtled to earth from their lofty perch as if a giant was astride the house, aiming limbs as spears. The smell of pine boughs on the basketball court reminded us of Christmas.

There are two feelings at work in the wake of such a storm. The first is relief that no one was hurt. The second is sadness, for the damage done to the venerable old trees. Grandfather had braced one stately maple many years ago, iron rods and cables holding limbs from which children's swings dangled. Mom called the tree expert this week. He shook his head: "Whole thing ought to come down." Mom looked around the yard for other accommodating limbs to relocate the swings. "I'm going to miss that tree," she said with a sigh, but in her mind, she was already adapting to Mother Nature's whims.

Monday, July 15. "She bowed. I want to make a broodmare out of her."

"But she never raced."

"Well, her father is by Northern Dancer, her mother by Sea-Bird. What will it cost to produce her first foal?"

"Fifteen, maybe twenty thousand—counting stud fee."

"What kind of mare could I buy for twenty thousand?"

"One who *raced*."

"And I'd have to pay upkeep on that mare. Why not just keep the mare I have? I *like* my mare."

To his last question, I had no answer. For a filly to become a successful broodmare, she must first have an owner who likes her.

Tuesday, July 16. Tony Portera tossed Shine On Sarah T-shirts around in the spacious Laurel paddock before the fourth race. The partners donned the casual garb and posed for photos while the small field of fillies circled. I had one eye in the camera lens and the other on the horses. This was a little too much action in the paddock, I thought, but what the heck, it's the fourth race on a thin Tuesday. Let's have a little fun. We're selling entertainment.

Moments later, after Sarah finished third, the partners all wanted to know: Did she run well enough to go to Saratoga? Trainer John Hicks nodded yes. One of the Golden Girls had read my last correspondence: "But we can't race on Lasix in New York!"

"Sarah'll be fine," John said. These folks are having the time of their lives with this filly. I am, too.

Wednesday, July 17. A groom on a local breeding farm was injured recently by a tough stallion. Most farms have encountered the "dangerous" stallion, the rogue to handle. A calculation is made by farm management: We know this stud is going to hurt someone, someday, but he is a sire worth having.

After we put down Lyllos several years ago, I hung his halter in my study. Since that day, whenever we look at stallion prospects, one of our first questions is: How's his temperament?

Thursday, July 18. "He's trying to buck shins," trainer Jerry Robb said about North Corridor. "I don't know how he'll run."

The first starter to represent Corridor Key broke on top, led by three turning for home—then the shins kicked in. Still, from our angle, he appeared to hold on for the win. The usher came to Hal Clagett's box to escort him to the winner's circle. The Tote board gave no clue. When the "Official" sign went up, our hopes went down. Lost by a short head. The usher made his apologies.

A male horse is merely a stallion until one of his foals wins a race. Then he is a sire. I had forgotten how exciting, and how elusive, this distinction can be.

Friday, July 19. A half-mile down Old Post Road, right off Route 40, is the old Havre de Grace race track. Four barns still stand, converted into machinery sheds for the National Guard, who commandeered the vacant plant in the nineteen-fifties as a parking lot for camouflaged trucks. You can still discern the roof lines of shedrows despite the make-over. Guard headquarters are in the old grandstand, an elegant, three-story, white facade completely out of place among the Quonset huts and low-slung barracks.

"The race train took thirty-six minutes from downtown Baltimore to Havre de Grace," racing writer Joe Kelly has told me. "They had a spur right there at the track. I saw Saggy beat Citation there in 1948. Citation would have won, but Hefty carried him wide. Ask your Dad. Hefty carried Citation wide, and Saggy won—the only time Citation lost as a three-year-old. The barns were right on the water, I mean *right on the banks* of the Chesapeake Bay. You couldn't ask for a prettier track than Havre de Grace."

The legend of Saggy in the slop. Race trains. Calumet glory. Barns on the Bay. Armed with a little oral history, you can almost hear the whoosh of ancient spirits galloping down the Old Post Road.

Saturday, July 20. "I'm on the lead. I'm gonna win. I'm flying. I'm way ahead of Housebuster. Safely Kept is nowhere. I'm gonna win. I heard those railbirds taunting me: 'What's *he* doing in here?' I guess they figure you can earn a half a mil' without talent, huh? My dad's Danzig. I was raised with Sunday Silence. I run my heart out every race. Now, I'm on the lead, and I'm gonna win—or die trying. Oh, no, here comes House. I can't beat House. Oh, he's the best. Who was I foolin'? But I'm second for sure.

"What? What? I'm falling! I've hurt myself!

"Where am I? I can't stand up! Yes, I can! I'm up! I'm running again! Why is that man grabbing my bridle? He looks scared to death. I'm all right. I just need to catch my breath. Was I second? What happened? Look! The winner's circle. Here comes House. Hey, you won, old buddy! You won! Here, right here. I'll move. I can move. There you go. All yours. Good work, old buddy. You're the best. My head's swimming. That sad song I heard in the shedrow is playing."

> *There's a red boat coming down the river.*
> *It's got numbers on its side*
> *and it's making big waves.*
> *Dad said red means trouble,*
> *numbers add up to nothing.*
> *I guess I never figured to be the one,*
> *to fade away so young.*

Sunday, July 21. The *front* page of the Sunday paper—not the sports page, mind you—carried a color photo of Bravely Bold breaking down. I know it's news. I know it's dramatic. I also know that when I talk to possible investors in a racing partnership, I'll need an answer when the person asks: "How come they're always breaking their legs? Every time I turn on the TV, or pick up the paper, they're carrying another horse off."

If Housebuster had won without incident, the Frank J. De Francis Memorial Sprint would not have made the front page.

Monday, July 22.
> Dear Denver Dan,
> Crabs are running good. Horses too. Big news is that the kid WALKED tonight, just four or five halting Young Frankenstein steps. All he needed was a little wall-to-wall to cushion his fall. We all clapped. He smiled and fell into my arms laughing. He's ten months on August 1st. I still can't believe he walked tonight.
> J. W. staying with Dad at Saratoga—two Walter Matthaus with shaving kits, hustling out to watch horses slip past in the morning mist. Valay Maid's half-sister sells at the Spa. First time since the fifties that a foal sired by our stallions has sold there. Ellen and me and JP III head to New Hampshire Friday. Lawn up to my knees. Gotta go. Gotta mow.

Tuesday, July 23. It was a hundred and four degrees in Baltimore today. At four-thirty this afternoon, the phone rang:
"Your horses are out," a woman said in a stern voice.
"Where are they?" my heart jumped. My mind immediately saw open gates, a car through the fence, horses on the highway. "Where are they out?"
"Out in the field. *In this heat.* I'm calling the Humane Society."
I breathed in relief. "They're *supposed* to be in the field, ma'am."
"I'm calling the Humane Society."
She hung up. She never identified herself. I felt as if I had been attacked. Steve Earl walked into the office:
"The studs are in. Just turned the foals out. Everybody's got water. Everybody's got hay. The mares are standing under the trees."
"Steve, we just got a call from some crank do-gooder bitching at us for having our horses out in this heat."
"What?"
"Top the water tubs off. If this lady shows up with the Humane Society, I want a full tub to dunk her head in."

Wednesday, July 24. At 11 P.M., Sandy called from the foaling barn. When I arrived, Trillora Gold was pacing the stall. I felt the foal's two front feet encased in the emerging placenta. Ten minutes and a dozen contractions later, the foal kicked her front feet through the placenta as her hind feet cleared the mare's vagina. The last foal of the year took her first breath.

I thanked Sandy for her vigilance. As I climbed into the car, my arms began to itch from the drying amniotic fluid, a sensation I've felt on cold nights in February, but never on a hot night in July. I stood on the steps of the house and looked back towards the fields—such round hills distinct in the light of the moon. When that filly first called for her dam, when a heretofore nervous maiden mare whinnied back, as a mother, in calm assuring tones, I had marveled anew at the subtle power of nature, as gentle as the full moon whose gravitational pull eased our new foal into this world. I could almost hear Mother Nature close the book on this season, on a goodnight story, sending us to bed as if we were children again.

Thursday, July 25. When the meet moves to Laurel, two-year-olds take over at Pimlico. This morning, we ducked in and out of shedrows, seeing creatures conceived at Country Life, as the pipeline reaches fruition. Smile Be Happy's groom lamented: "He'll stay with me till he gets good, then he'll go to the first string at Laurel." Clochal's groom advised: "Don't believe that work you saw in the *Form*. She's better than that."

Andrew galloped Edgar Allen Pro past the Discovery partners. He was flanked by Billy Santoro on Liz's Prospect. Together with Marshall Silverman on a gray two-year-old, the riders worked a three-man weave on horseback—first Edgar in the middle, then Liz, then the gray. Edgar was like a boxer doing road work, immersed in the rhythm of the exertion.

Allez Prospect, the lone three-year-old in Andy's stable, glided on the freshly harrowed track after the eight o'clock break, a bullet work. A misspelling on the claim form in December voided Allez Prospect's sale to another stable. Michael wanted to send a French-English dictionary to the trainer whose thirty-five thousand dollar mistake cost us a handsome Christmas bonus. An injury two days later put Allez on the shelf—unsold, still ours, damaged goods. All I could say to console Michael was that everything works out for the best. Today, Allez prepped for his return, and maybe the best is yet to come.

Tomorrow, we leave for vacation. There is much to look forward to, much to return to.

Friday, July 26. Some of the train's conductors are pulling the switch on the left, the one marked Vacation. Other conductors stay on track. The train company maintains service. Michael returns from his break, catches up on details he missed, previews the coming week. I tell him: "If John Hicks gets Sarah in for August 4th at Saratoga, call the partners right away. They'll need some lead time to make it up there. The blood work on Queenie's foal was good. Stay in touch with her owner. What've I forgotten?"

Goodbyes are hasty in the light rain that is falling. Mom ducks inside the car to kiss her grandson. Alice takes his little hand: "You be good, Mister Mister. You be good." The tiny pontiff acknowledges his audience, struggles against his car seat, hoping to be lifted free. We made it to Massachusetts. It poured all night. At first light, the portable crib rustled awake.

"It's like sleeping with a hamster," Ellen smiled.

Saturday, July 27. The Boston *Globe* carries Saratoga entries. The Test Stakes is today. Last year, Carnivalay's half-sister Go for Wand raced against Carnivalay's daughter Lucky Lady Lauren in the Test. Tomorrow, Citidancer's uncle, Tong Po, goes in the Jim Dandy Stakes. Next Sunday at the Spa, Shine On Sarah might run. There is a New York-bred two-year-old filly by Allen's Prospect, One Is Enough. She's ready to run at Saratoga. And, for the first time since Grandfather was alive, a home stallion will be represented at the select yearling sale.

The magic of this sport is that it renews itself, filling the holes left by Go for Wand as best it can. We have much to cheer for at the Spa this year, although we must be content to follow it through the *Globe* for the next ten days.

Sunday, July 28. "Forgiveness comes before penance," the white-haired preacher told vacationers at church on an island in a lake in New Hampshire. "Forgiveness frees the body from the paralysis of guilt, especially in the family setting."

Just when I was feeling unfettered from the farm, he had to go and mention guilt. Suddenly I was transported back, dealing with a dead foal, or an unhappy client, or any of a million tasks that pepper you throughout the breeding season. *Let it go. Just let it go.*

Tonight, the moon lifted itself over Red Hill, and I watched the moonbeam walk on the water towards the dock. Touched by this finger of light, I felt forgiven. It always has taken me a while to unwind.

Monday, July 29. Near this lake is the True Farm, in practice as well as in name. John True cleared its land in the seventeen-hundreds, when the sandy gravel driveway served as the stagecoach road. He built his barns out of red pine and oak. The barns have required little maintenance in two hundred years. Two black Percherons pick in the mountain pasture, amid white blossoms of Queen Anne's lace. A nearby camp hires Earle Chase to hook up the Percherons for weekly hayrides for camp children.

"I augment the farm's income best I can," Earle said. "Not many farms left in New Hampshire. Doesn't pay. Work's too hard."

Near the driveway are raised beds of vegetables, planted by Earle to implement the concept of Community Supported Agriculture. In other words, his neighbors pay him to grow their gardens.

"They pay one lump sum each spring, then they come in here in summer with their baskets, and ask for their buttercrunch lettuce, radishes, carrots, asparagus, tomatoes, spinach, whatever—all organically grown in raised beds. No fertilizer other than manure. No herbicides. Our neighbors, in essence, help us keep our farm a *farm*, a green space, undeveloped. That is the concept of Community Supported Agriculture. CSA started in Europe. California has a strong CSA program."

The True Farm passed out of the True family in the nineteen-twenties. Earle works for the farm's latest owner, a New York City resident. The Percherons pull wagons, Earle plants crops, the man in New York pays the property tax, and the True Farm makes it to another year on the effort and resourcefulness of its keepers. It is a true farm.

Tuesday, July 30. "His family stands some fine stallions," a relative introduces me to old friends of his. He can't pronounce Carnivalay, and he is vaguely aware that several of the stallions whose names he could pronounce are no longer with us. This unsettles him. He wants to ask about those early stallions, but I guess he figures why bring up old history. "Yes sir, some fine stallions," he repeats, but the old friends do not know how to follow up the introduction, and I just nod my head dutifully.

Wednesday, July 31. The hamster had a rough night. A new tooth is coming in. He was feverish all night, stood in his crib, and real tears fell. Two o'clock became three became four. Outside in the broad New Hampshire night sky, shooting stars were like matches being struck in the Milky Way. I wanted to pick the child up and carry him outside to distract him, but one good wail over the still water would awaken the entire camp. Near dawn, we all fell fast asleep. My last thought was how wonderful not to have to work tomorrow.

STONE PONIES

Thursday, August 1. Today is the deadline for submitting the Report of Mares Bred to The Jockey Club, listing the names of all the mares bred to each stallion this past season. How many foals can we expect from our efforts? The Jockey Club will answer: Approximately fifty-six percent of all mares bred will produce a foal who ultimately is registered.

Oh, some fancy stallions will be higher, but it is a constant in Mother Nature's equation: Mares Bred x .56 = Registered Foals. Been that way down through the ages. Old-time horsemen playing the new game of pinpoint palpations and HCG-hastened ovulation are stymied: "You mean we bred sixty mares and we're only gonna get thirty-three foals worth registering? Hell! We did that good back when we bred 'em on the second and fourth day of heat! Why do we even use vets?"

Approximately two out of five mares bred at Country Life this past spring will not be represented by a registered foal. Here's a sampling of the things that happen between conception and registration:

Mares pronounced in foal in spring will resorb their pregnancy by fall. Some mares—like Castalie, who was struck by lightning after being bred, and Frankie's Turn, who colicked after being pronounced in foal—won't even live to see the end of the breeding season. Foals will stand and nurse but fall and die, before the breeder gets around to sending in the Registration/Blood-typing kit. Pregnant mares will be given away as pleasure horses, to deliver live foals the following spring for new owners who do not care to register them as Thoroughbreds with The Jockey Club. Foals will be born, but estranged wives will withhold Stallion Service Certificates from their husbands until the divorce is final.

The Jockey Club statistics do not account for such situations. And though a bold disclaimer will accompany the registered-foal percentages issued by The Jockey Club two years after the August 1st Report of Mares Bred has been filed, many breeders construe those percentages as a reflection of a stallion's fertility. In actuality, The Jockey Club score is only about half the story, or—to be precise—fifty-six percent of the story. It is not a reflection of the number of mares pronounced in foal to a stallion during the breeding season: We average between eighty-five and ninety percent conception at the conclusion of a breeding season—nine out of ten mares, not five-and-a-half out of ten. The Jockey Club's calculations are merely a scorecard listing the total number of foals lucky enough to be born and survive under conditions that ultimately resulted in a registered Thoroughbred.

I don't want to dwell on this thought. It's dampening my vacation mood.

Friday, August 2. The greatest sire of the nineteenth century was Lexington. In the stud from 1856 through his death in 1875, Lexington sired five hundred and forty-three foals—in the days before ultrasounds and Regu-Mate, before shots of estrogen and artificial lighting. If five hundred and forty-three foals represent fifty-six percent of the mares Lexington covered, he was a busy boy. By extrapolation, in nineteen years at stud, he must have averaged fifty-one mares per season: Nine hundred and sixty-nine mares the total for his career.

Imagine the chagrin of the owners of the four hundred and twenty-six mares (forty-four percent) bred to Lexington who didn't have a registered foal to show for it—no chance at having an offspring such as Norfolk, Asteroid, Daniel Boone, Duke of Magenta, Harry Bassett, Kentucky, Kingfisher, Preakness. Double your depression if you bred a Glencoe mare and got no foal, for Lexington on Glencoe mares was the old-time equivalent of Mr. Prospector on Northern Dancer mares.

History is so uplifting. This has *never* been an easy game.

Saturday, August 3. In *The Stone Horse*, writer Barry Lopez discovers more than desert rocks arranged by American Indians three hundred years ago:

"I felt a headlong rush of images: People hunting wild horses with spears on the Pleistocene veld of southern California; Cortes riding across the causeway into Montezuma's Tenochtitlan (Mexico City); a short-legged Comanche, astride his horse like some sort of ferret, slashing through cavalry lines of young men who rode like farmers. A hoof exploding past my face one morning in a corral in Wyoming. These images had the weight and silence of stone."

The horse casts its spell over people of all ages: Hot–walkers on the track at sixteen, Michael's three-year-old son Philip on the pony at camp, my own son at ten months, captivated by the bulky black heads of Earle Chase's Percherons after the hayride, and, Lopez, staring at a stone horse, eighteen feet from poll to rump, eight feet from withers to hoof:

"I was not eager to move. The moment I did I would be back in the flow of time, the horse no longer quivering in the same way before me. A human being, a four-footed animal, the open land. That was all that was present—and a 'thoughtless' understanding of the very old desires bearing on this particular animal: To hunt it, to render it, to fathom it, to subjugate it, to honor it, to take it as a companion."

Sunday, August 4. "Shhh! Shhh! It's okay. Let's get out of this cabin before you wake up the whole world. Isn't this better? Strolling in your baby carriage down these pine lanes!"

"Eeee. Haaa. Mmmm."

"Let's play a game. You say the letter, I'll say the word. I'll teach you the foundation sires of the Thoroughbred breed. Let's make up an anagram we can both remember. Whaddya' say?"

"Eeee."

"Yes. *E* is for Eclipse, the line of the Darley Arabian. *Every Darn Arab....* "

"Haaa."

"Good. *H* is for Herod, from the Byerley Turk line. *Has Been To....*"

"Mmmm."

"Excellent! *M* is for Matchem, grandson of the Godolphin Barb. *Mecca, God Bless.* Now I've got my shortcut memorized: *Every Darn Arab Has Been To Mecca, God Bless.* Using the first letters of our anagram, I'll remember: Eclipse (Darley Arabian), Herod (Byerley Turk), Matchem (Godolphin Barb)— 'Every Darn Arab Has Been to Mecca, God Bless.' When's the test? I can't wait. Reducing complicated formulas to simple phrases is a trick they taught us in the Bar Review courses before the Bar Exam. It got me through."

"Kkkkk."

"Shhh! Don't confuse things. Enjoy your stroll. It's almost time for breakfast."

Monday, August 5. This morning's discourse at dawn found us at the gate to the True Farm. I recalled Earle Chase telling me about a brown bear who snatches sheep in nocturnal raids. It was barely daylight.

"Yep, he's got the whole White Mountain National Forest to roam, and he picks on my tiny sheep operation. Ate six last summer."

Farms everywhere have their own peculiar problems. I worry about vandals and bottle throwers, Earle worries about bears. I wondered how fast I could push a stroller with a bear in pursuit.

"Let's not work on our B's today, son. You say 'Baaa' and Mr. Bear might mistake you for a lamb pie."

Tuesday, August 6. There is something daunting about phoning home while on vacation. Home, where the lives of so many people are affected by how fast a horse runs, a perilous interdependency. You can't simply phone an answering machine. You *have* to know how Shine On Sarah ran at Saratoga: "She bled. Your father thought she broke down." You can tell from Mom's voice that Valay Maid did not fare well in the rich stakes at Monmouth. You're afraid to ask how Lucky Lady Lauren did in the Ballerina Stakes, only the third time a home-sired baby has competed in a Grade One race. (Xray was the first—he broke his cannon bone in the Champagne Stakes. Valay Maid was the second—she ran third after Go for Wand fell in the Breeders' Cup Distaff. Lucky Lady Lauren was the third—she was outrun, but unhurt, in the Ballerina.)

Tomorrow, you leave the lake and head for Saratoga, for a busman's holiday. The pages of summer reading books are no longer spellbinding. Horses as a business compete for attention. The vacation is over; the vocation continues.

Wednesday, August 7. The average price at the yearling sale tonight fell fifty percent from last year. Mr. Not Sold signed for the Corridor Key out of Valay Maid's mom. This rebuff did not discourage the filly's connections. "Let's put a syndicate together and race her." Most sellers scoured the grounds for interested underbidders, unable to assemble racing syndicates for entire consignments.

Earlier in the day, I had rummaged through the Lyrical Ballad Bookstore, and found a popular magazine of yesteryear: *Country Life*. Its pages were full of sports afield. My eyes stopped at an article entitled "Seduction at Saratoga," written in 1941 by Salvator, the *nom du plume* of racing historian John Hervey. The story was about the origins of the Saratoga yearling sales. William B. Fasig died in 1902. His partner, Ed A. Tipton, took over the business. Tipton took as his partner E. James Tranter.

"They were going ahead at a record-breaking pace when the 1910 Hughes anti-racing crusade succeeded in closing every race course in New York and paralyzed the sport through the greater part of the country. Tipton decided to 'get out while the getting was good.' Tranter, being younger and more optimistic, faced the future, confident that there was a golden lining to the clouds. The clouds did pass and the Fasig-Tipton Co. entered upon a new era of prosperity."

This year's dismal Saratoga yearling sale will reverberate back home, causing panic among mare owners anxious to rid themselves of breeding stock. I must remember the history lesson I learned in the magazine *Country Life*—the example of the fortitude of Mr. Tranter, who faced the future and came out on top. Maybe we'll take a piece of that Corridor Key filly. Maybe we'll have the last laugh.

Thursday, August 8. Secretariat's jockey Ron Turcotte from his wheelchair introduces another jockey, Pat Day. It is many moments before Day has the composure to speak, then he accepts the Hall of Fame induction on a stream of emotions, crediting family and fans, God and His creation the Horse. Hard-boiled horseplayers standing at the back of the tent wipe tears away, out of respect for a star certain enough of his craft to ride with teenage stamina, humble enough to recognize the next ride could be the last. After the ceremony, Dad slips a piece of paper into Day's hand. On it is written the phone number of a famous trainer undergoing treatment for alcoholism.

"I'll call him this afternoon," Day tells Dad. He will.

Fans linger under the tent, rubbing elbows with C. V. Whitney, who from the vantage of his ninety-three years assured the gathering that success in this game "is not easy." Cot Campbell had addressed the ceremony: "My father sold his business and entered racing in 1942, the year racing was declared a 'nonessential wartime activity.' Needless to say, my father's timing was not propitious."

It occurred to me, watching Turcotte and Day, Whitney and Campbell, that there is no shortcut for success in the racing game.

Friday, August 9. You don't even have to like horses to enjoy Saratoga. Today, Ellen drove us to the Saratoga National Historical Park, overlooking the Hudson River nine miles south of town—the battlefield where Gen. Benedict Arnold spurred his horse between volleys of British musket fire, rallying Colonial farmers to victory over the British Army in the first great triumph of the Revolutionary War. For three weeks, the British forces under Gen. John Burgoyne engaged American forces entrenched on Bemis Heights, a rocky defile over the river. On October 17, 1777, an exhausted Burgoyne proffered his sword in surrender, and the British Army stacked its arms on the banks of the Hudson.

The New England militiamen who had cleansed their wounds in the saline springs of the Saratoga region returned after the war, to farm the fertile plains and valleys. The land of the Mohawk soon belonged to the empire builders. Indian medicine springs became the site of the New World's most popular spa.

I've been coming to Saratoga since I was sixteen years old, when I lived in the longhouse near Barn Twelve, Jim Maloney's old barn. Until today, I had never appreciated the origins of this familiar town.

Saturday, August 10. Sudden Fiction. A term publishers use to define the *short* short story: Sudden. Without warning, from the Latin *subire*, to steal upon—unforeseen, swift, *sudden.*

Dr. George Pratt does not write fiction, but his style is sudden. In the National Museum of Racing, his words appear:

"The horse that walks around, eats grass, looks at the view and gives every appearance of tranquility was, in fact, designed by God to explode."

This morning, Ellen and I searched the recessed barns of the Oklahoma training track looking for Richard's Lass, an Allen's Prospect filly we had only heard about—last November, when she galloped in her racing debut, again when she won her second start by five lengths. She has been unheard of lately, a form of sudden fiction.

She *does* exist. She stood in her stall, a tattered plastic muzzle preventing her from revealing expression or munching hay on this, the day of a race. Half an hour before the Alabama Stakes today, Richard's Lass exploded out of the gate, racing a mile faster than any of the allowance fillies behind her. Now she'll set sail for the Maryland Million Oaks on September 8th. Her entry in the local showcase comes without warning, unforeseen. Suddenly, she is no longer a fiction.

Sunday, August 11. Saratoga doesn't sleep in August. At night, she stands under her trees behind the grandstand and catches her breath, her streets visible in the luminescent glow of TV monitors left burning blankly in the branches— embers from afternoon races, suspending action until the next day's card. Pavement reflects irregular ribbons of mica, crushed by spinning wheels of

four-in-hand carriages. Long after the procession of top hats and footmen has spun past, you can still hear the clop and whirr of shod horses and metal-rimmed wheels. A car's headlights catch orange in motion: A mounted police-man's horse, wearing Velcro reflector bandages.

Before dinner, we hold young Josh's hands as he circles the balcony in the sale pavilion named for Humphrey S. Finney. The boy is excited by the lights, by the exhibition of colorful oil paintings—animals drawn larger than he is. He utters his vowels, and they echo back from the burlap acoustical buffers dangling from the ceiling (an echo consignors must have heard last week). Long after dinner, the auctioneer's voice echoes out over the grounds, the New York yearling sale droning on into the restless night. Security guards at Oklahoma pace around their kiosk, and bands start up in the bars on Caroline Street. In a few scant hours, the first set of horses will gallop in the gathering light of the new day.

Life is a circle of constant activity at Saratoga in August.

Monday, August 12. Breakfast on the terrace? How 'bout Danish at the rail? Okay.

Impatiens are blooming. A horse's nose pokes through the flowers: It is Angel Cordero, Jr., clowning. On a red fire hydrant near the rail, a toddler clings, swinging to and fro: "Aaaa. Aaaa. Mmmm. Mmmm."

"El Senor," says the breakfast host over the loudspeaker. On the wide track, El Senor gallops past. Dishes rattle behind. Hooves beat. A green slate roof catches the morning sun. Ironwork tables, reliefs of all equine variety, dot the clubhouse apron. Red awnings cascade overhead. An artist concentrates at his easel. A gambler studies the *Form*. An entire seminar gathers at the paddock windows. The Big Red Spring bubbles from underneath a gazebo. A grandfather tucks his grandson into the car seat and says, "See you, sonny."

Seven hours down the road, Saratoga recedes into memory, like waves from a day at the beach.

Tuesday, August 13. "This is Peyton," the first message on the answering machine said last night. "I'm not happy about Follies' foal. Can't get the swelling down. I'll call you tomorrow. Hey, welcome back."

Our own foal by Allen's Prospect has a hind leg the size of a stovepipe. Trauma? Joint ill? Fracture? Who knows? One thing for certain, we *live* where we *work*. There is no easing back into farm life from a vacation.

Wednesday, August 14. Finney prepared the form Fasig-Tipton officials use when inspecting yearlings. His physical grading scale reads from Number Two to Number Eight, in this order: Wreck, Not Acceptable, Marginal (Needs comments), Acceptable, Desirable, Excellent, and Top. Fasig-Tipton officials Mason Grasty and Josh Taylor visited yesterday.

"Trainers are more forgiving than the agents who buy at our sales," Mason said, after examining our filly, the one Hall of Fame trainer Allen Jerkens selected as best in the yearling show. "An agent is going to say: She's a little lower in one heel than the other, travels a bit narrow up front, is light in the neck, and turns out slightly in the left fore."

I take it we didn't make the September select sale?

"You can argue with me if you'd like."

This is a business based on difference of opinion and strength of conviction. A blue ribbon filly to one man is a Marginal (Needs comments) to another. I thanked Mason for his candor, and took heart in Jerkens' simple comment the day he pinned the filly: "She just looks like she'll be a racehorse."

Thursday, August 15. "I don't buy it when folks at A.A. meetings say, 'You can lead a horse to water but you can't make him drink,' " Dad tells me on the drive to Laurel today, explaining in his own way how to avoid giving up hope that a fellow can be cured of, ironically, *drinking*. "You *can* make him drink if you make the water look attractive, sparkling—if you make it fun. You *gotta* keep going to meetings, but meetings have to be fun. Laughter is the best thing for a recovering alcoholic."

If Dad didn't have hope that he could lead a stubborn horse to water, *and* make him drink—something other than alcohol—then he would lose the gift he has for helping others.

Friday, August 16. "It's better than the old days," Dr. Pat Brackett said. She is one of the state veterinarians at the Maryland tracks, responsible for seeing that unsound horses do not compete, for their own welfare, and for the interests of the betting public. "We have perhaps one injury a month that requires euthanasia, and it's usually the sound horses who break down. The cripples take care of themselves. Bravely Bold, who broke down so terribly here last month, had run forty-two times. He was a very sound horse. He just had too much *class*—kept trying when he had no more to give. He broke his sesamoids in three places. And don't dream of some amputation fix-it procedure. These horses suffer terribly from the moment they break down. If you attempt to save them, they'll founder in the other foot, and you have to put them down anyway, after all that suffering."

Saturday, August 17. Keep turning left. Just *keep turning left.* Those are the mindless first instructions every hot-walker hears when the shank is proffered. But on a soft summer day in the shade of a Pimlico shedrow, as a bracing confusion of smells fill the air, as men on their hands and knees in deep straw slap liniments down long legs of young horses, hot-walking requires more than a sense of direction: It *demands* imagination. Under the roof of a forty-stall barn are eight or ten families, a neighborhood demarcated by the particular radio station—reggae here, rock there, Anne Murray in between.

Midway through every lap, as imagination takes full flight, the spell is broken by the lunging jaws of a colt rushing his webbing. *Wow,* that was close. A groom on a bucket blows saddle soap through holes in a halter: "Watch 'im!" he says, after the fact. Andrew stops us, puts his hand under the horse's belly. "You can put him away now. Nice job. Thanks."

My pleasure.

Sunday, August 18. Ben Rigdon is an auctioneer. Sometimes he works for horse vendors, sometimes he runs cars through at Manheim, or appraises estates he will be selling. He needs knowledge of many fields to be successful.

"Do you charge commissions?" he asked.

Not all the time. If I buy a horse and the horse is going to be boarded at Country Life, I don't charge.

"But you don't have a boarding contract that says the horse *has* to be boarded with you, do you?"

No, it's sort of a gentleman's agreement.

"So, if a guy moves his horses and you never charged him a commission, then you're out of luck. You selected nice horses for him but you've worked for free, didn't you?"

I suppose so.

"People *expect* to pay commissions. They feel they didn't get anything if you gave them your services for free."

Ben is right. A fellow for whom we bought, sold, and boarded horses, left us recently. This happens to trainers all the time. It doesn't happen on breeding farms too often. Nevertheless, it is a right of passage. I am sitting here kicking myself for not having charged commissions. Then again, he helped me get started. I don't want this experience to embitter me. I don't want to become, "No more Mr. Nice Guy."

Monday, August 19. I can't find his house. I've been driving around this seaside town two hours now. His phone is disconnected. No mail gets through. I want to tell him I am giving his mares away—I can't feed them *forever*. I know he has problems. We'll keep the weanlings, the board bill will be a wash. He wants them back, he can pay what they've cost us. That's how Grandfather did it. He carried guys through the Depression. Things happen to people. I *understand* that.

I pull up to the beach at sunset, and sit in the car near the boardwalk. Behind me is Asbury Park, a run-down, forgotten shore town. Hurricane Bob went through here today. The surf is very high. A tape is playing: *Independence Day*, Springsteen singing:

> *There's just a few people coming down here now,*
> *and they see things a different way.*
> *Soon everything we know*
> *will all be swept away.*

I look down the boardwalk. I see Madam Marie's fortune-telling booth. Behind me a white horse stands over the doorway to a cinder block tavern. The bar is called "The Stone Pony." I am in a very difficult business these days, and I don't like the metaphor of being swept away. I don't like being Sam Spade to find a guy. The last horse in my life *won't* be a stone pony, dammit. I'm gonna win a Derby with one of those weanlings, you just wait and see, Madam Marie.

Tuesday, August 20. Two things had to happen for Codys Key to have a chance in the first start of his career to win the Open Mind Stakes at Monmouth Park today. First, he needed a fast track, good footing. Second, the best New Jersey-bred in training, Thanks to Randy, needed to scratch.

"Randy Iffy for Minor Jersey Stakes," this morning's *Form* shouted.

Thanks to Bob, it rained and rained and rained. Thanks to Randy, Codys Key never had a shot. The second starter for Corridor Key made his racing debut in an added-money event. That's a reflection of his trainer's confidence.

"But you can forget the Maryland Million for him," Gary Contessa called to me after listening to the jockey's excuses. Two things didn't happen that needed to happen. Well, that's racing.

Wednesday, August 21. "The market for shares in stallion syndicates is pretty quiet. They don't appreciate like they used to. Now breeders wait until the last minute and buy no-guarantee seasons. Their accountants are telling them to sell their shares. Everybody must have the same accountant. I don't know what shares are worth anymore."

I've heard the maxim of rich men—that the time to buy is when everybody's selling—but darn if it doesn't take some courage to do that.

Thursday, August 22. Today is the Maryland Million preentry deadline. Said Hal Clagett: "North Corridor will be entered. We're holding his shins together, might run him four furlongs at Timonium to pick up enough earnings."

At Laurel this afternoon, Carlos Garcia said: "None of my horses are ready. I won't rush my babies. We'll be winning *later* this year, when breeders book their mares. That's when you want your stallions hot, no?"

Si! Senor.

Dad climbed down from the press box: "Valay Maid is in the Distaff, One's Not Enough in the Lassie, North Corridor in the Nursery. Richard's Lass and Ritchie Trail are in the Oaks. Pretty Amusing's entered, and Dum Crambo goes in the steeplechase. Maybe a few others. Not quite a horse in every race, but almost."

The cards were dealt today. I like our hand.

Friday, August 23. "Quick, come quick! Something's wrong with Shorty!"

When I arrived, Michael and Alice were kneeling over John King, who was down in the grass outside his house. I saw Shorty's arm wave them away. Well, he's alive.

"I found him here. He can't get up. I hope he didn't break his hip," Alice said.

His bag of groceries was scattered nearby. The taxi driver must have just pushed him out of the cab, to land in a drunken heap. Michael and I formed a fireman's lift and hoisted him into his house. He smelled as if he had just drunk his weight in beer. I was aware of the gaze of numerous cats as we put him on his bed. He was most grateful: "Mikey Mike...Alice...Thanks Alice...Thanks, *hear*?"

"Better let him sleep it off," I said quietly, and we closed the door behind us. We stood in the yard as Mike said: "I was worrying where we'd bury him."

The door creaked open. Shorty wavered in the door frame, expelling a few cats. We smiled at his unsteady but unharmed figure. Mike picked a basketball off the ground, aimed a sloppy hook shot at the hoop thirty feet away. It swished. Alice gasped in disbelief. This is a crazy place to live, sometimes. I've come to expect the unexpected, like a thirty-foot hook shot, like Johnny's crash landing this afternoon.

Saturday, August 24. Mr. Holly was his name. He talked so fast I couldn't understand him. But the Boy Scouts from the inner city whom he brought on summer weekends understood.

"Pay 'tenchen ta Mister Pons now boys. No runnin'!"

They poured from the wheel wells of station wagons, and we found baseball gloves and fishing rods and basketballs—their curiosity of horses exhausted after giggling at the fresh manure on the basketball court. Then one summer Mr. Holly just stopped coming. I have driven by the Boy Scout office near Druid Hill Park, but I haven't gone in to inquire about Mr. Holly. I miss him and his boys. I felt like we were helping those kids, just by opening up our farm in the quiet of summer to their energies.

Today, Mom hosted the picnic for the Baltimore Radio Reading Service, a volunteer group of city folks, mainly, who read books and newspapers over special radio transmitters to blind people. Children were everywhere. Watching the

kids playing, climbing on swings, running through the yard, I felt as if I was a cus-
todian of a treasured birthright—a lesson I learned from Mr. Holly.

Sunday, August 25. Follies' foal still has cellulitis in her fetlock. It blows
up at the least look. We hose off the poultice after removing the support bandage and the Vetrap. Then we hand-walk this stall-bound bundle of almost-
weanling nerves. She bounces on her front feet, disciplined against striking—
just like her daddy does on cool mornings, or when mares are presented for
breeding.

We walk in a large grassy paddock, up a gentle hill. She balks, waits to be
pulled on, lunges, bounces on both front feet simultaneously, not favoring either
hind leg, kicks back at her mother, strikes her mom's hocks. The filly winces in
pain at her exuberant mistake, takes two lame steps forward, walks it off. She
ain't well, but she's getting there, day by day—day-by-painstaking day.

Monday, August 26. Saw the trainers' standings from the recently concluded Laurel meeting today—all four hundred and forty-four trainers. I can't
believe that that many different trainers ran a horse at this particular meeting.

It is amazing how little money most trainers make—ten percent of their
horses' earnings representing, I suppose, the equivalent of a salary. Certainly
they don't make money off their day rate, as Uncle Sam and Workmen's Comp
get their share, and bad collections from tired owners eat any possible profit.
The *Form* prints these statistics, and every trainer who just had a bad meet feels
even worse, now that the whole world knows his in-the-money percentage. At
the Laurel meet, two hundred and eighty-two out of four hundred and forty-four
trainers didn't saddle the earners of ten thousand dollars. How many trainers
couldn't send a kid to kindergarten on ten percent of that?

Took my mind off this revelation when I stopped by Alice's garden to
admire the vegetables. Then the kid grabbed a tomato and ruined my shirt. That
was today.

Tuesday, August 27. "I'll need six copies," Hal Clagett informed track
photographer Jerry Frutkoff after North Corridor finished two lengths in front.
With reservation, supporters had gathered on the track.

"Mr. Clagett, please don't order the photos yet," I warned, recalling the
apologetic usher who escorted us to the winner's circle after the colt's first race,
a photo-finish loss. "It's not 'Official.' "

Timonium is the last of the "halfers," in a manner of speaking. A few years
ago, they lengthened the stretch (it's now almost a five-furlong track), but the
turns are still so tight that the jockeys draw their inside stirrups up a notch.
Number Six, North Corridor, had gunned to the lead, then dropped over to the
rail, as the jockey on the Number One horse stood straight up—just in case the
stewards forgot their binoculars. When he returned to weigh out, he grabbed the
phone. The "Objection" sign flashed.

In the balance hung North Corridor's Maryland Million eligibility. Without
the winning purse, he might not make the earnings cutoff for the one-hundred-

thousand-dollar Nursery Stakes. There were bigger problems in the world today—but not while that sign flashed, not for us. Suddenly, the Tote board quieted: Six stood atop One. Quite a dramatic first winner for Corridor Key.

Wednesday, August 28. The harmonica. I was all right until I heard the harmonica.

I laid him off because of horse farm economics, and because of some antics he pursued in his off hours. It is never easy to let someone go. Tonight, a young man suddenly unemployed and soon-to-be homeless sat on the steps of his loft apartment on the farm. He's never been much of a harmonica player. *Oh! Susannah* is the only song he knows. But tonight, he was following the bouncing ball of his own life, and he poured emotion into the vibrating reeds.

Don't be fooled by the power of music, I told myself—mindful, as any employer must be, of the boomerang problem of laid-off help: They often come back after flying an angry circle.

Thursday, August 29. Life on a family farm is a trip. Sometimes I fold my arms over my chest and smile at how ridiculous I feel. I swore as a very young man that I would *not* be dependent on horses for a living. I went to law school as an insurance policy on a career. Yet, here I am, with a field full of weanlings I can't sell for what Grandfather did fifty years ago, a partner in a farm that depends on horses.

Now I swear at the horse business: But I like it. I like dialing Louisiana Downs to see if Corridor Key got his second winner. (Nope. Naviator closed with a rush, though.) I like thinking about next weekend, when six of our babies race on Maryland Million Day. I like living on a farm. So who cares if the office isn't air-conditioned? (It was ninety-five degrees today.) Who cares if the heat has hatched out the yellow jackets who found a crack in the cedar shakes and dive-bombed us all day long? (Steve Brown will fix those guys tonight.) Who cares if every third caller wants you to sell his mare or share or yearling or weanling? (You'll have to speak up. Bertha's Tree Service is chain sawing limbs out of the tree I used to swing in.)

You do it because it is all you really know.

Friday, August 30. "Your father just called from Timonium," Mom phoned me at home late this afternoon. "He said Allez Prospect got plowed at the gate and lost all chance."

If you don't break good at Timonium, you don't have to worry about the tight turns: The race is over.

"Did you see yesterday's *Post*?" she continued. "Very hard words about the way racing in this state is being conducted."

Mom, I laid off two guys this week. At the moment, our racing stable is good business only for vets and vans. It's a hundred degrees today. *No*, I didn't see that story.

(One hour later.)

"Marty called from Monmouth—Codys Key won."

Great news, Mom. *Great.* Two winners in three days for a freshman stallion. That's the way to do it. Thanks for calling. Sorry about being such a drip earlier.

Saturday, August 31. At fifteen bucks a person, I didn't figure the two-day Horse Farm Tour would be much of a kickoff to Maryland Million Week. I was wrong. And I figured, somewhat snobbily, that no one would really be interested in horses—more likely to pitch a picnic blanket on the front yard, a free place to beat the city heat. I was wrong again. The folks who arrived in station wagons and vans, on bikes and on foot, were wonderful: Polite, curious, concerned about development, grateful to see farms they'd driven by for years, but never been invited to tour. They signed the Guest Register, paid the fifteen dollars for the benefit of Driving for the Disabled (allowing a person who can't walk to handle the reins of a carriage, to walk through fields again, with the help of a horse), and away we went.

There is something magnetic about horses, stone ponies in deserts will attest. These visitors felt the pull, standing on the hill overlooking the misty watershed, while the foals milled about in greeting, as if hosts at an equine cocktail party. By five o'clock, we were talked out. The last van drove out the lane. Duck Martin called from his Worthington Farm: "We were overwhelmed. We had a hundred and fifty people today. They're coming *your* way tomorrow."

By seven o'clock, summer was on the run, cool Canadian air rustling the first red leaves off the trees to land in the blue rapids of Winters Run. Let 'em come, Duck. It feels so awfully good to share this game with new friends. Tomorrow morning is likely to be brisk. Maybe I should have coffee ready for our early guests.

SAVORING SEPTEMBER

Sunday, September 1. "Do they know what they were bred to do?" the lady asked a leading question on Day Two of the Maryland Farm Tour. She was admiring a nursing foal, and I sensed that the Humane Society angle was coming.

"Yes, they do."

"Why?"

"Because for a thousand years, they've been bred to run. It's like terriers crawl down holes, and pointers point, and retrievers have a mouth so soft they won't leave teeth marks on your evening paper."

"But people don't *whip* them."

"It's not black and white like you think. Yes, every sport takes its toll on athletes. Have you ever seen two old football players walking together? Stiff as boards, most likely. If Maryland didn't have horse racing, you wouldn't have horse farms. Without horse farms, you'd have unbroken development. I couldn't raise hogs here in suburbia—people would complain about the smell. I couldn't raise enough cattle on a mere hundred acres. Soybeans are sprayed with herbicides that might contaminate your well. Erosion off a cornfield silts in the Bay. Horses are the *only* crop I can raise on this little piece of land, two miles from town, across the street from housing developments."

The lady's husband was enjoying this. He likes the sport, but she hates Saturdays at the races with him. Before this farm tour, he had exhausted his rationale. He didn't realize that twenty thousand Marylanders are employed in the horse industry, that racing's annual economic impact is five hundred million dollars, that horse farming's impact is another five hundred million, that the plank-board fences and open spaces of horse farms enhance Maryland real estate by a hundred million. It's a billion-dollar business for the state.

"So whaddya think honey? Can we go to the Maryland Million next week?" he asked her. He turned to me: "You got anything you like going that day?"

I didn't want to put him onto Richard's Lass, who either wins by a mile or loses by as much.

Monday, September 2. I had rocks in my head today, perhaps from calculating how much limestone the pastures needed. Twenty miles upstream, every mile of the boundary between Maryland and Pennsylvania once was marked by a rock, set in the seventeen-sixties by English mathematicians Charles Mason and Jeremiah Dixon. The rocks of the Mason-Dixon line became as sharp as a sword during the Civil War, cutting free states from slave-holding states, North from South.

A mile downstream, the myriad rocks of the Piedmont Plateau end at the fall line, where navigable waters begin, and fine sands of the Coastal Plain run to the sea. Our farm is on the fertile Piedmont, but our pastures need limestone to offset the acidity of land that was covered for thousands of years by hardwoods. Western Maryland, though, is a gently rolling ocean of land, underlain by thick layers of limestone, by the shells of eons of sea creatures.

The lands of Country Life, until 1933, did business as Rockland Farm. The big house is built from Port Deposit granite. Fieldstone foundations of hay barns are a testament to Amish artists. Walkways through yards are broad slabs of shale hauled from the bed of Winters Run. Yes, at rest on this Labor Day, I had rocks in my head.

Tuesday, September 3. One year ago, a circulation audit of *Daily Racing Form* reported a hundred and fifty thousand Saturday *Forms* printed, dropping back to ninety thousand printed for a routine Wednesday, Thursday, or Friday. About two hundred people work at the Hightstown, New Jersey, headquarters, and they often have only three hours from the time race tracks report their entries to the moment the presses run: Three hours to assemble a product upon which more than a hundred and fifty thousand Saturday horseplayers will rely on to wager millions of dollars. Race tracks call it the forty-eight-hour entry rule. Workers in Hightstown probably call it something else, when they must pare two days into a hundred and eighty minutes, simply so that every morning at Country Life, Dad can return from the Newstand and declare: "Look who's in for tomorrow!"

If the folks at Hightstown could see the pleasure they give Dad, they would know their efforts are not unappreciated. (He still calls the *Form* by its old name, the *Morning Telegraph*.) Propping his glasses on his eyebrows, he bares down on the Stewards Rulings. " 'Not a scintilla of evidence,' " he gleefully parrots the language of a decision. " 'Not a *scintilla* of evidence.' " He scans West Coast entries for runners by home stallions, guides his index finger through the fine print of morning workouts, gives weights for stakes ten days hence.

For an agonizing period this summer, the Newstand, citing decreased sales, quit carrying the *Form*. I took it as another tiny tumble in racing's fall from grace. Dad took it personally. He was miserable. He tried organizing a petition among the twelve other folks in Bel Air who buy the *Form* every day. The

Newstand sensed his aggravation, reinstated the *Form*. The folks in Hightstown could not have missed feeling the rush running through the universe when Dad once again could report a scoop: "Lucky Lady Lauren is headed for Arlington!"

Wednesday, September 4. Rain, rain, rain, rain. Such a lovely rain fell today, almost wiping out Wednesday night basketball until the decision was made to play wet—no lay-ups. The fields absorbed the soaking. Tomorrow Steve Brown will seed bare spots. We'll analyze soil samples, and the lime truck will put our fields in proper pH, with fertilizer to follow. September is here, the fall growing season at hand.

Thursday, September 5. An advertising agency in Baltimore inquired about using a mud room to shoot an ad for fur coats. Mom asked what's it worth to you. They said two hundred dollars. A tidy donation to the Radio Reading Service, Mom figured, so today the ad agency crew arrived: A security guard in a three-piece suit, two girls in slinky dresses, and thousands of dollars worth of whirring strobe lights and reflectors. Oh! And the furs. The cameraman started shooting at noon, went into the evening, looking for that perfect mix of propped-up saddles and snow boots, whips and racing silks, to sell a furry image to a tweedy clientele.

Meanwhile, in the office, another breeder wanted to sell her share, and the windswept foal's owner decided to give the baby away, and vet certificates

arrived for two mares pregnant in May but empty by the time the September 1st stud fee was due. A bit forlorn by the juxtaposition of furs and foals, I grabbed the *Form* and seized upon the cryptic notation of a black-type Belmont breeze: "Richard's Lass was *on the muscle.*"

Friday, September 6. "I hearken back sixty years," Hal Clagett began his presentation of an award to Dad at the Maryland Million breakfast, "to the early nineteen-thirties, when the great handicap horse Discovery carried the silks of Adolphe Pons. I jump ahead thirty years, to the nineteen-sixties, when I bred a mare to Correspondent, and relied on the expertise of Adolphe's sons Joe and John. I leap to the present, to the third generation—and when entries are taken shortly, I hope my colt by their freshman stallion is in the lineup for Sunday's Nursery Stakes."

The discovery by this correspondent that North Corridor would contest the Nursery was the result of six decades of a working relationship.

Saturday, September 7. "Do you ever get used to it?" racing newcomer Keith Dodd asked me today.

"No. Never."

He watched the horses load into the gate.

"It's the only part of the game I'm uncomfortable with, and I coached high school football. People get hurt in sports, too, you know."

"I know."

"Racing's a great game."

"Yes, it is."

"It takes a certain personality, though. It's not for everybody."

"No, it's not."

"The new owner's seminar this morning was very helpful to me. Anything by your stallions running tomorrow?"

"Five."

"Who do you like?"

Still, I hestitate to suggest Richard's Lass, even though her race at Saratoga last month gave me goose bumps.

Sunday, September 8. What's in a name? Apparently a lot, for a few babies by our sires. Sold as a yearling named Loaded Prospect, today she did business as One's Not Enough. When Maryland Lassie entries were taken, confusion arose as to her eligibility. Someone asked: "Was her name changed?"

Sold as a weanling named Carnival Hag, he is now Ace Eyes.

"My nightmare is that an ineligible runner might get to the post," fears Maryland Million assistant director Cricket Goodall. "Was *his* name changed, too?" Yes, he went from a Hag to an Ace. One's Not Enough bucked shins in the Lassie. Ace Eyes was second best in the Nursery. The worst-named Carnivalay colt ever, Dum Crambo, fenced smartly but finished fourth in the steeplechase.

Suddenly it was almost dark. ESPN had signed off at 6 P.M., fifteen minutes before the Oaks. Richard's Lass and Ritchie Trail were our last hopes.

BANG.

They broke right in front of us, tiny Jorge Chavez throwing the reins at Richard's Lass. The sudden fiction of her Saratoga race came true again. She won by fourteen lengths, the largest winning margin of any race in the six-year history of the Maryland Million. And steady Ritchie Trail closed for third— missing an Allen's Prospect Exacta by half a length.

What a weekend! Early this morning, Dad told us of Lucky Lady Lauren's win in the Matron Handicap at Arlington. By sundown tonight, we'd run one-three in the Oaks, second in the Nursery, first in the Matron. Four stakes placings in twenty-four hours.

Set 'em up, Uncle George. Crabcakes all around.

Monday, September 9. In the thick of a recession, it is no easy task to solicit corporate dollars. This morning, I wanted to write thank-you notes to Rich Wilcke and Jim McKay, to Jeff Huguely and Cricket, to everyone who carried the fund-raising effort to the board rooms of corporations, whose largess comprises a matching grant to the dollars voluntarily put up by the horsemen.

Yesterday, eleven winners represented eleven sires, and only one stallion, Deputy Minister, has left the state. We thus avoided a Windfields windfall— that departed breeding establishment often overpowering the mom-and-pop operations that are the backbone of Maryland breeding. And the day was good sport, eleven added-money events in which you knew the local participants and shared in their success. Why, the Maryland Million even featured a fifty-thousand-dollar steeplechase, and the front page of today's *Sunpapers* showed a color photo of four leaping horses, Dum Crambo's blaze face rising earnestly to the fences. There is a good feeling in Maryland racing circles today. We put on a first-class event, the envy of state breeding programs everywhere.

Tuesday, September 10. Attack it. Every day. It's two steps up, one step back. Don't let it be the reverse. Watch your costs, don't scrimp. Yes, lime the pastures, paint the gates. Keep the place up. A man wants his share sold? Sell it to a breeder who shares your confidence. A lady wants her mare sold? Get her a better one. Cull. Weed out. Quality, not quantity. If it's stormy, trim the sails, but meet the waves head on.

Wednesday, September 11. "Ace Eyes was not *registered* as a Maryland-bred," Georgia Dovell of the Maryland Horse Breeders told me today. "No one paid fifty dollars prior to May 31st of his yearling year. Now the fee is two hundred dollars. Carnivalay missed a stallion bonus for Ace Eyes' second in the Nursery."

Ouch! The old unregistered Maryland-bred trap. Ace Eyes fell through the cracks. Happens frequently when breeders sell their horses as weanlings or yearlings. But everybody loses: The breeder, the owner, the sire.

Breeder bonuses (ten-to-twenty percent of the purse) are paid on all races won by registered Maryland-breds, but owner's bonuses (about seventeen percent) are only paid if a registered horse wins for a twenty-five thousand dollar claiming tag or above. The stallion receives a bonus (five-to-ten percent) for all wins by his registered offspring. In Maryland, the philosophy is that the breeding and owning of *good* racehorses should be rewarded. Only a handful of annual stakes are restricted to Maryland-breds. Other states pay good money to bad horses in daily races restricted to state-breds.

Ace Eyes' owner will ante up the two hundred dollars hoping it comes back in owner's awards. But that sum also will benefit the breeder and sire as well—a form of unjust enrichment. Our stallion contracts inform the breeder that he is responsible for Maryland-bred registration. But too often, we can't find the breeder two years after he sold the weanling. Or the horse is a claimer running for less than twenty-five grand, and the owner refuses to pay the two hundred bucks for the sake of a stallion and a breeder he cares nothing about.

Today, we scoured Georgia's print-outs: Carnivalay Hag showed up as unregistered. Ace Eyes? Not on the list. The Jockey Club name-change had not caught up with Georgia's master list. This is a game where you can never know it all.

Thursday, September 12. From a California town named "Cool" came this letter:

"I thought no one alive but me remembered Camps Hale and Carson, the 10th Mtn. Div., and the mules. In the fall of 1943, some 300 head of 2-year-old mules were shipped to Carson for breaking and training. I was in the 30th Veterinary Gen'l Hospital and volunteered to learn how to break, train, and cargo-pack them. Later, I was assigned to the 254th Quartermaster mule pack train. We made many trips into the mountains to haul supplies to the 10th Mtn., mostly in subzero weather and for 20 days or so at a time. We used to laugh about the poor 10th Mtn. guys who had to walk and lead. We rode mules and either had a string of seven or eight mules or in the flat easy going areas herded them.

"The 254th was deactivated in 1944 in Fort Bragg, N. C. It was, I believe, the last QM mule pack train in the U.S. Army. The mules went from Elkins, W. Va., to Fort El Reno, Okla., and I don't think I've seen one since. I am now happily raising a T.B. baby each year on a one-horse ranch and losing my butt. Well, what can you expect at age 71?"

I've said it before: These Remount boys are everywhere.

Friday, September 13. "The hay shifted and I lost it on the shoulder of the road," the hay man phoned at lunch. "Think you could put a crew together, bring a couple of pickups, help me get this hay to you?"

Hay gets even more expensive when you have to handle it twice. Michael shrugged, reached for his gloves. He and Richard set off for north Harford County, to load a load of hay we'll be unloading here this evening.

Saturday, September 14. Back in May, Mom reserved a house at the beach. Through the heat of June and the flies of July, the thought of a deserted post-Labor Day seashore, the Gulf Stream still warming the waves, the pink sun rising over whitecaps, lifted our heads like the smell of pies cooling in the window. Mom and Dad are hosts for two weeks at off-season rates. Five children and nuclear families secure staggered three-day passes. One of the most wonderful natural resources of the mid-Atlantic region is the ocean, and we took advantage of it today.

Sunday, September 15. The beach house has cable TV, but the fellows in charge of programing assume that we're more interested in the West Coast pennant race than the East Coast horse race. Dad grabs the remote, anxiously awaiting the Woodward Stakes. He can only hold the scrambled feed for a millisecond. We see the hind end of a horse loading into the gate, then we're back to baseball. Dad zaps back and forth around the ABC station. We're going blind. It is almost subliminal, the fleeting image of silks, the blink of an eye that saw the field turn for home. Dad tries to hold the scrambled picture long enough to recognize a face in the trophy presentation. No luck. Not McAnally's smile. Not Lukas' hair. Who won? Who cares? I'm seasick at the ocean, exhausted after watching Dad watch the Woodward.

Monday, September 16. When we buried Stingo in the garden in July, my big fear was that he wouldn't stay buried. His canine compatriots search longingly for lost pals, and often retrieve them from entombment. For example, there was Alien, that needle-nosed numskull of a Dalmatian who met a fender in the Animal Hospital parking lot (fear of routine shots catapulting him into Route 1 traffic). Yes, there was Alien, again and again, his head in the paws of his ghoulish pals, Sarge and Stingo, a couple of Hamlets unable to grieve: "Alas, poor Alien, we knew him well."

Now Stingo has arisen, not literally, but reincarnated as the narrator, Col. John R. Stingo, in A. J. Liebling's *The Honest Rainmaker*, a ribald account of race track hustlers, first published in *The New Yorker* serials in 1952, but first known to me today.

Garrison Keillor baldly states in his introduction:

"A. J. Liebling was the wittiest American writer who ever lived."

There was no argument from Dad, who stumbled across an account of the 1930 Lawrence Realization Stakes between Questionnaire ("Why, Questionnaire *stood* here!") and Gallant Fox, the climactic event for Col. Stingo's Weather Control Bureau and the General Staff of the Rain Preventers. You see, Col. Stingo and Dr. Sykes set up a Detonatory Compound at the head of the Widener Chute, drained Consolidated Edison of thirty-two thousand kilowatts ("greatly annoying the good housewives on the Hempstead Plains busy with their can openers and electric stoves"), and the Rain Preventers delivered a bluebird day for the stakes. Honestly.

"I have three rules for keeping in condition," says Col. Stingo. "I will not let guileful women move in on me, I decline all responsibility, and above all, I avoid all heckling work."

For the corpulent Beagle of the same name, no finer epitaph could be written. Please, may he rest in peace.

Tuesday, September 17. Heat lightning burst over the farm tonight like bombs on Baghdad. A constant strobe effect flashed trees into high relief. The air seemed capable of spontaneous combustion. Gables on the barn roof jumped out from their eaves, a Victorian nightmare of gingerbread woodwork. And then it rained—rained so hard the water *bounced* off the dry ground, drops as big as hail.

Not the ideal drizzle for freshly seeded pastures, I muttered, with the resignation of a farmer who knows his precious grass seed is on its way to the Chesapeake Bay.

Wednesday, September 18. At 3 P.M., we heard what sounded like three very loud gunshots echo over the farm. Our blacksmith, Tom Rowles, showed up at the office minutes later.

"Some damn fool was driving down Old Joppa Road in a beat-up truck, turning the key on and off to make his engine backfire," Tom said. "*Three times* he did it. Richard was leading Corridor Key in when it happened. The horse reared, spun around, knocked Rich to the ground. His hat went flying, but Rich didn't let go. I swear, if I could of grabbed the kid in that truck…"

Thank goodness we always have a backup person on hand when we are leading the stallions. Ellen, meanwhile, had been helping the vet draw blood from the foals for registration purposes. Foals and needles aren't great friends anyway. They don't need a gun going off to throw their front feet skyward. Wow. We could have had a loose stallion, and a handful of injured employes, all thanks to some faceless clown who never gave us a passing thought as he passed dangerously through our lives.

Thursday, September 19. "Throwing those out?" I asked Cindy Deubler at the *Maryland Horse* today.

"Yes, we have the bound volumes."

With that, she loaded two bundles of old magazines into my arms. At home tonight, I read an old Stallion Register which featured a review of the highlights of 1968. The state breeding industry was thriving after five years of the Maryland Fund. Every burg and town stood a stallion. Chesapeake City captured Northern Dancer, his fifteen thousand dollar stud fee about to be ancient history. Sir Ivor won the International. Native son Joe Aitcheson led the nation's jump jocks, then stood like a native Indian for Peter Winants' camera, his high cheekbones perhaps the only unbroken bones he knew. A gray colt strode into the Glade Valley stud barn, and I read the first lines of Ogden Nash's poem for Peter Fuller:

Who'll run first in the Derby scrimmage? None, say I, but
Dancer's Image.

Vaughn Flannery's paintings were identified: *Chestnut Stallion Discovery; Daybreak at Old Pimlico; Interior Racing Barn, Sagamore.* The paintings perished in the hot flames of the old Pimlico clubhouse, but their warm colors were reproduced in this issue. The Stallion Register page for Rash Prince listed our telephone number as TE-8-5070. I haven't thought of that old number in years, maybe not since then, when I was in the eighth grade.

I got sand in my eyes reading late tonight.

Friday, September 20. Partly for sentimental reasons, but mostly to get a winner in a freshman stallion's first crop, we put Auntie Freeze's last foal into training. This family of grays goes back to the flea-bitten mare of our youth, The Heater. Her daughter was Heat Shield, by Saggy. Heat Shield's filly by Rash Prince was Heat Rash, the dam of Talc, a very useful New York sire. Heat Shield's filly by Uncle Percy was Auntie Freeze. Today at Pimlico, Auntie Freeze's last foal finished far back in a field of maiden claimers.

Up the road, you go, young man. Can't be sentimental these days. We need you to stick a win next to your Daddy's name on the Freshman Sire List, and you don't appear likely to do that here in Maryland.

Saturday, September 21. For a minute and ten and three-fifths seconds today at Pimlico, during the simulcast of the Ruffian Handicap from Belmont Park, I was afraid I'd be too excited to hold the kid in my arms. (In the Pimlico grandstand, he becomes a "stooper," kicking discarded pari-mutuel tickets with his red Padders, peering down in search of a Double Triple oversight, looking up only to brandish a half-smoked stogie.) Richard's Lass zoomed to the lead, in fractions identical to her Maryland Million Oaks. After six furlongs, though, the field of older mares blew past her like truckers on the Beltway once they've seen the Student Driver license plate.

"Close, but no cigars," I told the kid. "Let's blow this pop stand."

Sunday, September 22. Above me run the rusted tracks of the old hay fork, from the days when hay was stored loose, not in bales. Below me, the floor is worn smooth as a bowling alley from decades of farmers dragging straw across it. The south side of the barn is shelter for the horses; the north side is dug into the bank. Excavated stone is precisely stacked into walls two feet thick and two stories high. The poplar beams bear the marks of an adz swung a century ago. Today, while checking on hay inventory, I paused to admire the cathedral of craftsmanship that is the old bank barn.

Monday, September 23. In a letter to Charles Scribner, Hemingway pounded his publisher for suggesting a change in the draft of *For Whom the Bell Tolls:*

> *"Thanks too about the horse. I have never seen a horse*
> *myself so this is all hearsay but I am going to try and get*
> *near one sometime and see if I could find such mysterious*

*places as the Hock (which I always imagined to be inferior
white wine as drunk by the English) and the Gaskin which
for years I took to be a mis-spelling of the Gasket.*

*"Now take the Withers. I had always imagined this to be a
mis-spelling of the Whithers or the place the horse had just
come from. Don't tell me where they actually are. I want life
to hold some mysteries. My own experience with the horse
has been confined to his use as a bear bait."*

It gets worse, believe me. Ol' Hem must have been working through a
Sunday hangover. I have always admired the reality of the horses in his writings. I never imagined he had to defend his turf so zealously, nor with such an
acerbic pen.

Tuesday, September 24. A pall of anxiety prevails over the breeding business these days. When the last caller of the day has finished instructing me on
the disposition of his equine assets, I find humor to be the best balm; to wit,
Col. Stingo's assessment of Charles Evans Hughes, Chief Justice of the
Supreme Court, who, as governor of New York, abolished betting: "He is the
man who ushered in the Dark Ages of Hypocrisy. Prohibition was the inevitable
sequelae."

(Col. Stingo's patron saint was Gambrinus, the mythical Flemish king
reputed to have invented beer.)

Racing historian John Hervey observed in more serious tones: "During
1911 and 1912 not a single meeting was given in the Empire State. Only in
Kentucky and Maryland did racing maintain a precarious existence. Breeding
was also prostrate. With no market, ruin faced every stock farm. Hundreds of
animals were sold for workhorses. The condition was desperate. It required
great courage to face the situation."

Just about the time racing began recovering from Gov. Hughes' bill, World
War I started. In the decade between 1911 and the appearance of Man o' War in
1919, a generation of young horsemen passed through their prime. History
records the unfairness of fate, but Col. Stingo has an intoxicating flair for feathering the blow.

Wednesday, September 25. Today was a very difficult day on the farm.
Yet there is something about this land that fills me with resolve. I sat on the hill
in the rain, watching the lights of cars on Route 1. Across the highway, a yellow
sign screamed: "For Sale. Former Proposed Mall Site." I like that. We outlasted
the son of a gun who wanted to name his mall's restaurant: "Country Life
Cafe."

I walked home and in a field found Polly, a generally obnoxious Beagle
who, for the moment, had ceased her incessant barking. I petted her. When I
told her "Go home," I took my own advice, and immediately felt better. The sun
will be out tomorrow. That's good enough for me.

Thursday, September 26. A nine-year-old Maryland stallion named I Am the Game has been returned to training.

"Age doesn't break horses down," trainer King Leatherbury said. "Unsoundness does, and this horse was very sound."

Perhaps someday, we'll have a senior circuit for stallions coming out of retirement: "Precisionist Duels I Am the Game in Battle of Gray Beards." Until then, however, it is quite conceivable that I Am the Game will compete against his own offspring. I see the headline:

"Dad Beats Kid at Pimlico."

Friday, September 27. Despite the rush of Winters Run over the rocks as he patiently selects his dinner of fish, the Great Blue Heron senses us as we pad through the woods on a trail fifty feet above him. He is wary, lifts off, glides downstream, lands.

"You can drive past me on the pond," he seems to say, "but in the woods, I must be more careful."

An oak tree has fallen from the bank, its rootball revealing disease that hollowed out the trunk. Down it came, gathering an innocent beech in its arms. Sand builds up underneath the two fallen trees in the creek. An island is starting. So it was. Down through the ages. So it will be. Down through the ages— the heron, the stream, the trees, all in the constant motion of nature.

Saturday, September 28. What quality of the air makes a late September sunset so dramatic? It is as if the usher handed us 3-D glasses when he arrived with the cool air from Canada. That same air, though, carries the cries of a pair of weanlings, a sound as plaintive and beckoning as the call of the two geese who invaded the heron's domain tonight.

Sunday, September 29. Fasig-Tipton's Josh Taylor phoned from the sale grounds early this morning.

"An out-of-town buyer called, asked to scope the colt in your consignment. He won't bid unless you agree to the presale examination."

Oh, why this colt? He's only been with us four days. He's head shy as it is—broke his halter on the van ride down. He rears to the ceiling if you're too quick with a chifney bit.

The horseman in me says: "Don't stick a rubber tube down his nose. We're ten hours from selling him and, at the moment, he's in one piece. Mayberry comes all the way from California, and his policy is not to bother scoping or X- raying the cheaper yearlings. They grow out of most problems anyway."

The agent in me says: "I don't want a horse kicked back after the sale for failing an endoscopic exam. That's tough on the client we work for. Hard to find underbidders after the sale is over. Better to scope him now than later."

I called Josh back: "Go ahead. Just be careful."

The cynic in me says: "The absent buyer is wasting our time. We have yet to sell a horse at Timonium to a guy bidding over the phone."

Monday, September 30. Michael didn't get home last night until almost midnight. "I haven't been at a good sale at Timonium in years. I wanted to savor it."

While the general economy languished in deep recession, Timonium acted contrarian. Last night's select yearling sale jumped twenty percent in average. Cars were still pulling in an hour after the sale started. It looked as if the Marx Brothers had diverted traffic from busy York Road, knocking the "No" off the No Parking signs as they did in *A Day at the Races*. The absurd comedy continued today, for the non-select yearling sale, the ordinary manes and tails. Today's average jumped *forty percent.*

A man in Virginia bought a colt who last year would not have drawn a bid.

"The thing I liked about him," he said in an instructive drawl, "is he's a *gelding.*"

We told every looker about the cracked hoof on the filly. She topped our entire consignment today, selling for five times the average.

"I thought I could steal her," an underbidder explained. "I thought the crack would scare off everybody else."

Just when you think there isn't a person in the world who wants a horse, they sell like hot cakes. How do you figure? Who knows? Let's just savor it.

CONNECTED

Tuesday, October 1. New stallion halter. Mane pulled. Foretop straightened. Paddock push-mowed, not just bush-hogged. Photographer due this afternoon to take conformation photo of Citidancer, after his first season in the stud. Need low sun to catch light in stallion's eye, to lessen shadows. It's cloudy at 9 A.M., but clearing at noon. Neena says, "The sun will disappear. Scratch today." Blue sky at two. Change of heart.

At 4 P.M., the stallion is still damp from his bath. His first season has thickened his neck, deepened his girth. He is so very handsome.

"Push him back a step...now up a half...now back," Neena orders. "Hold it. Hold it. Hold it!"

Shutter clicks, again and again and again. Two hours into the shoot, she says, "I think we got it."

Wednesday, October 2. Young Josh's first birthday was yesterday. Surrogate uncle Marty Friedman bought him a wooden hobbyhorse, but vanning expenses from New Jersey were prohibitive.

"Splinters is bedded down in my garage," Marty said last night. "I'll ship her down carrier's convenience."

Splinters has no papers.

"She was sired either by The Axe II or Big Spruce," Marty guessed, "out of Myrtlewood or Miss Dogwood. Nevertheless, she was a hickory race mare, no matter who rode her—first Hedley, then Robert Woodhouse, finally Eddie Maple. Elliott Burch was her first trainer, then of course Woody Stephens got her. She won the Oaks and the Acorn, beat colts in the Wood, then she was retired to Forest Retreat.

"I think Splinters will be a fine first horse for your boy. But if she doesn't work out, let me know. I'd like to breed her to Woodman."

Thursday, October 3. Before I-64 was completed, Easterners traveling to

the Bluegrass region negotiated the old Route 60 from Lexington, Virginia, to Lexington, Kentucky. Stuck behind a tractor-trailer up the mountains of Gauley Gap, West Virginia, driving a Pontiac that used more oil than gas, we made our first trip to the horse capital of the world. Daniel Boone on *his* first trip to Kentucky made better time.

We would pop up for air in Charleston, West Virginia, but noxious fumes were vented from Union Carbide's plants along the Kanawha River—pollution so ubiquitous that a town named Nitro seemed perfectly appropriate. A short while later, as coal barges churned underneath us, the lights of the Ashland Oil Refinery—an industrial Disney World—signaled the border of Kentucky.

To a nineteen-year-old Marylander, raised on the lore of Bull Hancock and the possibility that the grass might really be blue, it was like imagining the inside of Madison Square Garden while standing in the rain on 34th Street.

"*This* is Kentucky?" I remember thinking.

Friday, October 4. Yesterday, I drove six hundred miles in ten hours. Today, behind a well-digging truck, I drove ten miles in one hour, crawling on the Paris Pike into Lexington. Time was not of the essence. I was lost in admiration for the limestone fences which ran unbroken for almost two miles, for the beauty of a meandering creek, for the Spanish tiles Joe Widener placed atop the guardhouse roof at Elmendorf Farm, for the mystery that lies over the hill at the entrance to Gainesway Farm, for the timeless quality of this rural boulevard, so innocently concealing the difficult times facing the Bluegrass region, indeed, facing Paris Pike herself. Her five miles of stone walls are jeopardized by plans to widen the road. Nevertheless, an hour on the Paris Pike is time well spent to the once-a-year visitor.

"Now, *this* is Kentucky!" I said to myself.

Saturday, October 5. At last night's Thoroughbred Owners and Breeders Association Awards Dinner, the Four Tops played for about fifty minutes, Levi Stubbs looking sixty but singing thirty, then they danced off stage as their announcer repeatedly proclaimed: "Ladies and gentlemen, the internationally famous Four Tops."

Most of the internationally famous horsemen left soon thereafter. A younger crowd took to the dance floor, as the supporting band played *THANK YOU Falettinme Be Mice Elf Again,* a funky seventies tune summing up our gratitude for the evening. The horse business for this generation is not what it was when the Four Tops were top. When you ask a fellow these days how he's doing, be prepared. Many of us tonight looked forty but acted twenty, thankful, for the moment, to be ourselves, again.

Sunday, October 6. Orange letters on the fuselage of the Kalitta Airline 727 stood out against the open sky of Bluegrass Field, as eighteen horses disembarked in single file down the ramps directly into waiting Sallee vans. This flight was the Tex Sutton Sunday Special from California, pampered equine athletes unloading like C.E.O.s climbing into limos. On a telephone in a nearby car, Oakwind Farm manager David Fiske listened to the call of the Prix de l'Arc de Triomphe, while his wife Martha at their home on the Russell Cave Pike held the receiver to the cable TV, catching the satellite signal from the race course in France. It's a small world these days.

"That's Tight Spot," Ted Berge said. "That gray horse is Itsallgreektome. This is a plane full of Breeders' Cup horses."

The Kalitta letters soon tilted skyward again. A scant five hours later, the plane was back at Bluegrass Field. Festin, who twenty-four hours earlier had skated away with the rich Jockey Club Gold Cup at Belmont Park, climbed down for the three-minute van ride to Keeneland, where moments earlier, Summer Squall had defeated Unbridled in a battle of classic winners. Now Festin stood at his stall door, halter off, yanking hay from his net, not a hair out of place.

"These horses might as well carry briefcases," Ted said.

Monday, October 7. The Midwest *Daily Racing Form* carries incredibly personal biographies of supporting players in the horse game.

"Who cares what a trainer's favorite movies are?" Teddy mused. "If they asked me that question, I'd say: '*Dr. Zhivago* and *Carwash.*' "

This morning, Ted's two-year-old daughter Meredith cried when the time came to leave for the baby-sitter. The baby-sitter, Peggy Brown, lives on a famous horse farm. She is an astute handicapper, and Keeneland is running now. "Don't book her bets," Ted warned. Meredith would not stop crying. Ted sat down next to her. I thought he was going to cry, too.

"Look, none of us want to go to work when Keeneland is running," he comforted her. Yes, the truth is, we all would rather have gone to the races, if only just to read bios in the *Form* all day.

Tuesday, October 8. "Good morning, Major! Why, I haven't seen you since the trees turned *last* fall. Where've you been?"

"Overindulging myself in Gambrinian delights with Col. Stingo, I'm afraid, while in search of the last wild horses on earth."

"Do tell!"

"We went to Mongolia, of course, to find the Przewalskis of Asia. Whereas Col. Stingo received his sobriquet by legislative fiat—that is, he is a 'Kentucky' Colonel—Col. Nikolai Przewalski, a Polish-born Russian officer, earned his stripes serving the czar. In 1870, Col. Przewalski shot a small horse in Mongolia, then shipped its skull and skin to St. Petersburg, where a biologist labeled the creature *Equus Przewalski,* and declared it the horse of the Stone Age."

"What does a Przewalski look like?"

"He's built much like your teaser, Jazzman—thickset and low-slung. The body is dun-colored, the mane dark, short and stiff, with no forelock. The ears, muzzle, and lower leg, are black. The nose, however, is encircled by an oatmeal-colored patch. A dark dorsal stripe runs the length of the spine. He looks —and is—pugnacious and irascible. I found him in Asia, but he is the prehistoric horse painted on the walls of the Lascaux Caves in France."

The Major shook slightly, as if a craving had overtaken him.

"I'd like to toast the horses of the Stone Age," he said, a dark twinge of obsession in his voice. "Any fermented mare's milk handy? If not, sherry will suffice."

Wednesday, October 9. "No, no, no, don't give her away. I've been fighting alligators up here in my real business. But I'll send you a check once a week until the board bill is current. I think she'll be a good producer. What's the earliest you can breed her next spring?"

Oh, to be an alligator.

Thursday, October 10. In the walking ring at Keeneland last week, I watched two grooms, one on either side of a prancing colt, leading him in tight circles. Back home, Corridor Key acts like a two-year-old colt these crisp mornings. Even a stallion will think twice if he feels restraint on both sides. This morning, the gray stallion tried to rear, then thought better of it when he felt two shanks. He did not get out of hand, or rather, hands.

Friday, October 11. "Swap these half-dozen mares for two good young ones. Entry fees and commissions would cost a thousand dollars for each horse in the December sale. Do you have six thousand dollars worth of horses? Doubtful. Sell some through the classifieds. Give a few away. Only enter the couple who'll bring a grand."

He and his wife work the farm themselves. He loves his horses for just *being* horses. I looked at the pregnant mares: How much in stud fees would have to be paid to get Stallion Service Certificates? Twice what the mares were worth, easily.

"Please, just resign yourself to taking a loss. It will be cheaper in the long run."

He looked at me blankly. I wondered whether I had overstated the case. No. If he intends to be in this business three years from now, he has to change course, now.

Saturday, October 12. This morning, I walked in unannounced on the Major. He quickly stashed his sherry bottle behind Capt. Hayes' *Veterinary Handbook for Horsemen.* He was embarrassed. My silence became his enabler.

"How did your man take the news yesterday?" he preempted the direction of the conversation. "Bet he didn't like it when you told him to get rid of his horses? Why should he be surprised? Mankind has never been kind to horses. Early man hunted the horse. Its flesh was the mainstay of prehistoric diets. There are arrows, in fact, in the drawings of the Lascaux Cave horses. In central France, a narrow outcrop called the Rock of Solutre, with cliffs on three sides, was a killing field. At the bottom of the cliffs, scientists have discovered the bones of countless horses, stampeded over the precipice by primitive hunting tribes. Man has always used horses to his advantage, and he has not been kind."

He shook from the effort of his diatribe. I could hold my tongue no longer.

"Major, you don't sound well. I'm concerned about you."

He would not look me in the eye, and he left the room without comment.

Sunday, October 13. We were *all* concerned about him, and have been for some time. We intervened. The Major entered Ashley, a nearby treatment center. It sounds easy; it wasn't. We met a groom there. He told us his story. He said when he was a young boy, he felt as if he'd been standing at the window, looking in. The other kids did not accept him, until he drank with them, that is, until he did drugs with them. He said the backstretch of a race track was an easy place to feel accepted.

The imagery of that young man at the window, yearning to be included with the group, struck us. We were moved by his recovery. We felt a spirit at work. Was "hope" the name of that spirit?

Monday, October 14. It is Columbus Day. No banks. No mail. Let's clean the office before one more cricket leaps through the door. Out went an avalanche of sales catalogues, phone books, and *Daily Racing Forms.* Win photos were hung: Richard's Lass and Ritchie Trail for Allen's Prospect, Lucky Lady Lauren and Valay Maid for Carny, even Per Quod, his group-winning rump checkerboarded in European style, for dear old Lyllos. It gave me great satisfaction to fiddle with the frames. No, too high. No, not straight.

A forecast of rain brought us to our knees to wax the floor quickly, to get furniture back inside. Tomorrow, we open for business in a brand-new frame of mind. Please, step into our office.

Tuesday, October 15. When John King waddles into Miss Nick's Candy Kitchen for an eighty-cent long-necked beer, the regulars swivel on their barstools to welcome him, as if he were Norm in the bar on the TV show *Cheers*.

"Shorty!" the chorus erupts.

He shuffles past the gallon jars of pickled meats and finds a booth, where he scratches the lottery tickets he just bought at High's.

"Johnny, this one has three matching numbers. Better hold onto it."

He does. Back at High's, the gal confirms it. Uncle John opens a savings account for him. "Mr. Astor," Mom calls him. When Shorty summons a cab for the trip to Miss Nick's, he speaks to a dispatcher also named Shorty.

"Hello, Shorty?" Shorty says. "This is Shorty."

The cab arrives momentarily to pick up John King. King John. Shorty.

Wednesday, October 16. A study of the cattle business concluded:

1) Breeders who employ a superior promotional program enjoy a notable increase in the price received for their products; 2) Purebred producers who breed the best performing bulls also excel at promoting cattle; and 3) For any significant price improvement to occur, promotion must be employed.

They are right. I hope we captured the moment in the conformation shot on October 1st.

Thursday, October 17. The first time we rode ponies into the Chesapeake Bay from the stables at Oakington, we swam so far out from shore that the ponies' backs dropped from underneath us, as we clung to their manes. In effect, we had turned ourselves over to the power of the little horses, taking it on faith that they could carry us back to safety.

People continue to turn themselves over to a higher power at Oakington, which in the last decade has been doing business as Father Martin's Ashley, an innovative alcohol and drug treatment center. Dad has referred a number of local horseplayers to Ashley through the Ryan Foundation. On any given day, you can find a kindred spirit.

"As the Major's extended family, there is much we can learn about this disease," said Dad, in so many words. "Ashley has an excellent family counseling program. We should all go for the three-day counseling sessions next week."

Friday, October 18. Brian Keelty, a horse owner for nearly ten years, a horseplayer for a decade prior, calls it, "The Ride of the Long Faces." It is that stretch of highway leading away from any race track after your horse has lost. Shine On Sarah ran desultorily today. On the ride home, Nanny the dog sat in the passenger seat, as Friday evening traffic crawled away from the metropolises of Baltimore and Washington. As we crossed the Gunpowder River into Harford County, we looked down into the streambed, bestrewn with boulders.

"Perhaps these huge rocks washed down in apocalyptic rainstorms, down from mountains young and high!" I told Nanny. "This is suspect terrain, Nanny. So wipe that long face away. Guess who's in tomorrow?"

Saturday, October 19. "I won with her daddy," jockey Marco Castaneda said as Carlos Garcia gave him instructions how to ride first-time starter Suspect Terrain, the gray filly circling anxiously in the Laurel paddock. Marco rode in Southern California, where Corridor Key ran. Now the colt is a freshman sire.

"Take her to the lead quickly," Carlos told Marco. "Don't let anyone bother her."

Carlos watched the filly leave the paddock, nodded his head when the pony boy took her from the groom: "Good, keep her walking. Don't let her think too much." He turned to me and said, "They see that crowd, they back up, come right back into the paddock sometimes."

At the start, the gray filly in the orange and blue silks of home stumbled, but Marco picked her head up, found himself on the lead. She won by five for Marco, just like her daddy.

Sunday, October 20. Corridor Key never looked as good as he did this morning, the sire of a homebred two-year-old who led from start to finish yesterday. When we turned him loose, he spun quickly, kicked his heels up, raining mud on us. One glop hit me right in the glasses. I looked ridiculous, as if I'd been in a pie fight, but I felt great.

Monday, October 21. "Oftentimes, the family is sicker than the alcoholic," Father Martin told us at Ashley on the first of our three days of family counseling. "The child of an alcoholic doesn't tell his classmates what his father or mother did last night. He just keeps it all in, and gets sicker. And he runs a high risk of growing up to become an alcoholic. Genetics is far more significant than any other factor in determining alcoholism."

I felt like the deer in *The Far Side* comic strip, a bull's eye painted on my back. I'm in the same business Dad was—the stallion business. I have the same worries, the same nagging paranoia that holds a horseman's dreams in check. What's to prevent me from having the same addiction as Dad?

Tuesday, October 22. On Day Two, we learned about the image business.

"A white horse is a favorite image of liquor advertisers," Father Martin said. "Look at a bottle of Courvoisier—a white horse gallops towards you. A can of Busch beer? Same horse. The identical message."

I thought about psychiatrist H. G. Baynes' account of two peasant farmers in England in the 1930s, sheltered from a storm as thunder exploded overhead:

"Did ye see the black horse in the sky?" one farmer called out to the other.

"Here," wrote Dr. Baynes, "we see the mythological moment, and the true mythological response. Observe that it is not the thunder as such, but rather the overwhelmingness, the impressiveness of the total experience, which brings up the image of the black horse from the archaic depths."

I see black horses in nightmares, white horses on billboards, and sometimes I am powerless over those images.

Wednesday, October 23. "You *were* affected. If you grew up in an alcoholic family, you grew up in a dysfunctional family. There is a bubble of denial that you must break. Al-Anon meetings will introduce you to other families similarly affected. 'Adult Children of Alcoholics'—A.C.O.A.—groups have sprung up in recent years. You don't have to carry this burden. There is help available."

We thought at first that we had come for counseling for the sake of the Major. Then we learned that Dad's alcoholism, in full cry during our most formative years, had affected us in ways we wanted to deny. By the end of Day Three, a healing process had begun. There was understanding, at last, for all of us.

Thursday, October 24. This morning, only two items on my desk seemed important. The first was an article called *Helping Hands*—a broad survey of organizations such as the Thoroughbred Addiction Council of Kentucky (TACK), the Winners Foundation in California, The Jockey Club Foundation, and the Ryan Foundation. Across the country, any segment of the horse industry can reach out for help.

The second was the article *Track Exclusions of Race Track Licensees as a Result of Drug and Alcohol Testing,* presented at the recent Equine Law Seminar at the University of Kentucky:

"Racing organizations should seriously consider a means to structure a substance abuse and rehabilitation program which would likely withstand legal attack, to enhance safety in horse racing and insure the continued integrity and public confidence in the sport."

I recalled hearing that soon Keeneland would be implementing procedures to drug-test its personnel. It's funny how your mind sees what it wants to see, once it centers down on a thought.

Friday, October 25. Today, we rotated six single mares into the rested back field, where shin-high grass has resisted the first frosts. Half an hour later, we added seven more mares. The thirteen girls pranced in confusing circles around each other—tails high, nostrils snorting—then instantly fell back into the two original herds of six and seven. They have stayed that way through sundown: two distinct clusters silhouetted against an orange October sky. Horses, like people, are comforted by support groups.

Saturday, October 26. Red leaves weighed down by the moisture of dense fog glide silently to ground. The mist blankets the farm to midday, and the temperature climbs into the seventies in this stretch of Indian summer, the finest part of the year.

Sunday, October 27. On a freshly mowed hay field off Black Rock Road in northern Baltimore County, a pride of over-the-hill athletes wheezed and strained through a touch football game. Radiating in all directions from the hay field was the spectacular landscape of the Piedmont Plateau, in full autumn glory, resplendent in the yellows of poplar, the reds of maple, the russet-brown of the turning oak leaf.

"Bring some beer back," a player called to a car.

"Bring something back for the guys who don't drink," another said. He saw the surprised look on my face. "Yeah, you knew me in college. *Moderation?* Whew! No way. But I don't want to end up like my dad. I had to stop. Besides, I'm standing on the most beautiful place on earth. How can you enhance this? Beer's not going to enhance this day for me."

I saw three of my good friends—three out of the twelve young men who had gathered to relive a bit of the glory days—and they were nodding their heads in agreement. My generation is coming of age during an enlightened period.

Monday, October 28. From an airplane descending at dusk, the Finger Lakes region of western New York state is bedecked in golden reflections, shimmering like sequins across the glacial plains. The fading sun reveals a strongly meandering river, the Genesee, the only significant north-flowing river in the northeastern United States.

Tonight, I began educating myself about the people who raise horses in this region, who are the Keepers of the Western Door among New York breeders—a proud and independent group known as the Genesee Valley Breeders Association. The GVBA inherited a legacy of farming from the agricultural Seneca Indians, the Keepers of the Western Door of the Iroquois Confederacy. They do not merely raise Thoroughbreds. They are stewards, presiding over a breathtakingly beautiful land.

Tuesday, October 29. Finger Lakes race track is a milking machine for a giant business concern. No need to raise purses so long as the New York State Thoroughbred Breeding and Development Fund subsidizes owners with bonuses of fifty percent of the purse. With downstate racing dark on Tuesdays, Finger Lakes management pours twelve races down the thirsty throats of simulcast bettors in New York City, who today handicapped the fine-tuned feature: "For 3 and up which have not won $6,605 since July 1." This main event was sandwiched in between three-thousand-dollar claiming events. A six-year-old bay gelding by Old Chronicle made his fifth start in five weeks today.

Finger Lakes runs from March to December, overweening Mother Nature's patience by six weeks on either side of opening and closing day. Nevertheless, a handsome paddock, a cozy clubhouse, an open-air grandstand, a sweeping view of the plains against a wide sky, and a family of dedicated horsepeople, makes the Finger Lakes experience most enjoyable.

Wednesday, October 30. Mrs. Martha Wadsworth (1864-1934) founded the GVBA in 1915. She established policies to improve the quality of *all* types of horses, not simply Thoroughbreds. Farmers were given mares, if they pledged to bred them to Association stallions—supplied by The Jockey Club and the U.S. Army Remount to the Lookover Stallion Station in nearby Avon. For forty-five dollars (a ten-dollar membership fee, a twenty-five-dollar delivery charge for the mare, and a ten-dollar stud fee), the farmer could breed a quality horse—a hunter, a workhorse, a racehorse, a show horse.

To Lookover, August Belmont II sent Wonder Boy, believed to be the largest Thoroughbred in the world at almost eighteen hands and more than fifteen hundred pounds. (I feel certain that Grandfather handled the details for Belmont.) Omaha, 1935 Triple Crown winner, left Claiborne for Lookover in 1943.

Mrs. Wadsworth's in-laws settled the Genesee Valley in the early eighteen-hundreds. James Wadsworth wrote of the region's advantages:

"The swales are timbered with elm, butternut, white ash; the upland with basswood, hickory, white oak, chestnut. (Goods) may be sent by land to Albany, by water to Montreal or from Ark Port down the Susquehanna River, in Arks to Baltimore."

I left for home today in a flying Ark, noting with irony the pilot's comments as we began the long descent toward Baltimore: "And down to our right, you'll see the Susquehanna River."

Connected. We are all connected through time. I must make a note when I meet Grandfather in the next world to ask him about Wonder Boy.

Thursday, October 31. The images of October are so diverse at this moment: Young Josh climbing the red sliding board on his first birthday; Splinters' arrival, Dad putting a dollar bill on the wooden horse's nose as his grandson rocks like a little Arcaro; the stone walls of the Paris Pike; Breeders' Cup horses on the airplane ramp; the trees of Keeneland in autumn; Suspect Terrain's gray coat beneath orange silks; the morning fog over the Bay at Ashley, clearing as if blown by the spirit; the Finger Lakes from the sky; Wonder Boy.

The horse business is not simply of dollars. It is of senses.

THE RISE AND FALL
OF NOVEMBER

Friday, November 1. The air near Lexington was thick with smoke at 5 A.M.
"It's just somebody burning leaves," Ted said. We were rushing to leave for
Louisville, to watch Breeders' Cup horses in morning drills.

"Leaves?" Ellen said. Ted shrugged, as if to say, what else could it be?

It was leaves all right, twenty-three thousand acres of them. The Daniel
Boone National Forest in Eastern Kentucky was on fire. The smoke had crawled
on cat feet through the Halloween night to darken our morning.

It is said that when Boone entered Kentucky, a squirrel could have jumped
from tree to tree the entire width of the state. Imagine such a wilderness. In the
eerie darkness, under the spell of smoke from ancient trees burning a hundred
miles away, Kentucky exerted a vague but powerful force. This is a state whose
raw beauty draws people to her. A small legion of Maryland horsemen have visited the Bluegrass and have never left. I would be one of them if it were not for
the family farm.

Saturday, November 2. Dawn of Breeders' Cup Day. Richard's Lass peers
from behind a mask of wool, a cowl made in Argentina keeping her face and
throat warm.

"Her groom spent the night with her," said her owner, Robert Perez, swinging his arms around his chest for warmth on this cold, dark morning. "He closed
the doors and slept in the stall with her."

In the light bulb's glow, the filly looked like a prizefighter, knuckles taped,
awaiting the gloves. Nine hours later, the heavyweights of her division chased her
chestnut rump and taunting yellow silks for seven furlongs before knocking her
out in the Breeders' Cup Distaff. She ran as fast as she could, as far as she could.

In one sense, it was exciting to watch the daughter of Allen's Prospect face
the best in the world. Yet watching Richard's Lass, and knowing she could not
last against such foes, made me wish she had picked a fight elsewhere today.

Sunday, November 3. In the Keeneland Library are leather-bound photo albums embossed with figures of horses—albums larger than coffee tables. Keeneland and Calumet, neighbors on the Versailles Pike, were born in the same era, and matured together, partners in each other's careers. This afternoon, the Keeneland auctioneer cried out the names of famous Calumet colorbearers. The once-proud stable was being dispersed. You could almost hear the leather book covers falling closed.

Monday, November 4. Not given to jewelry or pocketknives, Dad mysteriously, continuously, tripped the metal detector at Bluegrass Field as we prepared to leave Lexington. Finally, he reached into his raincoat and withdrew a shining horseshoe—a good luck memento from the Drug and Alcohol Conference he attended last week at Churchill Downs, sponsored by the University of Louisville.

"Can't call me 'Shoeless Joe' anymore," Dad declared, careful to hold the horseshoe up so the luck would not run out.

Tuesday, November 5. At the Keeneland Library yesterday morning, I researched Grandfather's role in August Belmont II's Turf empire, the Nursery Stud. In *Giants of the Turf,* Dan Bowmar wrote:

"On Dec. 9, 1924, Belmont complained of a slight pain in his right arm, and he told Adolphe Pons, his secretary for many years, to cancel his appointments."

Belmont died the next day. Joseph Widener, a creditor from Belmont's disastrous venture to dig a canal through the bedrock of Cape Cod, bought twenty-six yearlings and all the stallions and broodmares. Widener then engaged Grandfather and C. J. FitzGerald to arrange for a dispersal, and on Friday, May 15, 1925, the day before the Kentucky Derby, Belmont's breeding stock went on the block at the Nursery Stud on the Georgetown Pike. Wrote Bowmar:

"Special trains were operated from New York and Louisville, and 3,500 persons went to watch and eat burgoo. Fair Play (the sire of Man o' War), then 20 years old, topped the sale at $100,000. The late Walter J. Salmon opened the bidding at $75,000, but after Adolphe Pons, bidding for Widener, offered $100,000, the master of Mereworth Farm shook his head."

It seemed ironic that I would be at Keeneland for Calumet's dispersal, reading about Belmont's dispersal.

Wednesday, November 6. Struggled writing the script for the stallion video this morning. Ordered tapes of races and win photos from recalcitrant photographers at various tracks. The fellows shooting the video arrived early. I had yet to groom the most important characters in the production—the stallions.

"We're losing our light," the cameraman said politely.

As the dust flew from Carnivalay's coat, it dawned on me that film might not be my best medium.

Thursday, November 7. Clouds rolled in this morning and much-needed rain fell throughout the day. The aggravation of yesterday was replaced by the comforting realization that, just as in making hay, you have to roll tape while the sun shines.

Friday, November 8. The Meadowlands race track was David A. Werblin's inspiration: Give New York bettors a nighttime show in a glitzy, first-class pavilion, and betting will go through the roof. So Sonny Werblin built the Meadowlands on a swamp across the Hudson River from the largest concentration of horseplayers in the world. The Pegasus Room is the penthouse suite at this gamblers paradise: Gourmet food, hostesses who greet patrons by name, banks of mutuel windows, polished brass fittings.

Tonight, we joined Ray Sheerin, owner of Codys Key, at his Pegasus table, to cheer the son of Corridor Key in the New Jersey Futurity. Cody finished second. Our excitement, however, was tempered after we inquired of Mr. Werblin's health.

"Not good," said J. Willard Thompson, who trains the Elberon runners for Mr. Werblin. "Not good."

Saturday, November 9. Last week on a farm near Paris, Kentucky, we stood in the Jacob Aker family cemetery. The Akers farmed the land before and after the Civil War. Their headstones, and the court records Vanessa Berge researched when she and Ted bought the farm three years ago, provide insight into that era. Life itself was so precious that ages on headstones were measured in days as well as years:

Emma, Dau. of J. & R. Aker. Born Apr. 3, 1851. Died Nov. 18, 1859. Aged 8 yrs., 7 mos. & 15 days.

Life was precious, except perhaps for the slaves. The court records list the slaves as property, bought and sold, assets in the estate. Upon the elder Jacob Aker's death in 1841, his will read:

"My negro woman Patty shall become the property of said nephew Jacob Aker. Also my negroes Lewis, Lucy, Alfred, Daniel, Willis, Sophia, Betty, George, Mary, the above named ten negroes, to Jacob Aker and his heirs in fee simple forever, to dispose of them agreeable to his will and pleasure."

The value of these slaves? Patty $200, Lewis $700, Lucy $450, Alfred $450. Directly above the inventory of slaves was the notation: 11 Head Horses $500.

It seemed so incongruous that the lives of white people were measured in days, the lives of black people in dollars. The peculiar institution of slavery suddenly was no longer abstract to me.

Sunday, November 10. Pushing manure down the stallion barn aisleway this morning, I took a trip back in time. I was sixteen again, mucking stalls at Sylvia Hechter's farm every Sunday in exchange for a one-hour riding lesson from this legendary local instructor.

"Take your legs out of the stirrups," she said as I cantered circles in the indoor ring. "I'll teach you balance. Let the reins go. Now put your hands and feet up, *up* in the air—all your weight on your butt in the saddle."

I lost my balance. My flailing arms spooked the horse. I did a backwards somersault and landed face down, tanbark stuffed behind my glasses.

"Balance is everything," Mrs. Hechter said. "Let's try that again."

Monday, November 11. Today at Laurel, Rainbow Prospect was favored to win. He approached the wire four lengths on top. I held my breath. In his last race, the colt had led by three lengths with fifty yards to go—a sure winner. Then he suddenly slammed on the brakes and bounded like a bunny across the width of the track. Greg Hutton hung on, resumed race-riding, and ended up beaten a nose. Today, thank goodness, Rainbow broke his maiden uneventfully.

"Why didn't he run straight when he was ten-to-one?" a chagrined bettor groused.

Tuesday, November 12. "If you can see into the corner of the stall with the lights on," Dr. Bowman explained at a stallion management symposium today, "you probably have enough light to stimulate the mare into cycling."

Quick to edify were the panel of experts, who preferred a two-hundred-watt bulb.

"Fillies right off the race track often cycle quickly, sometimes within days," continued Dr. Bowman, in a tone whimsically unscientific, "probably because they are subject to much longer hours at the track, where lights come on at 5 A.M. for feeding, where a night watchman flicks the switch late in the evening to see that bandages are still on.

"When I interned at New Bolton, I saw broodmares cycling in February, even though the farms did not have a lighting program. Again, someone fed early, someone checked before bedtime. That little bit of light at routine hours stimulated the photosensors in the mares, and they began cycling."

On December 1st, barren and maiden mares here at Country Life are placed under lights. It is a royal pain through the cold of winter to maintain the lighting program. Some breeders, frustrated with the effort and expense, have abandoned lighting. That is one step up, two back, I remind myself.

Wednesday, November 13. Sold Suspect Terrain today. The headline in tomorrow's *Form* says: "Suspect Terrain Choice in Laurel Sprint."

If she takes a bad step, she goes from valuable racehorse to marginal broodmare—a risk we cannot afford. She will stay in Maryland. We have shares in her sire. We own her half-sister, in foal to Citidancer. We race horses in partnerships, and partners enjoy profits. Those considerations, plus the fact she vetted out, cinched the sale.

Thursday, November 14. For the first time in the modern history of this old farm, every stallion with foals of racing age sired a winner on the same card. It bothered me a little when Suspect Terrain walked into the winner's circle carrying someone else's silks, but I cannot second-guess the decision to sell her. Someday we will be in a position to take the risk, to refuse a sale, but we are young. It is not our time, yet.

Friday, November 15. Shareholders in freshman stallions monitor the earnings of two-year-olds. To be the leading freshman sire in a regional state such as Maryland does not ensure continued success. Travelling Music and Assault Landing once led this short list, yet are no longer here. It simply provides the market with some assurance that the sire is not a dud.

The distinction is not illusory. If you can book better mares in the sire's *fifth* breeding season—the first season after the two-year-olds have run—he has a second wind to his career, and you have a future. That is saying a lot these days. Yesterday, Corridor Key took the lead, thanks to Suspect Terrain.

Saturday, November 16. "The clocker has been in three John Waters movies," Andrew relates a morsel of race track trivia as he slides down the stair railing in the vacant Pimlico grandstand, after checking Edgar Allen Pro's workout this morning. "*Pink Flamingoes, Hairspray,* and...I forget the third movie."

Andrew spends half his waking hours either on a horse's back or on his own knees, attending the legs of his horses. Today, he is in a mild state of unfettered elation. There is a lightness to his spirit, and he is fun to be around. Edgar is no Einstein, but Andrew has patiently taught him how to run. Whether he can run *fast* has not been determined. This morning, it was enough that he worked well.

Sunday, November 17. Mrs. Hechter would be handcuffed by the liability issues facing today's riding instructors. Balancing on your butt bone is definitely not permitted on the Notice, Release, Assumption of Risk, and Indemnification form that is standard in riding stables. West Virginia, a beautiful state for riding, requires statutory compliance with a four-page release simply to ride a rented horse. (See Chapter 20, Article 4, Duties of Horsemen: Present to each participant, for inspection and signature, a statement explaining the responsibilities set forth in this article.)

Nowadays, liability—not balance—is everything.

Monday, November 18. Value is easy to define in a recession: Whatever it was last year, it should be cheaper this year. So today, we sent an announcement to our clients that we were reducing our stud fees for next season. After all, there is no commercial market in Maryland. Select yearlings average eleven thousand dollars. This means the breeder lost ten thousand producing that yearling. The race track is our sales ring: a one-mile oval where performance serves as a warranty, where runners who went unsold as yearlings are frequently claimed as two-year-olds for thirty-five thousand dollars, where allowance winners swap shedrows for sixty thousand, where budding stakes prospects set phones a-ringing.

People are tripping over each other getting out of the breeding business. With our lower fees, we hope some breeders will perceive value, stick around, breed a racehorse, sell it for a score, or win big pots in their own silks. This is how it was in Grandfather's day. After all, he started this farm in 1933, not in a mere recession, but in The Great Depression.

Tuesday, November 19. Today at Laurel, TV monitors promoted this weekend's Maryland Juvenile Filly Championship: "Can unbeaten Suspect Terrain become Maryland's top two-year-old filly?"

Her race footage rolled before they cut to a still shot, her steel-gray coat duplicated on dozens of monitors. Meanwhile, in New Orleans, Naviator prepares for the Old Hickory Stakes on November 30th at Fair Grounds, while the headline in tomorrow's *Form* shouts: "Codys Key a Lock in Meadowlands Feature."

It is not yet black and white whether the gray stallion will be a leading sire, but these are exciting times.

Wednesday, November 20. He knows the rules and walks a fine line to his paddock in the mornings. He bows his neck in the first step out of the stall. He tucks his chin into his chest and jigs in tight, dressage steps. His mouth works the chain shank as if he was ready to race again, a Munnings horse going to the post. His lips tighten across his teeth like Bogart. Then his eyes catch a movement. He stops, lifts his head, fixes on the distance. He is as still as a monument—a very big stone pony. The morning sun shines pink through the membranes of his flared nostrils. I feel as if I am watching his heart beat.

Thursday, November 21. It's a children's book come to life, this view from the lone bench at the gap where the barns at Laurel meet the track. The bench sits on a grassy knoll, underneath pines that follow the curve of the track's final turn. The grandstand anchors the horizon. Sunlight drops between pink clouds. The scene is slightly utopian, like an architect's sketch.

Suddenly, three red tractors dragging immense harrows execute a precise pirouette. In midstretch, panting horses returning from the previous race sidestep the harrows, while fresh runners for the next race bounce onto the track in single file, grooms yowling good-natured taunts to their opposition. Placing judges peer out from a station wagon stretching slowly around the turn, a giant red fox bolted to the roof.

A child will never see this side of a race track unless his parents are hotwalkers, yet the view from the bench would make a wonderfully educational children's book, a *Busiest People Ever* sequel, to seed those pretty clouds for another generation. The race track is so much more than merely a place to gamble.

Friday, November 22. "Your husband has holes in his toes from dragging his feet on buying a paint sprayer," Steve Brown chided Ellen today. "I could have this whole farm done if I had a sprayer."

He's right. I refused to borrow another farmer's sprayer because I knew it would break on us. We rented one, and it broke down *three times* in one week, but Steve whirlwinded around the farm, and the perimeter fences look great. Still, I keep making the same mistake, over and over. Proper equipment is necessary to run a farm. It is, in the long run, much cheaper than labor. I will get you a paint sprayer, Steve.

Saturday, November 23. "Suspect In 'Uncertain' Terrain Today," read the headline in the Washington *Post* racing column. Off two six-furlong sprints, she was asked to go more than a mile against seasoned fillies. When she tired after six furlongs, she did not step on her pedigree. Her mother is Allegedly, which *did* get stepped on once, by her mother, Princess Pout. The broken hind ankle kept Allegedly from racing, while her half-brother, Alleged, made headlines in Europe by twice winning the grueling Prix de l'Arc de Triomphe.

No, Suspect Terrain stepped on the calendar. The richest Maryland-bred juvenile filly race of the year occurred two weeks before the counterpart colt race.

Sunday, November 24. In the mild letdown following yesterday's race, we overlooked the entries at Aqueduct. It seems Richard's Lass used the Breeders' Cup Distaff as a prep for the Geisha Stakes. Sprinted one and one-eighth miles. Won by five. Uncle John clipped the chart from the New York *Post*.

"Maryland stakes don't turn any heads," he said, taking the historical perspective. "Need those New York wins."

Monday, November 25. When we bought Travelling Music, a major stakes winner as a two-year-old, we naively thought he would sire runners of similar ability. Wrong. Very *wrong*. I made a mistake about a horse, but I learned a lot about people from Sonny Werblin, for whom Trav raced. Once every autumn, I waited in the dark-paneled room on the eighteenth floor overlooking Madison Square Garden, of which Mr. Werblin was chairman. No visible doors led out of the waiting room. Instead, a panel would open, light would pour out, and his secretary Barbara would summon me. I had fifteen minutes ("Hold my calls," he told Barbara) to provide the stallion's majority shareholder with a complete update on his mares, on the coming season, on the status of Trav's runners. Mr. Werblin always wanted to know the score. Business was a game—a tough, but fair game.

Today, in the chapel at Rutgers University, the glee club of his alma mater sang Rutgers fight songs at his memorial service—bouncy irreverent lyrics about painting the town, and pushing the ball across the goal, playing the game with heart and soul, on the banks of the old Raritan. I was a water boy on the team that huddled in memory of the eighty-one-year-old sportsman today, but he had made me feel as if I was the most important water boy in the world. That was what I learned from Mr. Werblin.

Tuesday, November 26. Secret Halo raced one time, won by fifteen lengths, and cracked a hind sesamoid. Come spring, she'll visit Citidancer, just like her, a bit of a phenom. He won his first three races by wide margins, then cracked a front sesamoid. The two horses won four races by a total of fifty-one lengths.

Speed on speed is the stuff of dreams.

Wednesday, November 27. We did not want speed out of Secret Halo today. After her injury, doctors grafted a part of her rib onto her ankle. This afternoon, for the first time since her only start, she was free to run. Alice led her slowly around a small field until she settled into a quiet stride. When Alice unsnapped the shank, the filly put her head down and grazed, perhaps for the first time since she went into training two years ago. I am relieved to report that Secret is not speed crazy.

Thursday, November 28. Three Joes rode to Laurel today to cheer on Lucky Lady Lauren. The first one watched pro football on a transistor TV in the passenger seat, the second drove, the third peered from his car seat at the passing trucks.

"Brummm, brummm, brummm," J.P. Three trilled his lips. They were greeted at the paddock by a fourth Joe, who runs the track, and who reminded all his patrons: "Don't forget your pumpkin pie!"

L.L.L. ran fourth in the Thanksgiving Day stakes.

"Well, you gotta be in 'em to win 'em," the first Joe said philosophically. Pie in hand, three generations gave thanks for just being in the feature.

Friday, November 29. "I want to find a farm where I can settle down, make a living out of working with horses," the job applicant said. I hesitated to promise a young man he would have a job when seasonal mares go home in July. The fellow today has a college degree in agriculture, has worked for veterinarians, has his trainer's license, and doesn't smoke or drink. He's overqualified for the life of mucking stalls and leading horses in the rain, which is ninety percent of the springtime activity at a commercial breeding farm.

"We've got twenty horses in the mixed sale next weekend," I told him. "Let's see what you know."

Thus I bought some time on this decision.

Saturday, November 30. Deer move with impunity through our fields. A buck with a six-point rack walked within thirty feet of our back porch last week—a magnificent splendor to his careful gaze. He and his harem of does flow like Winters Run through the sanctuary of our farm. This afternoon, I was high atop the house, cleaning rain gutters of leaves. I looked downstream into the amphitheater of sound that is the watershed, where last week the buck's movement in the leaves betrayed him to our eyes. I gazed upstream to the open fields of Country Life. Invisible was the property line of the adjoining farm owner, upon whose land deer yank cornstalks straight out of the ground, and graze winter wheat.

Ellen called quietly up to me: "Look! Look!"

Four does cantered across our field. A long moment passed before the buck made his appearance—a rear guard, a sentry. The entire herd then paused at the five-foot wire fence. They gathered tight, then one by one, in single file, they leapt straight up and over from a standstill, and disappeared into the neighboring woodland. Such magnificent athletes. Oh, to have filmed this feat …

… *BOOM…BOOMBOOMBOOM* …

Dead silence followed the gunshots. So real was our sudden sense of loss. I felt as if I had turned my head at an execution. The lawyer in me searched in anger for a tenuous property claim to assert over the deer: "They are *ours,* too." Then the sound of water washing over the rocks of Winters Run climbed the hill to my ears. The last day of November, the first day of deer season, came to a disquieting close.

FOR GRANDFATHERS AND GRANDCHILDREN

Sunday, December 1. You can't believe your eyes sometimes on a farm.

A red-tailed hawk sweeps out of the sky, dive-bombs the barn. You see what he sees—a pigeon—but that pigeon is gone in the moment you realize he is the hawk's prey, lifted skyward in the talons of a raptor. You cannot believe your eyes, as one bird claims another.

A Great Horned Owl perches on a fence in midday during an eclipse. Who's fooling *whoooo*? A heron stands on the banks of the pond in December. You can almost see him shivering as he wraps himself inside the dull gray feathers of winter plumage, a skinny-legged old man. A flock of fifty Canada geese splash down onto the pond and splash back up in a display so breathtaking your eyes become camera shutters, and you see the droplets fall from their wings in your sleep that night. Migrating tundra swans shimmer white against the sky, while blackbirds paint the horizon on their journey south. The bird painter himself, John James Audubon, could not draw such broad strokes on the canvas of our imagination.

December is the month when the birds first stand out against the farm landscape. They are a sight to behold.

Monday, December 2. Trainer Bob Camac gave me this update on our racing partnership filly, Stable Gossip:

"I told the jock, 'Go an easy five-eighths.' He worked her in black type. She popped a splint. I'll fire it, give her thirty days. She won't race this year."

I typed the memo for the fifteen folks in the partnership: "Remember what I told you about two-year-olds? It's hurry up and wait."

Tuesday, December 3. The horse dentist is due here next week. Carnivalay acts silly merely if his halter gets over his ear, and yet he will permit a gentle, slightly built, once-a-year visitor to insert a steel speculum into his mouth to file down sharp teeth. This particular horse dentist has the demeanor of St. Francis of Assisi. I almost expect him to arrive with a squirrel on his shoulder, rabbits at his heels.

"I don't need any extra help," he said. "I'll be fine with one quiet man on the shank."

This fellow thinks like the horse thinks. He has a gift. Thus he is successful at his dangerous profession.

Wednesday, December 4. Gave away a weanling today. Her mother stepped on her hind ankle in August. It blew up and never returned to normal. If we took the filly to a sale, an out-of-state buyer might purchase her. She was bred here in Maryland. Can't earn state-bred bonuses for the sire if you race in Canada. Better to have given her away to a local trainer.

Thursday, December 5. It is love of family, and of this farm, that permits Uncle John, from the safety of the sidelines, to coach us, as if we were his sons. To score, sometimes we have to go to the air, downfield, into the distance. Sometimes we lose ground.

You should have seen that coming, he says when consequences first are in issue. Then he considers modern game conditions. *You see how that happened, don't you?* he says. *You must remember how that happened.* He dusts us off, puts his arm around us.

I want to say to him: *We're trying. We're trying our best.* But I stop myself: Close only counts in horseshoes. This is not a game. It's for real. It is then I appreciate his firm stance.

Friday, December 6. It is a Friday afternoon before a weekend full of possibilities.

"Codys Key is in at the Meadowlands tonight," Mike says. "Dad called from Laurel with Sunday's entries for the stakes. Only a five-horse field, including Suspect Terrain. Everybody gets a check."

A late phone call. A breeder says: "Yep, the table's set. Let's go on that share. Let's hope we have a good weekend."

Yes, let's hope.

Saturday, December 7. Fifty years ago today, Grandfather was certain his three sons would not be home for Christmas. Pearl Harbor had been attacked. For the ensuing four years, Grandfather managed the farm as best he could. By August of 1942, the help situation had become critical. All the young men had gone to war. A bill from the local paper read: "To adv. man wanted, 1 time, 50 cents."

Grandfather hired Bernard Moore, a shuffling Walter Brennan-looking fellow who was in the early stages of losing a lifetime battle with booze. Grandfather counted pennies hoping dollars would take care of themselves. He asked about the return of a hundred and fifty burlap bags to his feed man, who replied: "One hundred burlap sacks were returned for a credit of three dollars.

The forty-nine sacks returned the same day were cob corn sacks, which we do not give credit for." He sold his fifty-acre overflow farm three miles from here. He made hard choices in order to survive.

I never met Grandfather, but his persevering example is here, in his letters, to guide me through these difficult times.

Sunday, December 8. The December mixed sale at Timonium is a clearing-house. Every horse drew a bid today, but buyers didn't pay much. Their real expense comes later, in the time and effort making these small wagers pay off. The addresses of buyers reflected the far-flung Thoroughbred marketplace. Manitoba: Fifteen hours in a gooseneck. Pennsylvania: Look for her at Penn National next fall. Western New York: I'll breed this mare to my new stallion. New Hampshire: See you at the Rock. Maryland: Just needed a teaser. South Carolina: She'll make a perfect polo pony.

In midafternoon, a Post-it Notes on the bulletin board grabbed my attention: "Suspect Terrain beaten three-quarters of a length. Ran good. Dad." I hastily prepared Update Announcements for weanlings by Corridor Key, hoping trainers would sign their tickets. I prayed, "Please don't make polo ponies out of our weanlings."

Monday, December 9. "This sale's depressing."

"Why should somebody pay for something you don't want. If you wanted that horse, you'd keep it."

"Yeah, but a measly two thousand bucks for a racehorse?"

"It's fifty dollars a day to train a horse in Maryland. Think what you're saving."

"But look what I've got wrapped up in her!"

"That is sunk cost. Forget it. I got a barnful of stallions whose job it is to make babies—better babies than these. This is a game of constant culling."

"How do you win at this game?"

"You keep playing."

Tuesday, December 10. Betting handle is the umbilical cord that feeds the horse business, and the negative articles on Maryland's racing industry are constricting my oxygen supply. I am attached at the navel to the mercurial swings of betting handle. I felt light-headed and desperate back in the early nineteen-eighties, before the state lowered its takeout from four-point-five percent to point-five percent. I started to grow again. I felt another surge of oxygen when Sunday racing was approved, again when intertrack wagering was implemented.

Now, I'm dizzy again. Where's my air? Purses have been cut twice at Laurel recently. Handle has declined at *eight* consecutive race meetings. Please, somebody do something!

"Hey!" Rich Wilcke called across the parking lot of the *Maryland Horse* office. "We're having a press conference on Thursday to introduce the satellite wagering bill. Can you make it?"

I took a deep breath. Of course I can.

Wednesday, December 11. "Punch it up a little. It's a big deal, even to a clown like me who knows nothing about racing. Give it a strong read."

Okay (deep breath): "Saggy sired 1961 Kentucky Derby and Preakness winner Carry Back. That same year, Correspondent sired Belmont Stakes winner Sherluck. *The Triple Crown in 1961 belonged to Country Life sires.*"

"That's great. That's a wrap. Turn the mike off. Jeez, that was thirty years ago. Done anything lately?"

Thursday, December 12. In Pimlico's Sports Palace this morning, legislators were introduced to a well-appointed betting theater, with gracious amenities, rather than the New York City storefront parlors that their collective mind's eye associates with off-track betting.

"These teletheaters are a credit to the community," said the expert from Albany, as he passed around posters of an upscale restaurant for upstate New Yorkers.

The word "satellite" implies an entity that depends or relies on another entity. It might also imply a communications satellite. "Race tracks in the future will either be senders or receivers," explained track owner Joe De Francis. "We want Laurel and Pimlico to be senders."

This is a wired world, baby. Via satellite, Michael Jordan and the National Basketball Association reach eighty-eight countries—an international audience for the sport of basketball. We don't even reach the corners of this tiny state of Maryland with our sport.

Friday, December 13. Racing folks are a superstitious lot. Mrs. Jane Lunger wore the same shoes every time Go for Wand ran. Today at Laurel, a two-year-old by Go for Wand's half-brother squeezed a two-year-old by Allen's Prospect soon after the break. The filly fell hard, throwing her jock. Two races later, two babies by Allen's Prospect raced in tight quarters turning for home. Filex's Dream impeded Ritchie Trail and was disqualified from second.

Was it purely bad luck that four babies by the home stallions literally ran into each other today? Or was it simply Friday the 13th?

Saturday, December 14. There's a constricting element to working in the office of a family farm. It's close quarters. I've watched a parade of sisters, sisters-in-law, and my own wife, slowly go half mad from punching in computer minutiae about how a mare is teasing, from coding in credit card transactions for accountants' benefit, from sleuthing together a recent equine arrival's health records.

Alice is happily seeking a middle ground. Her mornings these days are spent outside with the horses. Her afternoons are spent inside, with the necessary, but never ending, paperwork. The physical exertion of working directly with the horses, not merely their computer code numbers, is great therapy.

Sunday, December 15. Ellen's mom mailed us an article about a Powhatan, Virginia, elementary school that uses ponies to help children with behavioral problems. The program, called therapeutic vaulting, teaches young students to ride backwards, to do scissors and around-the-worlds, to stand on a moving horse's back, while the horse gently circles on a lunge line. Children *earn* the right to participate by behaving. Occupational therapist Betsy Healey spearheaded the program after seeing positive results in riding programs for physically handicapped and autistic children. Said eight-year-old Kevin, who enjoys standing on the moving pony: "I feel gentle, fine, and great."

All I could think about was dear old Rollaids, our placid brain-damaged foal from two seasons ago. His new owners sent me a photo of him wearing a baseball cap. Their mentally handicapped daughter now has someone who needs her. The little girl and Rollaids both might be saying: "I feel gentle, fine, and great."

Monday, December 16. "Got your memo about the Fasig-Tipton sale closing today," Mr. Baylis said. "Why don't you just enter my three broodmares? We gave that two-year-old away after he finished last on Saturday. Yes, it's about time to sell the mares."

There go three year-round boarders. Phone rang again an hour later.

"Bad news," another client said. "My mare aborted Saturday night. I'll send you the vet certificate."

A live foal fee for the next spring thus was voided before the end of this fall. We've had worse Mondays, but not recently.

Tuesday, December 17. Bernie smiled as the final draft of the stallion video finished its advance screening: "Less than twenty minutes—nineteen minutes, fifty-eight seconds, to be exact. Saves on the dupes if it's less than twenty minutes."

Many prospective breeders are simply too busy to visit our farm. Video is the next best thing. In nineteen minutes and fifty-eight seconds, they can examine our four stallions and their progeny: Five minutes per stud. No beltways to fight. No lost Saturdays. No cold wind. No frozen hands on chain shanks. No muddy stallions. No walk 'em down and back, when all the studs want to do is run. It took seven weeks, a small fortune in Federal Express bills, and an embarrassing number of retakes, but it's done, thank heaven.

Wednesday, December 18. On the conference room wall in the editorial department of the Baltimore *Sun* is a photograph of a quintessential Maryland scene: Sagamore Farm, with its perfect white fences, its perfect red barns. At the conference table sat the shapers of public opinion. Recent editorials on off-track betting have not been favorable. The editors fear the Manhattan experience. They fear organized crime, video lottery, and slot machines. They fear the unknown.

I fear the known: That every breeding farm in Maryland relies on purses, and purses are not commensurate with the expense of raising a Thoroughbred. When asked why I supported OTB, I pointed to the photo of Sagamore. "Those fences have fallen down. Those barns are empty. That farm is deserted. I don't want that to happen to me. We need better purses. Purses come from handle. A fellow can't bet if he can't get to the track. Let us take our sport to him, rather than make him come to us."

Thursday, December 19. His crocodile tears gather his tiny eyelashes into wet triangles, and there is no consoling him. Even his favorite music (any music) fails to cheer him up. He is exhausted, and rests his head on my shoulder as I hold him. The music plays, The English Beat chanting, *"Feel love pumping through the doors of your heart."* But he has no bounce today, and the reggae beat is ineffective.

We stare at the horses on the hillside. What's a horse say? Come on, what's a horse say? He sighs in annoyance at his father. He is fifteen months old, and the flu has come to Maryland.

Friday, December 20. In 1934, Grandfather sold Airflame for twenty-five hundred dollars to Alfred Gwynne Vanderbilt. At two, Airflame set a world record and won his first four races. Vanderbilt's trainer, Bud Stotler, proposed to exchange three old mares for Airflame's dam, Flamante. Grandfather was offended, and his words leapt off the page he typed in response.

"I have always been honest with you and AGV and not gone after the almighty dollar just because it was AGV. Yet how is it that every time you deal with me you want to trade or exchange something? You know I have not much money and I must operate my farm to carry itself. Flamante has had five colts and all have won. She is only ten. Every one of her foals would bring me around three to four thousand dollars. She would certainly suit Discovery: Fair Play on Rock Sand blood. I am very fond of you, but …"

I tried to read on, but the handwritten second page was simply a series of pencil strokes and abbreviations.

"Shorthand," Mom explained. "August Belmont sent your grandfather to school to be his secretary. He learned shorthand."

That talented son of a gun. He knew shorthand. I never met a man who knew shorthand. *Never.*

Saturday, December 21. A family of elves busied themselves meeting the noon mail deadline—stuffing, stapling, stamping the outgoing Stallion Videos, which carried a cover letter shouting the December 31st deadline for Maryland Million nominations: Last chance for yearlings; miss it now, miss it forever. Rather Scrooge-like to hit folks with another horse-related bill at Christmastime, but we don't make the rules, we just follow them.

I cross-checked registrations, and in the process, discovered the names of next year's two-year-olds. Allen Paulson continues to name foals after aviation checkpoints. Could Belma or Cugar or Kyoko be the next Arazi? This exercise set my mind to dreaming. The happy ghost of Christmas Future danced about the office as I played Bob Cratchit, clerking my way through the holiday.

Sunday, December 22. "Why did you wait until this morning?" asked Thoreau. "Write while the heat is in you. The writer who postpones recording his thoughts uses a cooled iron."

Sorry. I had enough heat last night. The old wooden garage behind Andrew's house in the woods caught fire. It looked as if Andrew's house was ablaze. Debbie noticed it first. She ran into Uncle John's and Aunt Yvonne's house and dialed 911. Capt. Woodward took the call. He thought it was the main barn, that for sure he'd be drafting water out of the plastic hydrant on the pond—the one his crew practices on every summer. Uncle John called Dad, who flew out the lane in his car. Mom ran, *ran*, up the hill.

"It's only the garage," she heard, as she sighed a mother's relief.

Sparks landed in the crook of poplar trees a hundred feet high. Five fire trucks arrived. They doused Andrew's new pickup. Too late. Totaled. Andy had jumped in it and moved it as far as he could. He could have blown up. When the second story of the garage collapsed, down came a dozen bridles, two dozen feed tubs, anything that didn't fit in Andy's tack room. I wondered about "Contents Insurance." Then the hoses drenched everything. Capt. Woodward asked if any vagrants had snuck in to escape cold weather. Who can say? The dogs haven't barked.

"A most disturbing incident," said Thoreau. "Let us walk through the woods this afternoon, and assist in the aftermath."

Monday, December 23. Hit the highway today for Christmas in Virginia. Turned off I-95 at the Fredericksburg exit, headed south on old Route 1. Roadside plaques traced "Lee's Movements" along this same path. In the mist that rose over the Ni and the Po Rivers, I could almost hear the lonely fiddle refrain of the *Ashokan Farewell,* the signature song for the recent documentary entitled *The Civil War.* I thought about the book *Traveller,* written from the perspective of Lee's legendary war-horse—descriptions of the long night marches and the days of battle through which both men and horses passed.

The thought of what had transpired on this historic road a century ago made a long drive seem short, as the gentle fiddle played in my head.

Tuesday, December 24. A carriage ride over the lands of Goochland County on Christmas Eve: Deer dart into the woods, owls hoot, geese waddle off the bank unhurriedly, calves butt heads in play, and ponies gallop through a stand of cedars. Young Josh took it all in with rapture, as I held him in my lap, Mum Mum busy with the reins. What a lucky boy. What a lucky De De. Did he just say Daddy? No, it was De De, but that's close enough. Merry Christmas.

Wednesday, December 25. Back at the farm tonight for Christmas dinner, Alice presented me with a faded parchment—the plat of Rockland Farm, surveyed 1882. The woods were denoted by the tulip-shaped seedheads of poplar trees. After the four-hour drive back from Richmond, I was so anxious simply to walk in those poplar woods, to escape the current of people speeding along I-95. I fear I am becoming reclusive, too comfortable in front of a fireplace or alone in

the woods—that I am unrealistic about the future in a changing world. These are not the thoughts I wanted to have on a happy Christmas Day, but I-95 is only four miles away, and construction proceeds apace for a north-south interchange that will pour people into our backyards.

Perhaps Grandfather changed Rockland to Country Life because he was enamored of the name; it had meant home on his return each evening from New York City to Garden City, Long Island, where the train, to this very day, stops at a station called Country Life Press, for the nearby Doubleday publishing plant. I suspect Grandfather smiled when he changed the farm's name, how simple was his choice. I hope the pleasant name of our farm never becomes anachronistic.

Under the Christmas tree, the crackle of wrapping paper ceased for an instant as a present was admired by all. A grandchild pressed a button and the sound of a cricket filled the room: Someone had changed the Christmas carols to *The Civil War* disc. Crickets and fiddles silenced us. It was a magic moment for me, and for everyone else, too. We all felt so, well, so close.

Thursday, December 26. Do the facts foretell our future? Five years ago, four thousand mares were bred to Maryland stallions. Soon the number will be closer to two thousand. The breeding of Thoroughbreds in Maryland, an industry that peaked in the early nineteen-eighties, appears now to be something of a dinosaur on the crowded Eastern seaboard of the nineteen-nineties.

On the morning after New Year's Day, my neighbor, Marlene Magness, and myself—representing the last two active farms in the Route 1 development envelope—come face-to-face against the businessmen spearheading the push for a sewer system along our Route 1 road frontage. Land speculators hide in the trees. Architects draw plans for upgraded buildings. Engineers prepare to tunnel through the metamorphic rock under Winters Run.

This farm is itself something of a metaphor for so many other family farms. The big operations that once spread out over the Maryland landscape, Windfields and Sagamore, already have gone the way of the dinosaur. These farms were not cared for by an extensive family—the key to our survival. We are not extinct, but the family-owned Thoroughbred breeding farm is on Maryland's endangered species list.

Friday, December 27. A hickory branch tapped my shoulder. Startled, I turned to find Thoreau:

"I have just observed five deer grazing," he said. He waved his walking stick towards the pond. "The hunters you write about are not marksmen." With a sweeping gesture, he drew a circle around the woods: "A tunnel under precious Winters Run, you say? Well, treasure your heron. He will leave once his fish taste of a pumping station's overflow. I advise you to fight the proposal."

Lately, the perils of the horse business have consumed my attention. Thank you, Thoreau. I will make time for this pressing matter.

Saturday, December 28. Prepared the Fourth Quarter Statement for the Citidancer syndicate today.

"His first foal is due January 15th," I wrote in the cover letter. "Only three more years to wait."

Three more years before Citi's first crop of two-year-olds signal success or failure. The time it takes to prove a stallion often seems an eternity, but it seems like only yesterday I stood in Corridor Key's paddock with a video camera, shooting a truly homemade movie of which I made copies, one at a time, as prospective breeders inquired. Now the gray stallion's offspring take care of his promotion. Today's *Form* carried a headline on Codys Key: "Speed's the Key in Allegheny Stakes."

Even with such an immediate example at hand, I felt inadequate trying to reassure Citi's shareholders that, yes, three years can pass very quickly.

Sunday, December 29. Mild temperatures and a light rain greeted us this Sunday, and the stallions rolled leisurely in the mud near their gates, then grazed the morning away. My reflective mood was interrupted with concern for an old broodmare who stood stiffly in the warm rain. I've had my fingers crossed against the seeming inevitability of an aborted foal in the broodmare band. I lifted the mare's tail. There was no sign of a discharge. I clucked to her, and she moved off. Thank you, old girl. Thanks for holding on.

On the walk home, I startled Thoreau's deer, and they bounded down into the watershed, white tails bouncing through the woods, and the sound they made on the leaves was soon swallowed by the noise of the creek carrying away the rain.

Monday, December 30. After three years of recording farm life one day at a time, I am reminded of the foreword to *Blooded Horses of Colonial Days*, wherein Francis Barnum Culver wrote:

"It may be objected that the author has devoted more space to Maryland turf history than a fair apportionment of the subject warrants. His answer is, Maryland (holds) a supereminent place in the matters of the breeding of fine horses, and the records of this Province are rich in material."

The matter of apportionment aside, I have endeavored to chronicle our efforts to breed fine horses on this little hundred-acre province so rich in material. In so doing, perhaps a time capsule has been preserved: For grandfathers and grandchildren, for those not yet born, and most certainly for young Josh and his cousins Philip and Elizabeth, who might someday turn these pages with the same interest they now exhibit for Babar's travails.

Tuesday, December 31. A whole year seems to happen on the last day. In the morning, I am filling out Maryland Million nominations, tracking babies by our sires. I cannot quite believe that yearlings become two-year-olds tomorrow, yet their racing engagements are being mapped out this very day. I am reminded of Grandfather's secretarial chores in the days when two-year-old futurities were among the richest prizes of the racing season.

By late afternoon, a busy scene envelops me—a family farm in full swing: Michael is on the phone booking a mare, Alice is addressing the monthly board bills, Mom is logging in the mail, Uncle John is on a bank run, Ellen is updating vaccination charts, Andrew is tacking up the yearlings, Dad is on the line from Laurel.

"Won by a head," Dad says. Such a nice touch to finish off the year, even if it was merely a modest claiming event. A win's a win.

Two tractors—one hauling wood, one pulling the manure spreader—chug past the office. Young Josh will not sit still at the sound of the tractors, and in his new brown work shoes, he makes tracks toward the fields. I hold him at the gate as the noisy spreader throws straw into the air. What delightful fireworks are going off in his mind as the yellow straw floats to the ground!

Tonight at midnight, the watershed will echo with the sound of *real* fireworks, as neighbors celebrate the new year with a bang. Dogs will duck under beds, and horses will lift their heads. Another year of country life will be behind us, and another year will beckon.

THE END

Acknowledgements

This work contains brief excerpts from previously published material by other authors. Grateful acknowledgement is made for reference to the following copyrighted works:

Year One:

January 2: *Stud Farm Diary*, written in 1935. Copyright 1959 by Humphrey S. Finney, published by J. A. Allen & Co. Ltd., London, England. **January 14:** *The Mosquito Coast*, by Paul Theroux, copyright 1982 by Cape Code Scriveners. Published by Houghton-Mifflin, Boston. **January 26:** *Bonfire of the Vanities*, by Tom Wolfe, 1987 Bantam Books. **February 22:** *Lonesome Dove.* Copyright 1985 by Larry McMurtry. Published by Pocket Books, a division of Simon & Schuster Inc. **February 24:** *Registered Thoroughbred Names*, copyright 1990 by The Jockey Club, Lexington, Kentucky. **March 5:** *Rudyard Kipling's Verse Inclusive Edition 1885-1918*, published 1927 by Doubleday, Page & Company. Printed at the Country Life Press, Garden City, N.Y. **March 19:** *The American Turf Register and Sporting Magazine*, September 1834-February 1835, Numbers 1-6. **April 11:** *Horse Owners and Breeders Tax Manual.* Copyright 1989 by Thomas A. Davis, published by the American Horse Council, Washington, D.C. **April 18:** "Oh, Happy Day," 1952, by Don Howard Koplow and Nancy Binns Reed. **April 23:** *The Far Side Gallery 3*, by Gary Larson. Copyright 1988 by Universal Press Syndicate. **April 30:** *Rawhide*, theme by Ned Washington and Dmitri Tiomkin, sung over credits by Frankie Lane. **May 9:** *Manual of Equine Neonatal Medicine*, by John E. Madigan, D.V.M., M.S., Editor. Copyright 1987 by Live Oak Publishing. **May 18:** "Peg O' My Heart," words by Alfred Bryan, music by Fred Fisher, 1913. **June 9:** *Gunsmoke*, (character Chester Goode) CBS TV 1955-1975. **June 13:** *Poor Richard's Almanack*, by Benjamin Franklin. Copyright 1967 by Hallmark Cards, Inc. **June 19:** *Veterinary Notes for Horse Owners*, by M. Horace Hayes, F.R.C.V.S., published 1924 by Hurst and Blackett, Ltd., London, England.

July 20: "Who'll Stop the Rain," 1970, words and music by John C. Fogerty, by Creedence Clearwater Revival. **July 27:** *Misty of Chincoteague*, by Marguerite Henry, Macmillan Children Group, Inc. **July 30-31:** *Song of Horses*, arranged by Robert Frothingham. Copyright 1920 by Houghton-Mifflin Company. **August 1:** *I Want to Know Why*, by Sherwood Anderson. Copyright 1921 by B. W. Huebsch. Copyright renewed 1948 by Eleanor C. Anderson. **August 7:** *The Wizard of Oz*, by L. Frank Baum. Copyright 1982. Published by Henry Holt and Company. **August 8:** *All Creatures Great and Small*, by James Herriot. Copyright 1972 by James Herriot. **August 29:** *A Day at the Races* and "All God's Chillun Got Rhythm," 1937, words by Gus Kahn, music by Bronislaw Kaper and Walter Jurmann. Introduced film *A Day at the Races*, by Ivie Anderson and Duke Ellington and his Orchestra, with the Marx Brothers. **September 2:** *American Stud Book*. Copyright The Jockey Club. **September 19:** *The Thoughts of Thoreau*, by Henry David Thoreau (1817-1862). Copyright 1962 by Edwin Ray Teale. **September 24:** *The World According to Garp*, by John Irving. Rev. ed. 1990. Ballantine Books. **September 26:** *The One Minute Manager*, by Kenneth Blanchard, Ph.D., and Spencer Johnson, M.D. Copyright Blanchard Family Partnership. Published by Berkley Books. **October 2:** *Tesio, Breeding the Racehorse*, by Federico Tesio. Copyright 1958 by J. A. Allen & Co. Ltd., London, England. **October 10:** *Good Morning Vietnam*, directed by Barry Levinson, 1988. **October 18:** *The Noble Horse*, by Monique and Hans Dossenbach, 1987, Portland House, New York, N.Y. **December 20:** *The Great Gatsby*, by F. Scott Fitzgerald. Copyright 1925 by Charles Scribner's Sons. **December 24:** *Lee*, by Douglas Southall Freeman. Abridged 1982 by Richard Harwell. Macmillan. **December 25:** *A Christmas Carol*, by Charles Dickens. 1980 Edition. Buchaneer Books. **December 28:** "I Feel Like I'm Fixin' to Die Rag," by Country Joe McDonald and the Fish. From the album *Woodstock*. 1970. Atlantic Records.

Year Two:
(sources not previously cited in Year One)

January 3: *Blooded Horses of Colonial Days*, by Francis Barnum Culver. Copyright 1922, by Francis Barnum Culver. Press of Kohn & Pollock, Inc., Baltimore. *Racing in America 1866-1921*, by Walter S. Vosburgh. Copyright 1922 by The Jockey Club. The Scribner Press, New York. **January 25:** *Steel Magnolias*, directed by Herbert Ross, 1989. **February 1:** "What's the Matter Here?" by 10,000 Maniacs. From the album *In My Tribe*. Elektra/Asylum Records, 1987. **March 1:** *Robert Penn Warren Reader*, Random House, 1988. **March 7:** *The Other Side of the Mountain*, by E. G. Valens, HarperCollins, 1989 edition. **March 24:** "This Land is Your Land," words and music by Woody Guthrie (BMI) 1956. **April 3:** *The Jerk*, directed by Carl Reiner, 1979. **April 6:** *My Old Man*, by Ernest Hemingway (1898-1961). Copyright Charles Scribner's Sons, 1987. **April 10:** "A Lucky Guy," music and lyrics by Rickie Lee Jones. From the album *Pirates*. Easy Money Music (Ascap), 1981. **April 14:** "Take Me Home," sung by Phil Collins. From the album *No Jacket Required*. Atlantic Recording Corp. 1985. **April 20:** *The Anatomy of the Horse*, by George Stubbs (1724-1806). Published in 1766. Copyright by The Tate Gallery, 1984. **April 30:** "Still Crazy After

421

All These Years," words and music by Paul Simon. Warner Bros. Records, Inc. 1988. **May 19:** "What a Wonderful World," words by George Douglas, music by George David Weiss, 1968. **May 20:** *Weekend at Bernie's*, directed by Ted Kotcheff, 1989. **June 7:** *John Muir, In His Own Words*, compiled and edited by Peter Browning. Copyright by Peter Browning, 1988. **June 14:** *September*, by Rosamunde Pilcher. Copyright by Robin Pilcher, Mark Pilcher, Fiona Pilcher, and Philippa Imrie, 1990.

July 10: *Winnie-The-Pooh*, by A. A. Milne. 1981 Edition. Dell. **August 2:** *The History of Thoroughbred Racing In America*, by William H. P. Robertson. Copyright 1964 by W. H. P. Robertson. **August 24:** "Raindrops Keep Falling On My Head," words by Hal David, music by Burt Bacharach. Copyright 1969. **September 1:** *Zelig*, directed by Woody Allen, 1983. **September 2:** *Out of Africa*, by Isak Dinesen, Random House, 1984 Edition. **September 3:** "If I Had a Boat," by Lyle Lovett. From the album *Pontiac*. Copyright 1987 Michael H. Goldsen, Inc. MCA Records (Ascap). **September 5:** *Racing in Art*, by John Fairley. Copyright 1990 by John Fairley. *Racing Days*, Photographs by Henry Horenstein, text by Brendan Boyd. Copyright 1987 by Horenstein and Boyd. Published by Viking Press. *The Body Language of Horses*, by Tom Ainslie and Bonnie Ledbetter. Copyright 1980 by Tom Ainslie Bonnie Ledbetter. *Thoroughbred, A Celebration of the Breed*, photography by John Denny Ashley, written by Billy Reed. Copyright 1990 by John Ashley and Billy Reed Enterprises. Published by Simon and Schuster, New York. **September 6:** *The Look of Eagles: Racing Story*, by John Taintor Foote. 1970 Edition, Saifer. **September 11:** *Bury My Heart at Wounded Knee: An Indian History of the American West*, by Dee Brown. 1971, HR&W. **September 22:** *Radio Days*, directed by Woody Allen, 1986. **September 29:** *Dr. Strangelove, or How I Learned to Stop Worrying and Love the Bomb*, directed by Stanley Kubrick. 1962. **October 1:** *The Best of Damon Runyon*, selected by E. C. Bentley, Circle Books Edition. Published 1945.

Year Three:

January 5: *Rocky Horror Picture Show*, 1973. **January 11:** *The Horse in Motion*, by Leland Stanford. Copyright 1881. Published by James R. Osgood and Company 1882 Boston. *Thoroughbred Racing Stock*, by Lady Wentworth. 1938 George Allen & Unwin, London, England. *British Sporting and Animal Drawings*. Copyright 1978 by Judy Egerton and Dudley Snelgrove, by the Tate Gallery. **February 15:** *To Have and Have Not*, by Ernest Hemingway. Copyright 1937 by Ernest Hemingway. Published by Charles Scribner's Sons. **February 25:** *One Flew Over the Cuckoo's Nest*, by Ken Kesey. 1963, NAL-Dutton. **March 2:** *A Sand County Almanac*, by Aldo Leopold. Copyright 1949 by Oxford University Press, Inc. **March 4:** "This Must Be the Place (Native Melody)," by Talking Heads. From the album *Speaking In Tongues* 1983 Sire Records Company. **March 23:** "Blue Sky," by The Allman Brothers Band. From the album *Eat A Peach*. Copyright 1972 Polygram Records, New York. **March 24:** *Endangered and Threatened Species of the Chesapeake Bay Region*, by Christopher P. White. Copyright 1982 by Tidewater Publishers. **April 4:** "Eternal Father Strong To Save," by William Whiting (1825-1878) and John B. Dykes (1823-1876). **April 7:** *Stone Work, Reflections on Serious Play and Other Aspects of Country Life*, by John Jerome. Copyright 1989 by John Jerome. Published by the Penguin Group. **April 28:** *In Suspect Terrain*, by John McPhee. Copyright 1983 by John McPhee. **May 26:** "Pancho and Lefty," sung by Willie Nelson and Merle Haggard. From the album *Half Nelson*, 1985 CBS Records, Inc. **June 6:** "Sun Comes Up, It's Tuesday Morning," by The Cowboy Junkies. From the album *The Caution Horses*. Copyright 1990 BMG Music Canada Inc. **June 17:** *The Things They Carried*, by Tim O'Brien. Copyright 1986 by Tim O'Brien. **June 30:** *At Play in the Fields of the Lord*, by Peter Mathiessen. Copyright 1965 by Peter Mathiessen.

July 9: *Sophie's Choice*, by William Styron. Random House, 1979. **July 20:** "Powderfinger," by Neil Young. From the album *Neil Young and Crazy Horse*, 1991 Reprise Records. **August 3:** *The Stone Horse*, by Barry Lopez. Copyright 1987 by Barry Holstun Lopez. Appeared in *Best American Essays* 1987. Published by Ticknor and Fields. **August 19:** "Independence Day," by Bruce Springsteen. From the album *The River*. 1980 Columbia Records. **August 28:** "Oh! Susannah," by Stephen Collins Foster, 1848. **September 16:** *The Honest Rainmaker, The Life and Times of Colonel John R. Stingo*, by A. J. Liebling. Copyright 1953 by A. J. Liebling. North Point Press, San Francisco, 1989. **October 5:** "THANK YOU Falletinme Be Mice Elf Agin," by Sly and the Family Stone. Written by Sylvester Stewart. Copyright 1970 by Sylvester Stewart. From the album *Sly and the Family Stone Greatest Hits*. Epic Records. **November 5:** *Giants of the Turf*, by Dan Bowmar III. Copyright 1960 by *The Blood-Horse*. **December 19:** "Doors of Your Heart," by The English Beat. From the album *Wha-ppen*, 1981 IRS A&M Records. **December 23:** "Ashokan Farewell," by Jay Ungar. Copyright 1984. Flying Fish Records. From *The Civil War*, a film by Ken Burns, 1990 Elektra Entertainment. *Traveller*, by Richard Adams. Copyright 1988 Edward Casson Promotional Services Limited. Published by Dell.